# Evaluation of Systems' Irregularity and Complexity:  Sample  Entropy, Its Derivatives, and  Their Applications across Scales and Disciplines

# Evaluation of Systems' Irregularity and Complexity: Sample Entropy, Its Derivatives, and Their Applications across Scales and Disciplines

Special Issue Editor

**Anne Humeau-Heurtier**

MDPI • Basel • Beijing • Wuhan • Barcelona • Belgrade

**MDPI**

*Special Issue Editor*
Anne Humeau-Heurtier
University of Angers
France

*Editorial Office*
MDPI
St. Alban-Anlage 66
Basel, Switzerland

This is a reprint of articles from the Special Issue published online in the open access journal *Entropy* (ISSN 1099-4300) from 2017 to 2018 (available at: https://www.mdpi.com/journal/entropy/special_issues/Scales_and_Disciplines)

For citation purposes, cite each article independently as indicated on the article page online and as indicated below:

LastName, A.A.; LastName, B.B.; LastName, C.C. Article Title. *Journal Name* **Year**, *Article Number*, Page Range.

**ISBN 978-3-03897-332-4 (Pbk)**
**ISBN 978-3-03897-333-1 (PDF)**

Photo cover image courtesy of J.-L. Heurtier.

# Contents

# About the Special Issue Editor

**Anne Humeau-Heurtier** received her PhD degree in Biomedical Engineering in France. She is currently a full professor in Engineering with the University of Angers, France. Her research interests include signal and image processing, mainly multiscale and entropy-based analyses. Main applications concern the biomedical field. She was guest editor for several Special Issues in the journal *Entropy*.

*entropy*

MDPI

*Editorial*

# Evaluation of Systems' Irregularity and Complexity: Sample Entropy, Its Derivatives, and Their Applications across Scales and Disciplines

**Anne Humeau-Heurtier**

Laboratoire Angevin de Recherche en Ingénierie des Systèmes (LARIS), University of Angers, 62 avenue Notre-Dame du Lac, 49000 Angers, France; anne.heurtier@univ-angers.fr

Received: 8 October 2018; Accepted: 9 October 2018; Published: 16 October 2018

**Keywords:** sample entropy; multiscale entropy; fuzzy entropy; multivariate data

Based on information theory, a number of entropy measures have been proposed since the 1990s to assess systems' irregularity, such as approximate entropy, sample entropy, permutation entropy, intrinsic mode entropy, and dispersion entropy to cite only a few. Among them, sample entropy has been used in a very large variety of disciplines for both univariate and multivariate data. However, improvements to the sample entropy algorithm are still being proposed because sample entropy is unstable for short time series, may be sensitive to parameter values, and can be too time-consuming for long data.

At the same time, it is worth noting that sample entropy does not take into account the multiple temporal scales inherent in complex systems. It is maximized for completely random processes and is used only to quantify the irregularity of signals on a single scale. This is why analyses of irregularity—with sample entropy or its derivatives—at multiple time scales have been proposed to assess systems' complexity.

This Special Issue invited contributions related to new and original research based on the use of sample entropy or its derivatives. The papers published in this Special Issue can be divided into two categories. First, some papers present new applications of sample entropy or its derivatives. Second, some papers propose improvements to sample entropy or its derivatives.

In addition to these articles, Sun et al. [1] performed a systematic review to summarize the complexity differences of important signals in patients with asthma. They obtained the overall trend of entropy change in the physiological signals related to asthma, evaluated the potential of using entropy of biological dynamics as new clinical indices, and discussed possible strategies for future research. The authors mention that entropy of heart rate variability (HRV), airflow, center of pressure, and respiratory system impedance are lower in patients than in healthy people, whereas the entropies of respiratory sound, airway resistance, and reactance are higher in patients than in healthy people. This might be explained by the unstableness of local respiratory tract and the increase in breathing difficulty.

## 1. Applications of Sample Entropy or Its Derivative

In the biomedical field, Liao et al. used multiscale entropy to study complexity of skin blood flow in type 2 diabetic patients with peripheral neuropathy and in healthy controls [2]. Skin blood flow was recorded at the first metatarsal head using laser Doppler flowmetry in response to locally applied pressure and heating. Indeed, quantification of skin blood flow responses to loading pressure and thermal stresses may be a reasonable way to assess the risk of diabetic foot ulcer. Multiscale entropy was computed using a modified sample entropy algorithm. The results showed that during reactive hyperemia and the biphasic response induced by local heating, the modified sample entropy in diabetics presents only small changes compared to baseline but undergoes significant changes in controls. Moreover, during baseline and skin blood flow responses—except for the pressure loading period—the modified sample entropy at small scales exhibits different transitions between the two groups. The findings support the use of nonlinear measures of skin blood flow responses induced by mechanical and thermal stresses to assess the risk of diabetic foot ulcer.

Simons et al. studied electroencephalogram (EEG) of Alzheimer's disease (AD) and age-matched control subjects [3]. For this purpose, they used fuzzy entropy. Their results showed that fuzzy entropy is lower in Alzheimer's patients than in control subjects for electrodes T6, P3, P4, O1, and O2. Moreover, fuzzy entropy led to better results than approximate entropy or sample entropy when diagnostic accuracy was computed with the receiver operating characteristic curves. However, the authors note that the results are dependent on the input parameters used in the fuzzy entropy computation.

Ruiz-Gomez et al. aimed to detect AD and its prodromal form (i.e., mild cognitive impairment, MCI) from healthy controls through an EEG-based methodology [4]. For this purpose, after a data preprocessing step, the authors used a combination of spectral measures and nonlinear methods, with both frequency (spectral features) and time domain (nonlinear features: Lempel–Ziv complexity, central tendency measure, sample entropy, fuzzy entropy, and auto mutual information) analyses applied to EEG recordings. Furthermore, to avoid redundant features sharing similar information, an automatic feature selection stage based on the fast correlation-based filter (FCBF) was used. Finally, different multiclass classifiers were tested: logistic discriminant analysis (LDA), quadratic discriminant analysis (QDA), and multilayer perceptron neural network (MLP). The results showed that both AD and MCI elicit changes in the EEG background activity, including a slowing of EEG rhythms, alterations in the frequency distribution of the power spectrum, a complexity loss, a regularity increase, and a variability decrease. Spectral and nonlinear features allowed the brain abnormalities associated with AD and MCI to be characterized. Moreover, the brain activity in AD patients is less complex, more regular, and less variable than in MCI and healthy control subjects.

In their work, Kumar et al. proposed a method for an automatic diagnosis of myocardial infarction (MI) using ECG beat with a flexible analytic wavelet transform (FAWT) method [5]. For this purpose, using lead-2 ECG signals, first, a preprocessing step was performed to remove the baseline wandering and other noise present in the ECG signals. Then, ECG signals were segmented into the beats. Furthermore, these beats were decomposed up to the 24th level of decomposition using FAWT. Sample entropy was computed from each sub-band signal, which was reconstructed from the corresponding coefficients of the FAWT-based decomposition. The computed features were subjected to random forest (RF), J48 decision tree, back propagation neural network (BPNN), and least-squares support vector machines (LS-SVM) classifiers to separate the ECG beats of MI and normal classes. The authors found that lower frequency sub-band signals show higher sample entropy values for normal ECG beats than MI ECG beats and that higher frequency sub-band signals extracted from normal ECG beats have lower values of sample entropy. Finally, the method used by the authors achieved 99.31% accuracy using LS-SVM classifier with radial basis function kernel.

Fazan et al.'s work is related to the analysis of the changes in physiological complexity entailed by physical training [6]. This work was performed through the study of HRV time series recorded in rats divided into two groups: rats that performed medium intensity training and a sedentary group. HRV signals were recorded five days after the experimental protocol. The analysis of the HRV time series was done with different algorithms: multiscale entropy, multiscale dispersion entropy, and multiscale SDiffq. Multiscale SDiffq is a measure of entropic differences, and differences of entropy between time series and its uncorrelated version, i.e., surrogate data, are used to represent the complexity. From the latter, three quantities ($q$-attributes) were computed: SDiffqmax, $q$max, and $q$zero (SDiffqmax represents the maximum value for SDiffq in the range of $q$, whereas $q$max and $q$zero represent the $q$-value where SDiffq finds its maximum and zero values, respectively). The results showed that the significant difference was only found between trained and sedentary rats in the mean $q$max at long time scales. Physical training therefore increased the system complexity. The authors also conclude that multiscale SDiffq is an alternative tool for characterizing the complexity of HRV time series as it can add information in some situations where multiscale entropy is not accurate enough.

Shi et al. studied the within-subject changes of HRV during walking with a regular speed of 5 km/h on a treadmill (the Walk protocol) compared to those in a resting seated position (the Rest protocol) [7]. The analysis of the time series was performed with eight different entropy measures: approximate entropy, corrected approximate entropy, sample entropy, fuzzy entropy without removing local trend, fuzzy entropy with local trend removal, permutation entropy, conditional entropy, and distribution entropy. The authors also explored the potential effects of nonstationary linear or very low-frequency trend. From their results, the authors note the importance of stating whether detrending has been performed in studies and, if so, which process has been performed.

Bakhchina et al. were interested in the "mind–body" relationship and explored how the system organization of behavior could be reflected in the irregularity of the heart rate, as measured by sample entropy [8]. Thus, they proposed a model explaining HRV in relation to neuronal processes in the brain. For this, they studied the sample entropy of the heart rate in different conditions. The results revealed that irregularity of the heart rate reflects the properties of a set of functional systems subserving current behavior, with higher irregularity corresponding to later-acquired and more complex behavior. This showed that the dynamics of functional systems supporting current behavior is reflected not only in activity of the brain but also in the activity of the rest of the body. The authors finally conclude that sample entropy of the heart rate can be used as a new tool to study psychological processes and organization of behavior.

Ye et al. proposed a method for the recognition of driving fatigue [9]. They used sample entropy associated with kernel principal component analysis to recognize driving fatigue. The framework was tested on EEG data. Using support vector machine for classification, a driving fatigue state recognition model was constructed. The results were compared with the ones given by sample entropy alone, fuzzy entropy, and combination entropy. The authors showed that their approach significantly improved the classification recognition rate compared with the traditional sample entropy, fuzzy entropy, and combination entropy. Moreover, fuzzy entropy and combination entropy associated with kernel principal component analysis give worse results than those obtained with sample entropy and kernel principal component analysis.

In another field of application, Lin et al. [10] proposed a structural health monitoring (SHM) system based on multiscale cross-sample entropy (MSCE) to detect damage locations in multibay three-dimensional structures. Through MSCE, the degree of dissimilarity between the response signals of vertically adjacent floors was used to localize damage for each bay analysis. Moreover, a damage index was proposed for rapidly and efficiently diagnosing the damaged floor, axis, and bay in the structure. The work of the authors shows the feasibility and further potential of the proposed SHM system for the detection and localization of damage in large and complex structures.

In a completely different context, Yin et al. analyzed the complexity of carbon market and an illustration was performed on pilot carbon markets in China [11]. Because of the short length of the time series used, the analysis was done through the use of the modified multiscale entropy proposed by Wu et al. in 2013. The results showed an overall low complexity in those carbon markets, far smaller than that in the European carbon market. Furthermore, the complexity of the carbon market (except Chongqing) was found higher in small time scales than in large scales. Moreover, complexity level in most pilot markets increased as the markets developed, showing an improvement in market efficiency.

## 2. Improvements of Sample Entropy or Its Derivatives

Sample entropy has the drawback of necessitating a long computational time. This is why Manis et al. presented three algorithms to compute sample entropy quickly [12]. The first algorithm is an extension of the kd-trees algorithm, customized for sample entropy. The second one is an extension of an algorithm initially proposed for approximate entropy—the bucket-assisted algorithm—customized

*Entropy* **2018**, *20*, 794

for sample entropy. The last one is the most rapid for specific values of *m*, *r*, time series length, and signal characteristics.

Looney et al. proposed an analysis dealing with multivariate sample entropy [13]. The authors first revisited the embedding of multivariate delay vectors. They also proposed a new multivariate sample entropy algorithm. Their results showed the improved performance of this new algorithm over existing work for synthetic data and for classifying wake and sleep states from real-world physiological data. Moreover, they showed that synchronized regularity dynamics are uniquely identified via the new multivariate sample entropy analysis.

Chen et al. proposed the hierarchical cosine similarity entropy (HCSE) to overcome some limitations of the multiscale sample entropy, such as undefined entropy value for short time series [14]. HCSE takes both lower and higher frequency components into consideration. The algorithm proposed by the authors is composed of three steps. First, the hierarchical decomposition is used to decompose the time series under study into subsequences. Second, the sample entropy is modified using Shannon entropy rather than conditional entropy. Moreover, angular distance is used instead of Chebyshev distance. Third, the complexity of each subsequence is quantified by the modified sample entropy. An application of HCSE is shown first on synthetic signals and then on the classification accuracy of real ship-radiated noise, which is the main signal source of passive sonar for underwater target detection and recognition. The results show the superiority of the new algorithm over traditional multiscale entropy.

Azami et al. proposed to assess the impact of coarse-graining in multiscale entropy estimations based on both sample entropy and dispersion entropy [15]. Thus, the computation of multiscale entropy relies on two steps: (i) a coarse-graining approach, which is a combination of moving average filter and downsampling process; (ii) computation of the sample entropy for each scale factor, i.e., for each coarse-grained time series. A low-pass Butterworth filter was proposed as an alternative to moving average. Thus, the authors compared existing and newly proposed coarse-graining approaches for univariate multiscale entropy estimation. Among others, their results show that the downsampling may lead to increased or decreased values of entropy depending on the sampling frequency of the time series. The authors also concluded that downsampling within the coarse-graining procedure may not be needed to quantify the complexity of signals, especially for short ones. Moreover, the authors showed that dispersion entropy leads to more stable results than sample entropy in the estimations based on coefficient of variation values and ensures that the entropy values are defined at all temporal scales.

Fuzzy entropy is a derivative of sample entropy that has shown to give better results than sample entropy in some situations; fuzzy entropy presents a stronger relative consistency and shows less dependence on data length than sample entropy. However, fuzzy entropy still has some drawbacks as it depends on the number of samples in the data under study: The shorter the signal, the lower is the precision of the fuzzy entropy values. This is why Girault et al. proposed a new fuzzy entropy measure that presents better precision than the standard fuzzy entropy [16]. This is performed by increasing the number of samples used in the computation of the entropy measure without changing the length of the time series. For this purpose, the constraint of the mean value in the comparison of the patterns is removed. Moreover, not only translated patterns, but reflected, inversed, and glide-reflected patterns are also considered. The new measure (so-called centered and averaged fuzzy entropy) was applied to synthetic and biomedical signals (fetal heart rate time series). The results showed that the centered and averaged fuzzy entropy leads to more precise results than the standard fuzzy entropy does.

**Acknowledgments:** I express my thanks to the authors of the above contributions and to the journal *Entropy* and MDPI for their support during this work.

**Conflicts of Interest:** The author declares no conflict of interest.

*Entropy* **2018**, *20*, 794

## References

1. Sun, S.; Jin, Y.; Chen, C.; Sun, B.; Cao, Z.; Lo, I.L.; Zhao, Q.; Zheng, J.; Shi, Y.; Zhang, X.D. Entropy Change of Biological Dynamics in Asthmatic Patients and Its Diagnostic Value in Individualized Treatment: A Systematic Review. *Entropy* **2018**, *20*, 402. [CrossRef]
2. Liao, F.; Cheing, G.L.Y.; Ren, W.; Jain, S.; Jan, Y.-K. Application of Multiscale Entropy in Assessing Plantar Skin Blood Flow Dynamics in Diabetics with Peripheral Neuropathy. *Entropy* **2018**, *20*, 127. [CrossRef]
3. Simons, S.; Espino, P.; Abásolo, D. Fuzzy Entropy Analysis of the Electroencephalogram in Patients with Alzheimer's Disease: Is the Method Superior to Sample Entropy? *Entropy* **2018**, *20*, 21. [CrossRef]
4. Ruiz-Gómez, S.J.; Gómez, C.; Poza, J.; Gutiérrez-Tobal, G.C.; Tola-Arribas, M.A.; Cano, M.; Hornero, R. Automated Multiclass Classification of Spontaneous EEG Activity in Alzheimer's Disease and Mild Cognitive Impairment. *Entropy* **2018**, *20*, 35. [CrossRef]
5. Kumar, M.; Pachori, R.B.; Acharya, U.R. Automated Diagnosis of Myocardial Infarction ECG Signals Using Sample Entropy in Flexible Analytic Wavelet Transform Framework. *Entropy* **2017**, *19*, 488. [CrossRef]
6. Fazan, F.S.; Brognara, F.; Fazan Junior, R.; Murta Junior, L.O.; Virgilio Silva, L.E. Changes in the Complexity of Heart Rate Variability with Exercise Training Measured by Multiscale Entropy-Based Measurements. *Entropy* **2018**, *20*, 47. [CrossRef]
7. Shi, B.; Zhang, Y.; Yuan, C.; Wang, S.; Li, P. Entropy Analysis of Short-Term Heartbeat Interval Time Series during Regular Walking. *Entropy* **2017**, *19*, 568. [CrossRef]
8. Bakhchina, A.V.; Arutyunova, K.R.; Sozinov, A.A.; Demidovsky, A.V.; Alexandrov, Y.I. Sample Entropy of the Heart Rate Reflects Properties of the System Organization of Behaviour. *Entropy* **2018**, *20*, 449. [CrossRef]
9. Ye, B.; Qiu, T.; Bai, X.; Liu, P. Research on Recognition Method of Driving Fatigue State Based on Sample Entropy and Kernel Principal Component Analysis. *Entropy* **2018**, *20*, 701. [CrossRef]
10. Lin, T.-K.; Laínez, A.G. Entropy-Based Structural Health Monitoring System for Damage Detection in Multi-Bay Three-Dimensional Structures. *Entropy* **2018**, *20*, 49.
11. Yin, J.; Su, C.; Zhang, Y.; Fan, X. Complexity Analysis of Carbon Market Using the Modified Multi-Scale Entropy. *Entropy* **2018**, *20*, 434. [CrossRef]
12. Manis, G.; Aktaruzzaman, M.; Sassi, R. Low Computational Cost for Sample Entropy. *Entropy* **2018**, *20*, 61. [CrossRef]
13. Looney, D.; Adjei, T.; Mandic, D.P. A Novel Multivariate Sample Entropy Algorithm for Modeling Time Series Synchronization. *Entropy* **2018**, *20*, 82. [CrossRef]
14. Chen, Z.; Li, Y.; Liang, H.; Yu, J. Hierarchical Cosine Similarity Entropy for Feature Extraction of Ship-Radiated Noise. *Entropy* **2018**, *20*, 425. [CrossRef]
15. Azami, H.; Escudero, J. Coarse-Graining Approaches in Univariate Multiscale Sample and Dispersion Entropy. *Entropy* **2018**, *20*, 138. [CrossRef]
16. Girault, J.-M.; Humeau-Heurtier, A. Centered and Averaged Fuzzy Entropy to Improve Fuzzy Entropy Precision. *Entropy* **2018**, *20*, 287. [CrossRef]

*entropy*

MDPI

*Review*

# Entropy Change of Biological Dynamics in Asthmatic Patients and Its Diagnostic Value in Individualized Treatment: A Systematic Review

Shixue Sun [1,†], Yu Jin [1,†], Chang Chen [1], Baoqing Sun [2], Zhixin Cao [3], Iek Long Lo [4], Qi Zhao [1], Jun Zheng [1], Yan Shi [5,*] and Xiaohua Douglas Zhang [1,*]

1    Faculty of Health Sciences, University of Macau, Taipa, Macau, China; yb77639@umac.mo (S.S.); yb67647@connect.umac.mo (Y.J.); yb67646@connect.umac.mo (C.C.); qizhao@umac.mo (Q.Z.); JunZheng@umac.mo (J.Z.)
2    State Key Laboratory of Respiratory Disease, the 1st Affiliated Hospital of Guangzhou Medical University, Guangzhou 510230, China; sunbaoqing@vip.163.com
3    Beijing Engineering Research Center of Diagnosis and Treatment of Respiratory and Critical Care Medicine, Beijing Chaoyang Hospital, Beijing 100043, China; 18301564184@163.com
4    Department of Geriatrics, Centro Hospital Conde de Sao Januario, Macau, China; drloieklong@yahoo.com.hk
5    Department of Mechanical and Electronic Engineering, Beihang University, Beijing 100191, China
*    Correspondence: yesoyou@gmail.com (Y.S.); douglaszhang@umac.mo (X.D.Z.)
†    These authors contributed equally.

Received: 16 March 2018; Accepted: 23 April 2018; Published: 24 May 2018

**Abstract:** Asthma is a chronic respiratory disease featured with unpredictable flare-ups, for which continuous lung function monitoring is the key for symptoms control. To find new indices to individually classify severity and predict disease prognosis, continuous physiological data collected from monitoring devices is being studied from different perspectives. Entropy, as an analysis method for quantifying the inner irregularity of data, has been widely applied in physiological signals. However, based on our knowledge, there is no such study to summarize the complexity differences of various physiological signals in asthmatic patients. Therefore, we organized a systematic review to summarize the complexity differences of important signals in patients with asthma. We searched several medical databases and systematically reviewed existing asthma clinical trials in which entropy changes in physiological signals were studied. As a conclusion, we find that, for airflow, heart rate variability, center of pressure and respiratory impedance, their entropy values decrease significantly in asthma patients compared to those of healthy people, while, for respiratory sound and airway resistance, their entropy values increase along with the progression of asthma. Entropy of some signals, such as respiratory inter-breath interval, shows strong potential as novel indices of asthma severity. These results will give valuable guidance for the utilization of entropy in physiological signals. Furthermore, these results should promote the development of management and diagnosis of asthma using continuous monitoring data in the future.

**Keywords:** entropy; irregularity; asthma; physiological signal; individualized treatment

## 1. Introduction

Asthma is a chronic respiratory disease with an increasing incident rate (8% in 2009 vs. 7% in 2001, globally) and its typical symptoms include breathing constriction, reduced oxygen intake, limitation of activity, and even life-threatening respiratory failures that require immediate intervention [1]. It is reported that asthma affects 234 million people around the world and places a heavy burden on patients and their families [2]. The foremost purposes of asthma therapy are to achieve and maintain the control of symptoms, reduce respiratory impairments, and prevent exacerbation [3]. Appropriate management of asthma can help patients live a relatively high quality of life [2]. The lung function is the key indicator of potential risk assessment, thus continuous monitoring of lung function plays a pivotal role in the treatment and care of asthmatic patients.

Besides the classical measurement of static lung function tested at particular time points, increasing attention is being paid to the study of pulmonary function dynamics in expectation of discovering novel indices that can reflect and predict fluctuation of respiratory function during a longer period [4,5]. For this reason, nonlinear analysis methods are widely used to study the complexity of biological data collected from continuous monitoring devices. One of the meaningful characteristics of these dynamic biological data is the irregularity measured by entropy. Currently, several extended concepts of entropy, such as approximate entropy (ApEn) [6], cross-approximate entropy (Cross-ApEn) [7], sample entropy (SampEn) [8] and multi-scale entropy [9,10] are being applied in the quantification of inner irregularity of physiological signals and classification of different disease phases or severities. Entropy has been applied in the classification of different biological parameters of various diseases [10–12]. Research articles have been published on the patterns of entropy change of electroencephalogram (EEG) or electrocardiogram (ECG) signals of epilepsy [13], ECG segments or sound signals of obstructive sleep apnea syndrome, heart rate variability (HRV) of cardiovascular diseases, respiratory signals in chronic obstructive pulmonary disease (COPD), and so on [11,14–16].

In the area of asthma research, several clinical trials investigating the entropy of various physiological signals associated with asthmatic disease have already been carried out [17–24]. These signals include airflow, heart rate variability, center of pressure and respiratory impedance, respiratory sound, airway resistance, and so on. Do all these signals have the same trend of changes in entropy? In what types of signals does entropy show high potential as novel indices of asthma severity? To answer these questions, a systematic review on all available key literatures in this field has yet to be conducted. Therefore, we present this systematic review in which we obtain the overall trend of entropy change in the physiological signals related to asthma, evaluate the potential of using entropy of biological dynamics as new clinical indices and discuss possible strategies of future research. These results will give valuable guidance for the utilization of entropy in physiological signals measured by health monitoring devices as well as for research on complexity in asthma. Furthermore, these results should promote the development of management and diagnosis of asthma in the future.

## 2. Materials and Methods

### 2.1. Search Strategy and Inclusion Criteria

Databases of EMBASE, PubMed, and Google Scholar were searched for papers on the application of entropy analysis in asthma until May 2017. Key words used for our search included "asthma", "asthmatic", "respiratory tract diseases", "respiratory disorders", and "entropy". The search in these databases resulted in 812 articles and 40 articles remained after removal of duplicates and exclusion of records based on title and abstract. The Preferred Reporting Items for Systematic Reviews and Meta-Analyses (PRISMA) workflow is shown in Figure 1.

Trials included in the final discussion met the following criteria: (1) patients with asthma were studied; (2) entropy of asthma associated physiological signals was analyzed; and (3) study results were published. Conference abstracts were excluded for the uncertainty of their reliability as they had not been strictly peer-reviewed. After the search, 11 trials meeting the criteria were reviewed.

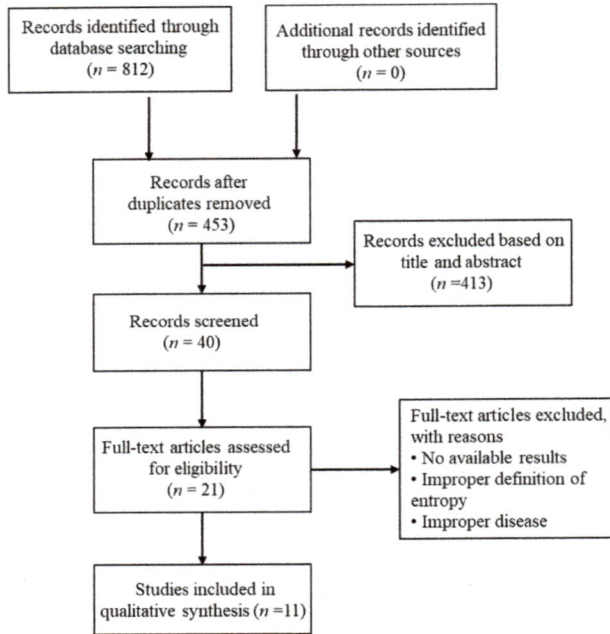

**Figure 1.** PRISMA flow diagram.

## 2.2. Information Extraction and Quality Evaluation

The following information, if possible, was gathered from the reviewed studies: study type, study design, subject number, age, gender ratio, pulmonary function, physiologic signals, entropy types and results.

Jadad's GRADE (grading of recommendations assessment, development, and evaluation) scale was used to evaluate the quality of the publications [25]. Overall quality of the articles was assessed based on their study type, study design (randomization and blinding methods), and description of participants' outcome. The strength of the recommendation to conduct our systematic analysis using these studies was finally graded as very low, low, moderate, or high (Table 1).

**Table 1.** GRADE analysis: applied entropy to physiologic parameters related with asthma.

| OutcomeMeasure | N (Arms) | Risk of Bias | Limitation of Study | Inconsistency | Indirectness | Imprecision | Effect Size | Quality of Evidence |
|---|---|---|---|---|---|---|---|---|
| Airflow | 128(3) | No | No obvious limitations | No | No indirectness | No | Significant | High |
| HRV | 24(1) | No | No obvious limitations | No | No indirectness | No | Significant | High |
| entre of Pressure | 39(1) | No | No obvious limitations | No | No serious indirectness | No | Significant | Moderate |
| Respiratory sound | 51(3) | No | Limitation in study design and data collection | No | No serious indirectness | No | Significant | Low |
| Respiratory impedance | 74(1) | No | No obvious limitations | No | No indirectness | No | Significant | High |
| Airway resistance | 186(2) | No | Limitation in study design and data collection | No | No indirectness | No | Significant | Moderate |

## 2.3. Brief Introduction of Entropies

The concept of entropy was proposed for the first time in 1865 by Rudolf Clausius. Physicist Ludwig Boltzmann deduced the physical nature of entropy which can be associated with internal disorder in 1870. In 1948, Shannon extended the concept of entropy in statistical physics to the process of channel communication, thus creating the discipline "information theory". Since Shannon's extension of the initial thermodynamic concept of entropy, the application of entropy was gradually expanded to the area of statistics [26], information theory [27,28], physical system [29,30], and then physiological dynamic system.

Shannon's definition of "entropy" is also known as "Shannon entropy" or "information entropy" [31], which is calculated as:

$$S(p_1, p_2, \ldots, p_n) = -K \sum_{i=1}^{n} (p_i \log p_i)$$

In the formula, $i$ stands for all possible samples in the probability space, $p_i$ is the probability of occurrence of the sample, and $K$ is a constant. Kolmogorov entropy calculates the change rate of Shannon entropy of a system. The calculation of Kolmogorov entropy is rather complicated. Thus, the lower bound of Kolmogorov entropy is commonly used [28]. For the analysis of short and noisy time series, Pincus [6] introduced a family of measures termed approximate entropy as approximate estimation to the lower bound of Kolmogorov entropy. Then, sample entropy was introduced to eliminate self-matches effect in approximate entropy. Sample entropy [8] has the advantage of being less dependent on time series length and showing relative consistency over a broader range of parameter values. In the 2000s, the multiscale entropy (MSE) was promoted by Costa et al. [10] to represent the complexity of a signal in different time scales. MSE relies on the computation of the sample entropy over a range of scales. In the MSE algorithm, coarse-grained time series, which represent the system dynamics on different scales, are analyzed with the sample entropy algorithm. For scale one, the MSE value of a time series corresponds to the sample entropy value.

## 3. Results

In the studies resulted from our database search, various asthma associated physiological signals have been collected and analyzed. The signals we studied include airflow, heart rate variability, center of pressure, respiratory impedance, airway resistance, respiratory sound, etc. (Table 2).

**Table 2.** Studies on Entropy Comparing Healthy subjects and Asthma Patients.

| Physiologic Signals | Study (Year) | Study Type | Entropy Method | Location | Number of Subjects | Age in Years as Mean ± SD or Range | Gender Ratio (M/F) | Pulmonary Function | Entropy Result | AUC |
|---|---|---|---|---|---|---|---|---|---|---|
| Airflow | Veiga et al., 2010 | Observational | ApEn | Brazil | Control 5, NE 5, Mild 5, Moderate 6, Severe 5 | Control 47.6 ± 19.7, NE 33.2 ± 8.5, Mild 49.2 ± 14.7, Moderate 54.3 ± 7.8, Severe 61.4 ± 6.7 | N/A | FVC, FEV$_1$, FEF$_{25\text{-}75\%}$, FEV$_1$/FVC, FEF/FVC | lower in asthmatic patients | No |
| | Veiga et al., 2011 | Observational | ApEn | Brazil | Control 11, NE 11, Mild 14, Moderate 14, Severe 12 | Control 54.4 ± 15.1, NE 34.9 ± 10.3, Mild 51.1 ± 13.5, Moderate 54.2 ± 10.7, Severe 60.5 ± 12.5 | N/A | FVC, FEV$_1$, FEF$_{25\text{-}75\%}$, FEV$_1$/FVC, FEF/FVC | lower in asthmatic patients | Yes |
| | Raoufy et al., 2016 | Observational | SampEn | Iran | Control 10, CAA 10, UAA 10, UNAA 10 | Control 27.6 ± 5.3, CAA 30.8 ± 9.8, UAA 31.1 ± 7.2, UNAA 32.7 ± 8.1 | N/A | N/A | lower in asthmatic subjects | Yes |
| HRV | Garcia-Araujo et al., 2014 | Observational | ApEn SampEn Shannon | Brazil | Healthy 10, Asthma 14 | Healthy 31 ± 8.7, Asthma 28 ± 8.5 | Healthy: 10/0, Asthma: 11/3 | FEV$_1$, FVC, FEV$_1$/FVC, VO$_2$ | lower in asthmatic patients during respiratory sinus arrhythmia maneuver | No |
| Center of pressure | Kuznetsov et al., 2014 | Observational | SampEn | USA | Healthy 18, Asthma 21 | Healthy 9.87 ± 2.77, Asthma 20.04 ± 1.85 | Healthy: 3/15, Asthma: 6/15 | N/A | lower in asthmatic patients | No |
| | Jin et al. 2008 | Observational | SampEn | Singapore | Control 7, Asthma 7 | N/A | N/A | N/A | SampEn is effective for wheeze detection. | No |
| Respiratory Sound | Aydore et al., 2009 | Retrospective | Renyi | USA | 7 (COPD & asthma) | 50 ± 17 | 4/3 | N/A | The Renyi entropy of wheeze signal has a uniform distribution | No |
| | Mondal et al., 2014 | Retrospective | SampEn | India | Normal 10, Abnormal 20 | N/A | N/A | N/A | higher in asthmatic subjects | No |
| Respiratory Impedance | Veiga et al., 2012 | Observational | ApEn | Brazil | Control 12, NE 12, Mild 20, Moderate 18, Severe 12 | Control 52.7 ± 16.4, NE 35.2 ± 9.9, Mild 51.8 ± 13.8, Moderate 53.2 ± 14.2, Severe 60.5 ± 12.5 | N/A | FVC, FEV$_1$, FEF$_{25\text{-}75\%}$, FEV$_1$/FVC, FEF/FVC | higher in asthmatic patients | Yes |
| Airway Resistance | Gonem et al., 2012 | Observational | SampEn | UK | Control: 30, GINA4: 33, GINA5: 33 | Control: 47.0 ± 2.2, GINA4: 51.0 ± 2.3, GINA5: 56.5 ± 1.9 | Control: 12/18, GINA4: 16/17, GINA5: 15/18 | FEV$_1$, FEV$_1$/FVC | higher in asthmatic patients | No |
| | Umar et al., 2010 | Observational | SampEn | UK | Control: 27, Asthma: 66 | Control: 54.1 ± 1.4, Asthma: 48.4 ± 2.2 | Control: 9/18, Asthma: 31/35 | FEV$_1$ | higher in asthmatic patients | No |

**Abbreviations:** NE: normal to exam; ILD: interstitial lung disease. FVC: Forced Vital Capacity; FEV$_1$: Forced Expiratory Volume for the first second; FEF: Forced Expiratory Flow FEF$_{25,75\%}$: FEF between 25% and 75%; CAA: controlled atopic asthma; UAA: uncontrolled atopic asthma; UNAA: uncontrolled non-atopic asthma; VO$_2$: maximal oxygen consumption; GINA: Global Initiative for Asthma; AUC: area under the Receiver Operating Characteristic curve of (ROC).

## 3.1. Airflow

The pathophysiology of asthma is characterized by chronic airway obstruction which gradually affects the entire tracheobronchial tree. Therefore, asthma patients may have altered airflow pattern, control of respiration, and entropy in respiratory dynamics [25]. There are studies aiming to characterize the complexity of respiratory patterns using nonlinear dynamical analysis [32–34].

Veiga et al. [21] conducted an observational clinical trial on the relationship between increased airway obstruction and changes of ApEn in the airflow dynamics of asthmatic patients. Totally, 26 subjects were involved in this study: 5 healthy subjects compared with 21 asthmatics patients with different levels of airway obstruction (5 normal spirometry exam (NE), 5 mild, 6 moderate, and 5 severe). It was found that there was a significant decrease in asthmatic patients in ApEn of airflow ($p < 0.002$) which was significantly correlated with the airway obstruction indices of Forced Expiratory Volume in the first second (FEV1) ($R = 0.60$; $p < 0.001$). Viega et al. then conducted an enlarged study [35] with 62 subjects: 11 healthy controls and 51 asthmatic patients with different levels of airway obstruction (11 NE, 14 mild, 14 moderate, and 12 severe). Significant decrease of ApEn was observed in asthma groups ($p < 0.02$) which was significantly correlated with FEV1 (%) ($R = 0.31$; $p = 0.013$). Receiver Operating Characteristic (ROC) curve was plotted, which shows that the area under the curve (AUC) for ApEn of airflow reached acceptable values for clinical use in discriminating healthy subjects from patients with mild, moderate, and severe airway obstruction (AUC > 0.8).

Raoufy et al. [36] used four nonlinear analysis methods (sample entropy, cross-sample entropy, largest Lyapunov exponents, and detrended fluctuation analysis) to compare the respiratory complexity pattern in asthmatic patients with that of healthy subjects. Forty subjects were enrolled in this study: 10 healthy subjects and 30 patients with different types of asthma (controlled atopic asthma (CAA), uncontrolled atopic asthma (UAA), and uncontrolled non-atopic asthma (UNAA), with 10 patients for each group). The metrics of inter-breath interval (IBI) and lung volume (LV) were continuously recorded and analyzed using nonlinear analysis methods. Significant decrease of SampEn in IBI ($p < 0.001$) and LV ($p = 0.002$) was observed in the asthma group than in the healthy control, while no significant difference was found in SampEn of LV between the patients in UNAA group and healthy subjects. Cross-sample entropy between IBI and LV was further calculated, which showed increased synchronization in UAA and UNAA groups ($p < 0.001$). ROC curve was drawn and SampEn showed a high clinical potential as a complexity index in differentiating asthmatic patients from healthy people (AUC = 0.95, 95% CI: 0.89–1.02).

## 3.2. Heart Rate Variability

Cardiac function is often highly influenced by breath difficulty in asthma patients. The increased obstruction during respiration gradually causes imbalance in the autonomic regularity of heart rhythm. One clinic index used to measure the variation of cardiac function is heart rate variability (HRV) which assesses the variation in the R-wave intervals.

Garcia-Araujo et al. [19] studied the HRV in the supine, seated positions and during the respiratory sinus arrhythmia maneuver (M-RSA) in 14 asthmatic and 10 healthy subjects. Nonlinear analysis methods including ApEn, SampEn, and Shannon entropy were used to evaluate the complexity/irregularity of HRV. It was found that, in the asthma group, the patients had a higher SampEn in the supine position than in the seated position ($p < 0.05$). During the M-RSA, a significantly lower value ($p < 0.05$) of ApEn was observed in asthma group than that in control group. The correlation between airway obstruction and HRV entropies also showed potential as biological markers of cardiovascular function impairment due to asthma.

## 3.3. Center of Pressure

Long-term asthma may lead to alteration in motor function and decrease in postural control. Postural control is clinically evaluated by the magnitude of variability of the center of pressure (COP) during standing in different conditions. Kuznetsov et al. [37] conducted research to investigate the ability of postural control in 21 young asthma patients and 18 age-matched control participants. The results showed no significant difference in the task completion between asthma and control group, while SampEn of $COP_{AP}$ (anterior-posterior COP data, collected by a Bertec force plate) was lower in the asthma group ($p = 0.04$).

### 3.4. Respiratory Sound

The respiratory sound signals of asthmatic patients carry important information might be used for discriminating different disease stages. The complexities of respiratory sound signals are higher in patients with lung function impairment than in healthy people because of the existence of additional auxiliary signals in unhealthy lungs. However, the complexity of respiratory sound dynamics makes it difficult to access the disease status using the baseline entropy measures. Recently, some novel algorithms have been proposed to improve the accuracy and sensitivity in distinguishing different types of respiratory sounds.

Jin et al. [38] conducted an observational study to test a novel statistical algorithm. It uses SampEn histograms of the respiratory sound signals to identify wheeze segments, a type of lung sound featured with periodic waveforms at a dominant frequency above 100 Hz and with a duration of over 100 ms. The sound signal recordings were collected from seven healthy and seven asthmatic subjects as well as from two lung sounds databases. Sample entropy of the sound data was calculated followed by mapping a histogram of mean distortion of the SampEn. This algorithm had an overall accuracy of 97.9% when detecting high-intensity wheezes and an accuracy of 85.3% when identifying low-intensity wheezes.

Aydore et al. [24] tested their algorithm for dividing wheeze and non-wheeze segments within respiratory sounds. The respiratory sound signal recoding from patients with asthma and COPD was collected in hospital's database. The sound signals were divided into wheeze and non-wheeze segments based on the calculation of their four features: kurtosis, Renyi entropy, f50/f90 ratio and mean-crossing irregularity. Then, all data were projected into four-dimensional feature space applying Fisher Discriminant Analysis and then was tested using Neyman–Pearson hypothesis. In the training dataset, their method had a final classification rate of 95.1% and, in the testing dataset, leave-one-out approach of the method resulted a correction rate of 93.5%.

Mondal proposed and tested a new classifier consisting of four statistical parameters: kurtosis, skewness, lacunarity, and sample entropy [24,39]. The experimental sound data, which was collected from various resources, contained 120 cycles of recordings from 10 normal and 20 abnormal individuals. The pathological problems being involved in this experiment contained asthma, COPD, and interstitial lung disease (ILD). Extreme learning machine and support vector machine networks were used to evaluate the efficiency of this method on discriminating abnormal sound signals from normal ones. The results, which were five-fold cross-validated, showed that the accuracy, sensitivity and specificity of this algorithm were 92.86%, 86.30%, and 86.90%, respectively.

### 3.5. Respiratory Impedance and Airway Resistance

The increase in the airflow obstruction in patients with asthma causes modification of system resistance during respiration. The forced oscillation technique (FOT) is the most commonly used method in clinic to collect physiology metrics of system impedance in respiratory tracts, such as airway resistance, airway reactance, and respiratory impendence (Zrs) [40]. It should be noted that there is relationship between airway resistance and respiratory impedance as described below. FOT applies external force oscillations on the spontaneous breathing, measures the pressure–flow patterns and determines lung mechanical parameters. Que et al. revealed that when the force oscillation has a frequency near 5 Hz, the impedance has a resonance and the magnitude of system impedance is approximately equal to the airway resistance [4]. Impulse oscillometry is a kind of FOT which utilizes impulse-shaped external pressure oscillation to estimate the airway resistance or the real part of respiratory impedance. The detected resistance is actually a component of the pressure drop over the entire respiratory system in phase with flow and normalized with flow at certain frequencies. Because the chest wall also has resistance, technically impulse oscillometry measures total respiratory resistance instead of airway resistance. However, when oscillation frequency is above 5 Hz, most of the detected resistance comes from the airways.

Veiga et al. [22] conducted a trial to study the relationship among the severity of airway obstruction in asthmatic patients, their approximate entropy of respiratory impedance (ApEnZrs) and entropy of the density of respiratory impedance recurrence period (RPDEnZrs). Totally 74 subjects were enrolled including 12 healthy subjects and 62 asthmatic patients with no (NE, $n = 12$), mild ($n = 20$), moderate ($n = 18$), and severe ($n = 12$) airway obstruction. Respiratory impedance was measured at 5 Hz using forced oscillation technique. The resulting signals of pressure (P) and airflow (Q) were used to calculate the respiratory impedance (Zrs = P/Q). It was found that both ApEnZrs and RPDEnZrs decreased significantly ($p < 0.002$ and $p < 0.00001$, respectively) with the increase of airway obstruction. The association was accessed using Pearson's correlation which showed that ApEnZrs and RPDEnZrs presented inverse correlation with Zrs ($p < 0.0001$). ROC curve indicates that ApEnZrs presents AUCs < 0.80 in all the study groups while RPDEnZrs reached acceptable value for diagnostic use (AUCs > 0.80).

Gonem et al. [20] studied the pattern of SampEn in airway resistance of asthmatic patients and its correlation with the frequency of asthma symptom exacerbation. This study involved 66 patients with severe asthma (33 on Global Initiative for Asthma (GINA) treatment step 4 and 33 on GINA treatment step 5) and 30 healthy people with matched demographics. A Jaeger MasterScreen Impulse Osillometry (IOS) system was used to record airway resistance and reactance in a frequency range of 5–35 Hz. Data were collected both at baseline and following bronchodilator administration. The results showed that, compared with healthy subjects, asthmatic patients had increased SampEn of airway impedance. For example, median SampEn of reactance area at baseline was 0.42 in healthy group, compared to 1.05 in GINA 4 group and 1.19 in GINA 5 group ($p < 0.0001$). The results also showed that, among all the 6 parameters studied in this trial, SampEn of R5–R20 (resistance at 5 Hz minus that at 20 Hz, a measure of the diversity of bronchial tree obstruction) was the only variable independently associated with frequent exacerbations (defined as two or more exacerbations in the previous year) ($p = 0.016$, odds ratio = 3.23:0.1).

Umar et al. [41] investigated entropy changes in airway resistance of 66 patients with severe asthma (GINA Stage 4/5) and 27 controls. IOS system was used to measure the airway resistance in the 5–35 Hz range, both at baseline and after salbutamol inhalation. The results showed that the SampEn of airway resistance in severe asthma group increased significantly compared to control group ($p = 0.007$ and 0.009 at baseline and post bronchodilation, respectively), and was correlated significantly with the exacerbation frequency (r score = 0.3 both at baseline and post bronchodilation).

### 3.6. Summary

For the airflow, respiratory system impedance, heart rate, and center of pressure, the entropy of these signals are lower in an asthmatic patient when compared with a healthy subject. The loss of entropy in airflow, HRV, and respiratory impedance associate with the increase in airway obstruction. ApEn and SampEn of Airflow, and RPDEnZrs have shown acceptable potential as clinical indices when evaluated using the ROC curve (AUC area > 0.8) (Figure 2). For the respiratory sound and airway resistance and reactance, the entropy increases in asthmatic subjects. This general result discussed here is illustrated in Figure 3.

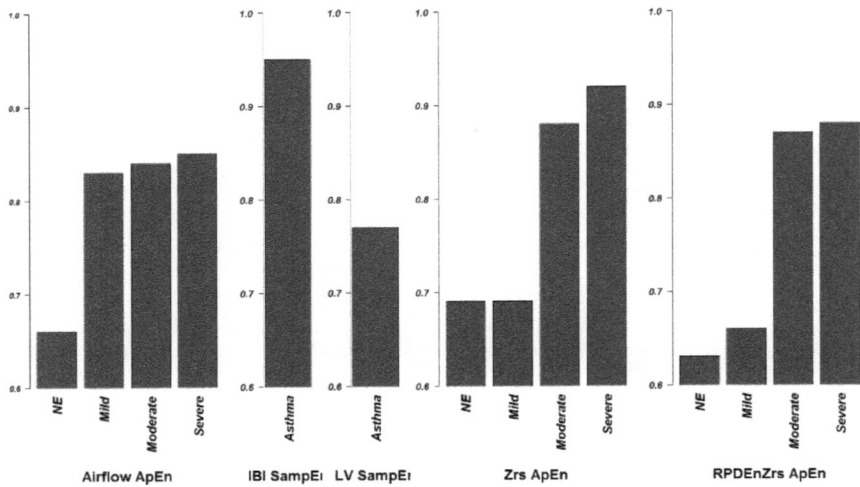

**Figure 2.** AUC area of Receiver Operating Characteristic curve. AUCs of some trials were calculated to evaluate the diagnostic ability of entropy for the severity of asthma. An index has an AUC value over 0.8 is considered good enough, and a value over 0.9 is considered excellent. SampEn of IBI has an excellent performance in distinguishing asthma patients from healthy people, ApEnZrs also performs excellently in distinguishing severe asthma patients. Besides, the ApEn of airflow has a good performance when distinguishing healthy subjects from patient with mild, moderate or severe asthma, ApEnZrs has a good performance in identifying moderate asthma patient, RPDEnZrs also performs well in distinguishing patients with moderate and severe asthma.

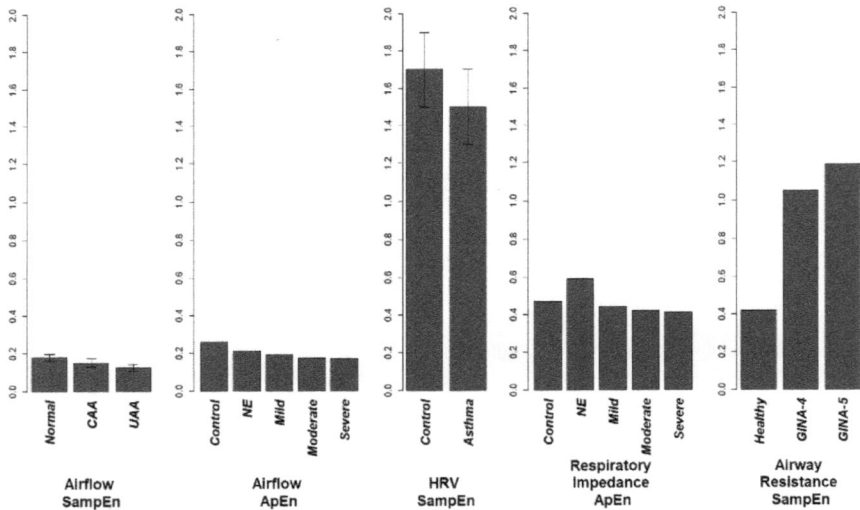

**Figure 3.** Complexity change in asthma. Entropy results are extracted and drawn to the same scale. Entropies of airflow, HRV and respiratory system impedance decrease associated with the disease progression. On the contrary, SampEn of airway impedance had a 2–3-fold increase in severe asthma patients. This contradiction might be explained by the way in which airway resistance is measured.

## 4. Discussion

With the rapid development of large data storage technologies and mobile network technologies, continuous monitoring of physiological signals is becoming increasingly important in the diagnosis and treatment of asthmatic patients [42,43]. Based on the continuous monitoring data, scientists hope to establish novel clinical indices for diagnosis and prognosis of asthma. Different from existing clinical examinations, these new indices will be able to classify patients into more precise disease types, so that more individualized treatment can be prescribed. Furthermore, the trends of some of these indices fluctuate prior to that of asthma symptoms, so that prevention can be addressed to avoid possible exacerbation.

Nonlinear analysis methods have been introduced to study the characteristics of these continuous dynamic data [39,40]. Recently, several studies have been conducted to explore the regularity of physiological signals in asthmatic patients [21,35–37]. To explore the pattern of complexity changes in asthma related physiological signals, we conducted this systematic review on available literature articles.

The quantified complexity of physiological signals obtained by entropy analysis provided novel perspective on the changes in asthma pathophysiology. Moreover, this review may contribute to the clinical evaluation of asthmatic patients and prediction of disease progression. From the summarized results, we can see that the pattern of entropy change in asthma may differs in different physiological signals. The entropy of physiological signals of HRV, airflow, center of pressure, and respiratory system impendence (Zrs = P/Q) are lower in a patient than in a healthy person, whereas the entropies of respiratory sound, airway resistance and reactance are higher in patients than in healthy people (Table 2).

The patterns of entropy in the physiological signals of airflow, HRV, and respiratory system impendence (Zrs) are similar to the results in previous studies on cardiovascular and COPD patients. This can be well explained by the widely accepted hypothesis proposed by Goldberger [44], who considered complexity as a kind of variability that "generates scale-invariance (self-similarity) and long-range organization", said that the disease states are correspondingly characterized by a "loss of complexity". The physiological signals of airway resistance and reactance, although detected in a similar method as that for respiratory system resistance, had an opposite patterns of entropy change. The reason for this inconsistency is currently unclear, but there may be several possible contributing factors. Firs, the method of impulse oscillometry in the published study is not as reliable as the standard forced oscillation technique using sinusoids. Second, the airway resistance is calculated differently from the impedance of entire respiratory system. As to respiratory sounds, the higher complexity in asthmatic subjects may be explained by the inclusion of sound signals which are more complex or heterogeneous in nature, such as wheeze or crackle. These highly irregular sound signals are products of the impaired respiratory tracts associated with the asthma progression [39].

ROC analysis is performed in some of these trials. The area under ROC is an index commonly used to evaluate whether the sensitivity and specificity of a new index are high enough to ensure reasonable type 1 and 2 error rates. Sample entropy of IBI has the highest AUC of 0.95 when comparing asthma and healthy subjects, however it did not have a high enough score when discriminating controlled from non-controlled asthma patients, or non-atopic from atopic asthma (AUC < 0.8). Approximate entropy of airflow, ApEnZrs, and RPDEnZrs had acceptable AUC values when comparing healthy subjects to asthma patient with moderate or severe airway obstruction. ApEn of airflow had an AUC > 0.8 when discriminating patient with mild airway obstruction to healthy people. None of the indices have an acceptable AUC value when comparing healthy subjects to asthmatic patient with normal spirometry exam. This is reasonable because, by definition, NE patients have normal lung function test; thus, the complexity of their biological dynamics is still unaffected.

Although the entropy of respiratory sounds is generally accepted to be higher than that of a healthy person, the characteristics of the complicated sound waveforms make it difficult to identify abnormal sound signals from normal ones only based on the calculation of single complexity feature such as entropy. Two of the newly developed machine learning algorithms are based on entropy and other nonlinear analysis methods and have demonstrated high ability to discriminate variant types of sound signals with acceptable sensitivity and specificity. More algorithms such as neural network has been developed with outstanding performance in data analysis of various biological signals, such as detection of sleep apnea syndrome [45], arrhythmias [46] and sputum [47,48], and classification and evaluation of EEG waves [49]. Therefore, we suggest future investigation for establishing new diagnostic indices might be conducted with the help of novel deep learning algorithms.

Our review demonstrates the potential for entropy as a novel index of individualized diagnosis and prognosis prediction for asthma severity. However, before being fully developed as a valid tool in clinical practice, more investigations need to be done. First, the overall pattern of entropy changes should be explored in different statuses of diseases as we did in this review. There are other types of entropies, such as MSE and phase entropy computed using higher order spectra. We may need to explore the utility of these types of entropy in asthma related physiology signals. Second, analysis methods of entropy should be further optimized for different physiological signals. Per our review, practicable complexity analysis method has not been well developed for respiratory sound signals. For many signals, entropy results did not reach acceptable AUC values in every study group. Furthermore, the analytic results of entropy vary due to individual differences. Therefore, standardized value ranges of healthy subjects based on their vital biological information (sex, age, height, weight, etc.) are required as the foundation of the discrimination and classification of disease status. Besides, to evaluate the predictive power of entropies, we suggest longitudinal studies to reveal long-term time relations between the change of entropies and that of patients' conditions.

In this paper, we systematically reviewed existing studies on entropy changes in asthma associated physiology signals and discussed their possible role in facilitating individualized medical sciences. Due to the limitation of our inquiry, certain articles might have been missed if they: (1) were published very recently; (2) were not included in the databases we searched; or (3) were written in a language other than English.

## 5. Conclusions

We reviewed existing research articles and summarized the entropy changes of physiology metrics associated with asthma. We found a significant correlation between complexity changes of physiological signals of asthmatic patients and the severity of asthma. For most signals we studied, such as airflow, HRV, COP and respiratory system impedance, their entropies decrease with the exacerbation of asthma. For respiratory sounds and airway resistance, their entropies increase in asthma patient. This might be explained by the unstableness of local respiratory tract and the increase of breath difficulty. ROC analysis demonstrated that entropies of several physiology signals have excellent sensitivity and specificity (AUC > 0.8 or even 0.9) when they were used to discriminate different severities of asthma. For future work, we suggest more clinical trials should be conducted to probe the entropy analysis for diagnosis and management of asthma. Besides, we also advise researchers to pay attention to the application of deep learning algorithms which has already shown extraordinary capability in physiological data analysis. The summarized result obtained through this review should give valuable guidance for further research on complexity patterns of physiological signals in asthma and provide basis for the use of entropy as a clinic indicator aiming for more individualized diagnosis and patient care.

**Author Contributions:** All authors conceived of the idea. S.S. and Y.J. reviewed the articles and analyzed data. S.S., Y.J. and X.D.Z. drafted the manuscript. All authors reviewed and commented on the manuscript and approved the final draft.

**Acknowledgments:** This work was supported by University of Macau through Research Grants SRG2016-00083-FHS and FHS-CRDA-029-002-2017 and FDCT (Fundo para o Desenvolvimento das Ciências e da Tecnologia de Macau) grant FDCT/131/2016/A3.

**Conflicts of Interest:** The authors declare no conflict of interest.

# References

1. Pocket Guide for Asthma Management and Prevention. Available online: http://ginasthma.org/wp-content/uploads/2016/01/GINA_Pocket_2015.pdf (accessed on 12 March 2017).

2. World Health Organization. Asthma. Available online: http://www.who.int/mediacentre/factsheets/fs307/en/ (accessed on 12 March 2017).

3. Pijnenburg, M.W.; Baraldi, E.; Brand, P.L.; Carlsen, K.H.; Eber, E.; Frischer, T.; Hedlin, G.; Kulkarni, N.; Lex, C.; Makela, M.J.; et al. Monitoring asthma in children. *Eur. Respir. J.* **2015**, *45*, 906–925. [CrossRef] [PubMed]

4. Que, C.L.; Kenyon, C.M.; Olivenstein, R.; Macklem, P.T.; Maksym, G.N. Homeokinesis and short-term variability of human airway caliber. *J. Appl. Physiol.* **2001**, *91*, 1131–1141. [CrossRef] [PubMed]

5. Frey, U.; Brodbeck, T.; Majumdar, A.; Taylor, D.R.; Town, G.I.; Silverman, M.; Suki, B. Risk of severe asthma episodes predicted from fluctuation analysis of airway function. *Nature* **2005**, *438*, 667–670. [CrossRef] [PubMed]

6. Pincus, S.M. Approximate entropy as a measure of system complexity. *Proc. Natl. Acad. Sci. USA* **1991**, *88*, 2297–2301. [CrossRef] [PubMed]

7. Pincus, S.; Singer, B.H. Randomness and degrees of irregularity. *Proc. Natl. Acad. Sci. USA* **1996**, *93*, 2083–2088. [CrossRef] [PubMed]

8. Richman, J.S.; Moorman, J.R. Physiological time-series analysis using approximate entropy and sample entropy. *Am. J. Physiol. Heart Circ. Physiol.* **2000**, *278*, H2039–H2049. [CrossRef] [PubMed]

9. Costa, M.D.; Goldberger, A.L. Generalized multiscale entropy analysis: Application to quantifying the complex volatility of human heartbeat time series. *Entropy* **2015**, *17*, 1197–1203. [CrossRef] [PubMed]

10. Costa, M.; Goldberger, A.L.; Peng, C.K. Multiscale entropy analysis of biological signals. *Phys. Rev. E Stat. Nonlinear Soft Matter Phys.* **2005**, *71*, 021906. [CrossRef] [PubMed]

11. Jin, Y.; Chen, C.; Cao, Z.; Sun, B.; Lo, I.L.; Liu, T.-M.; Zheng, J.; Sun, S.; Shi, Y.; Zhang, X.D. Entropy change of biological dynamics in COPD. *Int. J. Chron. Obstr. Pulm. Dis.* **2017**, *12*, 2997–3005. [CrossRef] [PubMed]

12. Zhang, X.D.; Zhang, Z.; Wang, D. CGManalyzer: An R package for analyzing continuous glucose monitoring studies. *Bioinformatics* **2018**, *1*, 3. [CrossRef]

13. Sharma, R.; Pachori, R.; Acharya, U. An integrated index for the identification of focal electroencephalogram signals using discrete wavelet transform and entropy measures. *Entropy* **2015**, *17*, 5218–5240. [CrossRef]

14. Volterrani, M.; Scalvini, S.; Mazzuero, G.; Lanfranchi, P.; Colombo, R.; Clark, A.L.; Levi, G. Decreased heart rate variability in patients with chronic obstructive pulmonary disease. *Chest* **1994**, *106*, 1432–1437. [CrossRef] [PubMed]

15. Corbo, G.M.; Inchingolo, R.; Sgueglia, G.A.; Lanza, G.; Valente, S. C-reactive protein, lung hyperinflation and heart rate variability in chronic obstructive pulmonary disease—A pilot study. *COPD J. Chron. Obstr. Pulm. Dis.* **2013**, *10*, 200–207. [CrossRef] [PubMed]

16. Chen, C.; Jin, Y.; Lo, I.L.; Zhao, H.; Sun, B.; Zhao, Q.; Zheng, J.; Zhang, X.D. Complexity change in cardiovascular disease. *Int. J. Biol. Sci.* **2017**, *13*, 1320–1328. [CrossRef] [PubMed]

17. Da Luz Goulart, C.; Simon, J.C.; De Borba Schneiders, P.; San Martin, E.A.; Cabiddu, R.; Borghi-Silva, A.; Trimer, R.; Da Silva, A.L.G. Respiratory muscle strength effect on linear and nonlinear heart rate variability parameters in COPD patients. *Int. J. Chron. Obstr. Pulm. Dis.* **2016**, *11*, 1671–1677. [CrossRef] [PubMed]

18. Dames, K.K.; Lopes, A.J.; de Melo, P.L. Airflow pattern complexity during resting breathing in patients with COPD: Effect of airway obstruction. *Respir. Physiol. Neurobiol.* **2014**, *192*, 39–47. [CrossRef] [PubMed]

19. Garcia-Araujo, A.S.; Pires Di Lorenzo, V.A.; Labadessa, I.G.; Jurgensen, S.P.; Di Thommazo-Luporini, L.; Garbim, C.L.; Borghi-Silva, A. Increased sympathetic modulation and decreased response of the heart rate variability in controlled asthma. *J. Asthma* **2015**, *52*, 246–253. [CrossRef] [PubMed]

20. Gonem, S.; Umar, I.; Burke, D.; Desai, D.; Corkill, S.; Owers-Bradley, J.; Brightling, C.E.; Siddiqui, S. Airway impedance entropy and exacerbations in severe asthma. *Eur. Respir. J.* **2012**, *40*, 1156–1163. [CrossRef] [PubMed]

21. Veiga, J.; Faria, R.C.; Esteves, G.P.; Lopes, A.J.; Jansen, J.M.; Melo, P.L. Approximate entropy as a measure of the airflow pattern complexity in asthma. In Proceedings of the 2010 Annual International Conference of the IEEE Engineering in Medicine and Biology Society (EMBC), Buenos Aires, Argentina, 31 August–4 September 2010; Volume 2010, pp. 2463–2466.

22. Veiga, J.; Lopes, A.J.; Jansen, J.M.; Melo, P.L. Fluctuation analysis of respiratory impedance waveform in asthmatic patients: Effect of airway obstruction. *Med. Biol. Eng. Comput.* **2012**, *50*, 1249–1259. [CrossRef] [PubMed]

23. Veremchuk, L.V.; Yankova, V.I.; Vitkina, T.I.; Nazarenko, A.V.; Golokhvast, K.S. Urban air pollution, climate and its impact on asthma morbidity. *Asian Pac. J Trop. Biomed.* **2016**, *6*, 76–79. [CrossRef]

24. Aydore, S.; Sen, I.; Kahya, Y.P.; Mihcak, M.K. Classification of respiratory signals by linear analysis. In Proceedings of the Annual International Conference of the IEEE Engineering in Medicine and Biology Society, EMBC 2009, Minneapolis, MN, USA, 3–6 Septtember 2009; pp. 2617–2620.

25. Jadad, A.R.; Moore, R.A.; Carroll, D.; Jenkinson, C.; Reynolds, D.J.M.; Gavaghan, D.J.; McQuay, H.J. Assessing the quality of reports of randomized clinical trials: Is blinding necessary? *Control. Clin. Trials* **1996**, *17*, 1–12. [CrossRef]

26. Chakrabarti, C.G.; De, K. Boltzmann entropy: Generalization and applications. *J. Biol. Phys.* **1997**, *23*, 163–170. [CrossRef] [PubMed]

27. Shannon, C.E. Communication theory of secrecy systems. *Bell Labs Tech. J.* **1949**, *28*, 656–715. [CrossRef]

28. Grassberger, P.; Procaccia, I. Estimation of the kolmogorov entropy from a chaotic signal. *Phys. Rev. A* **1983**, *28*, 2591. [CrossRef]

29. Gonzalez Andino, S.L.; Grave de Peralta Menendez, R.; Thut, G.; Spinelli, L.; Blanke, O.; Michel, C.M.; Seeck, M.; Landis, T. Measuring the complexity of time series: An application to neurophysiological signals. *Hum. Brain Mapp.* **2000**, *11*, 46–57. [CrossRef]

30. Slomczynski, W.; Kwapien, J.; Zyczkowski, K. Entropy computing via integration over fractal measures. *Chaos* **2000**, *10*, 180–188. [CrossRef] [PubMed]

31. Shannon, C.E.; Weaver, W. The mathematical theory information. *Math. Gazette* **1949**, *97*, 170–180.

32. El-Khatib, M.F. A diagnostic software tool for determination of complexity in respiratory pattern parameters. *Comput. Biol. Med.* **2007**, *37*, 1522–1527. [CrossRef] [PubMed]

33. Bates, J.H.; Davis, G.S.; Majumdar, A.; Butnor, K.J.; Suki, B. Linking parenchymal disease progression to changes in lung mechanical function by percolation. *Am. J. Respir. Crit. Care Med.* **2007**, *176*, 617–623. [CrossRef] [PubMed]

34. Venegas, J.G.; Winkler, T.; Musch, G.; Melo, M.F.V.; Layfield, D.; Tgavalekos, N.; Fischman, A.J.; Callahan, R.J.; Bellani, G.; Harris, R.S. Self-organized patchiness in asthma as a prelude to catastrophic shifts. *Nature* **2005**, *434*, 777–782. [CrossRef] [PubMed]

35. Veiga, J.; Lopes, A.J.; Jansen, J.M.; Melo, P.L. Airflow pattern complexity and airway obstruction in asthma. *J. Appl. Physiol.* **2011**, *111*, 412–419. [CrossRef] [PubMed]

36. Raoufy, M.R.; Ghafari, T.; Darooei, R.; Nazari, M.; Mahdaviani, S.A.; Eslaminejad, A.R.; Almasnia, M.; Gharibzadeh, S.; Mani, A.R.; Hajizadeh, S. Classification of asthma based on nonlinear analysis of breathing pattern. *PLoS ONE* **2016**, *11*, e0147976. [CrossRef] [PubMed]

37. Kuznetsov, N.A.; Luberto, C.M.; Avallone, K.; Kraemer, K.; McLeish, A.C.; Riley, M.A. Characteristics of postural control among young adults with asthma. *J. Asthma* **2015**, *52*, 191–197. [CrossRef] [PubMed]

38. Jin, F.; Sattar, F.; Goh, D.Y. Automatic wheeze detection using histograms of sample entropy. In Proceedings of the 30th Annual International Conference of the IEEE Engineering in Medicine and Biology Society, EMBS 2008, Vancouver, BC, Canada, 20–25 August 2008; Volume 2008, pp. 1890–1893.

39. Mondal, A.; Bhattacharya, P.; Saha, G. Detection of lungs status using morphological complexities of respiratory sounds. *Sci. World J.* **2014**, *2014*, 182938. [CrossRef] [PubMed]

40. Oostveen, E.; MacLeod, D.; Lorino, H.; Farré, R.; Hantos, Z.; Desager, K.; Marchal, F. The forced oscillation technique in clinical practice: Methodology, recommendations and future developments. *Eur. Respir. J.* **2003**, *22*, 1026–1041. [CrossRef] [PubMed]

41. Umar, I.; Desai, D.; Corkill, S.; Shelley, M.; Singapuri, A.; Brightling, C.; Siddiqui, S. The use of impulse oscillometry (IOS) to study fractal scaling and sample entropy in airway resistance time series in severe asthma. *Thorax* **2010**, *65*, A129. [CrossRef]

42. Kvedar, J.C.; Fogel, A.L.; Elenko, E.; Zohar, D. Digital medicine's march on chronic disease. *Nat. Biotechnol.* **2016**, *34*, 239–246. [CrossRef] [PubMed]

43. Elenko, E.; Underwood, L.; Zohar, D. Defining digital medicine. *Nat. Biotechnol.* **2015**, *33*, 456–461. [CrossRef] [PubMed]

44. Goldberger, A.L. Fractal variability versus pathologic periodicity: Complexity loss and stereotypy in disease. *Perspect. Biol. Med.* **1997**, *40*, 543–561. [CrossRef] [PubMed]

45. Kaguara, A.; Myoung Nam, K.; Reddy, S. A Deep Neural Network Classifier for Diagnosing Sleep Apnea from ECG Data on Smartphones and Small Embedded Systems. Available online: https://www.researchgate.net/publication/273633242_A_deep_neural_network_classifier_for_diagnosing_sleep_apnea_from_ECG_data_on_smartphones_and_small_embedded_systems (accessed on 11 April 2018).

46. Acharya, U.R.; Fujita, H.; Lih, O.S.; Hagiwara, Y.; Tan, J.H.; Adam, M. Automated detection of arrhythmias using different intervals of tachycardia ECG segments with convolutional neural network. *Inf. Sci.* **2017**, *405*, 81–90. [CrossRef]

47. Niu, J.; Shi, Y.; Cai, M.; Cao, Z.; Wang, D.; Zhang, Z.; Zhang, X.D. Detection of sputum by interpreting the time-frequency distribution of respiratory sound signal using image processing techniques. *Bioinformatics* **2018**, *34*, 820–827. [CrossRef] [PubMed]

48. Shi, Y.; Wang, G.; Niu, J.; Zhang, Q.; Cai, M.; Sun, B.; Wang, D.; Xue, M.; Zhang, X.D. Classification of sputum sounds using artificial neural network and wavelet transform. *Int. J. Biol. Sci.*. in press. [CrossRef]

49. Li, K.; Pan, W.; Li, Y.; Jiang, Q.; Liu, G. A method to detect sleep apnea based on deep neural network and hidden markov model using single-lead ECG signal. *Neurocomputing* **2018**, *294*, 94–101. [CrossRef]

# entropy

MDPI

*Article*

# Application of Multiscale Entropy in Assessing Plantar Skin Blood Flow Dynamics in Diabetics with Peripheral Neuropathy

Fuyuan Liao [1], Gladys L. Y. Cheing [2], Weiyan Ren [3,4], Sanjiv Jain [5] and Yih-Kuen Jan [3,4,*]

[1] Department of Biomedical Engineering, Xi'an Technological University, Xi'an 710021, China; liaofuyuan1024@163.com
[2] Department of Rehabilitation Sciences, the Hong Kong Polytechnic University, Hong Kong 999077, China; gladys.cheing@polyu.edu.hk
[3] Rehabilitation Engineering Lab, Department of Kinesiology and Community Health, University of Illinois at Urbana-Champaign, 1206 South Fourth Street, MC-588, Champaign, IL 61820, USA; renweiyan03@163.com
[4] Beijing Advanced Innovation Center for Biomedical Engineering, School of Biological Science and Medical Engineering, Beihang University, Beijing 100083, China
[5] Department of Physical Medicine and Rehabilitation, Carle Hospital, Urbana, IL 61801, USA; sanjiv.jain@carle.com
* Correspondence: yjan@illinois.edu; Tel.: +1-217-300-7253

Received: 22 January 2018; Accepted: 12 February 2018; Published: 15 February 2018

**Abstract:** Diabetic foot ulcer (DFU) is a common complication of diabetes mellitus, while tissue ischemia caused by impaired vasodilatory response to plantar pressure is thought to be a major factor of the development of DFUs, which has been assessed using various measures of skin blood flow (SBF) in the time or frequency domain. These measures, however, are incapable of characterizing nonlinear dynamics of SBF, which is an indicator of pathologic alterations of microcirculation in the diabetic foot. This study recruited 18 type 2 diabetics with peripheral neuropathy and eight healthy controls. SBF at the first metatarsal head in response to locally applied pressure and heating was measured using laser Doppler flowmetry. A multiscale entropy algorithm was utilized to quantify the regularity degree of the SBF responses. The results showed that during reactive hyperemia and thermally induced biphasic response, the regularity degree of SBF in diabetics underwent only small changes compared to baseline and significantly differed from that in controls at multiple scales ($p < 0.05$). On the other hand, the transition of regularity degree of SBF in diabetics distinctively differed from that in controls ($p < 0.05$). These findings indicated that multiscale entropy could provide a more comprehensive assessment of impaired microvascular reactivity in the diabetic foot compared to other entropy measures based on only a single scale, which strengthens the use of plantar SBF dynamics to assess the risk for DFU.

**Keywords:** multiscale entropy; regularity; skin blood flow; diabetic foot ulcers

---

## 1. Introduction

Diabetic foot ulcer (DFU) is a common complication of diabetes mellitus [1,2] and a major cause of hospitalization and non-traumatic lower-extremity amputations among people with diabetes [3]. The yearly and lifetime incidences of DFU are estimated to be about 2% and 15–25%, respectively [1], and the amputation rates in diabetics were reported to be 10–30 times higher than in the non-diabetic population [4,5]. Since treatment of DFUs is challenging and its economic burden is high [1,2], prevention of DFUs is highly important and has been recognized as a priority of diabetes healthcare [2].

The formation and development of DFUs involve a number of risk factors, among which peripheral neuropathy and peripheral arterial disease are crucial factors [1,2]. Diabetic peripheral neuropathy induces a series of pathologic alterations in the foot such as a loss of protective sensation for detecting mechanical stresses and/or trauma, foot deformities may result in elevated plantar pressure, and dryness of the skin that contributes to skin breakdown [6]. These alterations increase the risk of trauma and subsequent ulceration [6]. Also, it has been found that the majority of foot ulcers involve tissue ischemia [3], which is thought to be a major factor of the development of DFUs [6]. In the diabetic foot, impaired vasodilatory response to repetitive plantar pressure during walking is a main cause of plantar tissue ischemia [7], while impaired vasodilatory response to elevated temperature of the foot can aggravate tissue ischemia, because elevated skin temperature increases the metabolic demands of local cells and tissues, thus requiring an increase in skin blood flow (SBF) to meet the metabolic demands. Therefore, quantification of SBF responses to loading pressure and thermal stresses may be a reasonable way to assess the risk of DFU [7,8].

Traditionally, SBF response to mechanical stress is quantified using time-domain parameters such as normalized mean blood flow (divided by basal blood flow) during mechanical stress and measures of hyperemia [7,9]; SBF response to local heating is quantified using normalized first peak, nadir, and second peak (divided by basal blood flow) [7,9]. Moreover, wavelet-based spectral analysis has been utilized to investigate the underlying mechanisms of the responses [7]. It has been found that blood flow oscillations (BFO) in the human skin contain six characteristic frequency components in the frequency interval 0.005–2 Hz [10,11]. Two components with higher frequencies are originated from cardiac activity (0.4–2 Hz) and respiration (0.15–0.4 Hz), respectively. The other four components are associated with the myogenic activity of vascular smooth muscle (0.05–0.15 Hz), the neurogenic activity of the vessel wall (0.02–0.05 Hz), nitric oxide-related endothelia activity (0.0095–2 Hz), and nitric oxide-independent endothelia activity (0.005–0.0095 Hz), respectively. Jan et al. [7] investigated SBF responses at the first metatarsal head of diabetics induced by pressure loading and local heating, and showed an attenuated myogenic component during reactive hyperemia and attenuated metabolic, neurogenic, and myogenic components in response to local heating compared to healthy controls.

Although time-domain parameters provide direct features of the SBF responses, and wavelet analysis provides a mean for characterizing the state of the regulatory mechanisms of SBF during the responses, they are unable to characterize the nonlinear features of BFO [12], which are the structural features of BFO, e.g., complexity and self-similarity, rather than the magnitude of variability. There is evidence that altered nonlinear properties of physiological signals are an indicator of pathologic changes in the physiologic system [8,13–15]. In our previous study [8], we utilized a modified sample entropy method [13] to quantify the regularity degree of SBF in diabetics, and showed promising results. However, because we used a fixed parameter, i.e., the time delay between neighboring data points of the sequences to be compared, we were unable to gain insight into how the regularity degree of BFO changed with time scales, which is likely associated with the homogeneity degree of the combination of characteristic frequency components. Therefore, the objective of the current study was to investigate the regularity degree of SBF responses at the first metatarsal head of diabetics induced by loading pressure and thermal stress at multiple scales and how regularity degree changed at various scales. We hypothesized that the transition of regularity degree of BFO could reflect microvascular dysfunction in diabetics.

## 2. Methods

### 2.1. Participants and Data Collection

Eighteen people with type 2 diabetes and peripheral neuropathy (13 men and 5 women) and eight healthy controls (four men and four women) were recruited into this study. The demographic data of participants and experimental protocols were presented in our previous publication [8]. Briefly, the diabetic subjects had a mean age (standard deviation, SD) of 48.5 (9.4) years, body mass index (BMI) of 28.3 (7.1) kg/m$^2$, duration of diabetes of 15.2 (5.1) years, and HbA$_{1c}$ level of 7.8 (0.9) %. Each of them suffered from peripheral neuropathy and had a history of foot ulcers. The healthy controls had a mean age

(SD) of 21.8 (2.4) years and BMI of 25.8 (3.3) kg/m$^2$. This study was approved by a university institutional review board (IRB #14707).

The experiments were conducted in a research laboratory with the room temperature being maintained at 24 ± 2 °C. Prior to any test, the subject was acclimated to the room temperature for at least 30 min. Then, the subject lay in a supine position and underwent two experiments. The first experiment was aimed to examine SBF response to locally applied pressure at the first metatarsal head. SBF and skin temperature were measured with a sampling rate of 32 Hz using a Laser Doppler flowmetry (PeriFlux 5001, Perimed, Ardmore, PA, USA), and a probe with heating function (Probe 415–242, Perimed). This protocol included a 10-min basal measurement, followed by a 3-min period during which a 300 mmHg pressure was applied to the probe via a computer-controlled indenter [7], and a 17-min recovery period. Figure 1A shows SBF responses in a diabetic subject and a healthy control; Figure 1C shows normalized SBF during the loading period and normalized peak hyperemia in two groups.

The second protocol was aimed to examine SBF at the first metatarsal head in response to local heating. This protocol included a 10-min basal measurement, a 30-min heating period during which the skin was heated to 42 °C in 2 min and the temperature was maintained at that level, followed by a recovery period lasting 10 min. Figure 1B shows SBF responses in a diabetic subject and a healthy control. This response was quantified using three indices: first peak (P1), nadir, and second peak (P2) divided by basal blood flow. Because SBF exhibited a plateau during the later period of heating (Figure 1B), the mean value of SBF during the last 10 min was defined as P2. Figure 1D shows the results of normalized P1, nadir, and P2 in two groups.

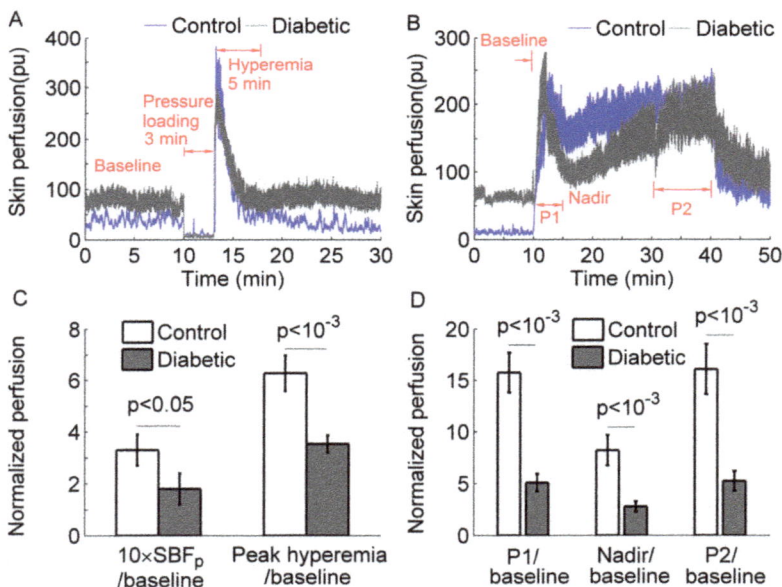

**Figure 1.** (**A,B**) Skin blood flow (SBF) responses to a loading pressure of 300 mmHg (**A**) and local heating (**B**) at the first metatarsal head of a diabetic subject and a healthy control. pu, perfusion unit. (**C**) Normalized SBF (divided by basal SBF) during the loading period and normalized peak hyperemia in two groups. (**D**) Normalized SBF during P1, nadir, and P2 in two groups. Data are represented as mean ± standard error. The differences in SBF between two groups were examined using Mann–Whitney U tests.

## 2.2. Sample Entropy and Its Derivatives

Sample entropy ($E_s(m,r,N)$) is defined as the negative natural logarithm of the conditional probability that two sequences of $m$ points within a tolerance $r$ remain within the tolerance at the next point [16]. A smaller (larger) value of $E_s$ indicates a higher degree of regularity (irregularity). It has been demonstrated that $E_s$ depends on the relationship between the frequency of the dominant oscillations and the sampling rate [13]. Oversampling may lead to misleading results, i.e., the obtained $E_s$ value does not reflect the regularity degree of the dominant oscillations [13]. To address this problem, we recently developed a modified sample entropy algorithm [13]. Its procedures are presented briefly as follows. Figure 2 illustrates a main procedure of the algorithm.

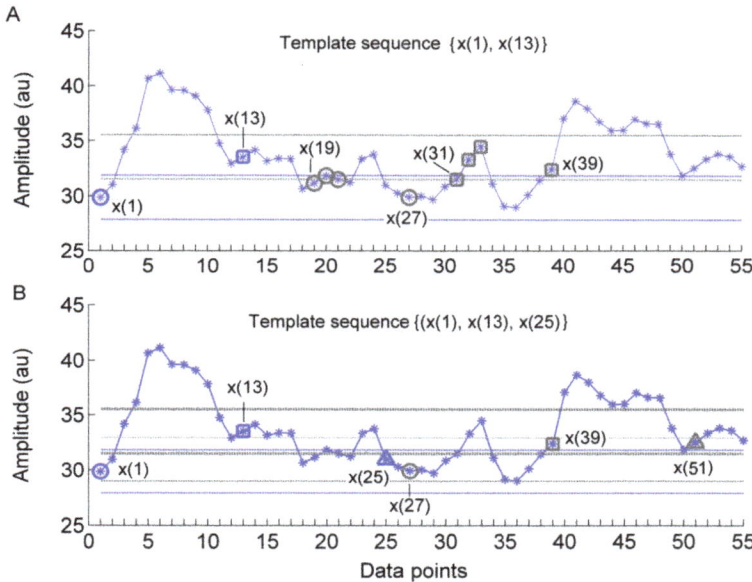

**Figure 2.** Illustration of a main procedure for calculating the modified sample entropy ($E_{MS}$) in the case of $m = 2$, $\tau = 12$, and $r = 0.2 \times SD$ (the standard deviation of the series). (**A**) For a $m$-component template sequence $\mathbf{x}_m^\tau(i) = \{x(1), x(13)\}$, i.e., $i = 1$, there are four sequences $\mathbf{x}_m^\tau(j) = \{x(19),x(31)\}$, $\{x(20),x(32)\}$, $\{x(21),x(33)\}$, and $\{x(27),x(39)\}$, i.e., $j = 19$, 20, 21, and 27, satisfying $d[\mathbf{x}_m^\tau(i), \mathbf{x}_m^\tau(j)] < r$, $|j - i| > \tau$. This procedure is repeated for the next template vector $\mathbf{x}_m^\tau(i) = \{x(2),x(14)\}$, i.e., $i = 2$, and so on. The dotted horizontal lines around data points x(1) and x(13) represent $x(1)\pm r$ and $x(13)\pm r$, respectively. (**B**) The above procedure is repeated for all $(m + 1)$-component sequences $\mathbf{x}_{m+1}^\tau(i)$, e.g., $\{x(1), x(13),x(25)\}$.

For a time series $\{x(i), i = 1, \ldots, N\}$, consider the $m$-point sequences:

$$\mathbf{x}_m^\tau(i) = \{x(i + k\tau), 0 \le k \le m - 1\}, 1 \le i \le N - m\tau, \tag{1}$$

where $\tau$ is a lag. The condition $1 \le i \le N - m\tau$ ensures that $\mathbf{x}_{m+1}^\tau(i)$ exits for $i = N - m\tau$. The distance between two sequences $\mathbf{x}_m^\tau(i)$ and $\mathbf{x}_m^\tau(j)$ is defined as

$$d[\mathbf{x}_m^\tau(i), \mathbf{x}_m^\tau(j)] = \max\{|x(i + k\tau) - x(j + k\tau)|, 0 \le k \le m - 1\}, |j - i| > \tau. \tag{2}$$

For a given sequence $\mathbf{x}_m^\tau(i)$, let $n_i$ be the number of $\mathbf{x}_m^\tau(j)$ satisfying $|j - i| > \tau$ and $n_i^m(r)$ the numbers of $\mathbf{x}_m^\tau(j)$ satisfying $d[\mathbf{x}_m^\tau(i), \mathbf{x}_m^\tau(j)] < r$, $|j - i| > \tau$, where $r$ is a tolerance, usually being set to be proportional to the SD of the time series. The constraint condition $|j - i| > \tau$ is aimed to reduce influence of the correlation on entropy estimation [13]. Thus, $C_i^m(r) = n_i^m(r)/n_i$ represents the probability that any sequence $\mathbf{x}_m^\tau(j)$ is within $r$ of $\mathbf{x}_m^\tau(i)$, and $C^m(r) = \sum_{i=1}^{N-m\tau} C_i^m/(N - m\tau)$ represents the probability that any two sequences $\mathbf{x}_m^\tau(i)$ and $\mathbf{x}_m^\tau(j)$ are within $r$. Likewise, $C^{m+1}(r)$ represents the probability that any two sequences $\mathbf{x}_{m+1}^\tau(i)$ and $\mathbf{x}_{m+1}^\tau(j)$ are within $r$. The modified sample entropy is defined as:

$$E_{ms}(m, r, \tau) = \lim_{N \to \infty} -\ln \frac{C^{m+1}(r)}{C^m(r)}, \tag{3}$$

which is estimated by:

$$E_{ms}(m, r, \tau, N) = -\ln \frac{C^{m+1}(r)}{C^m(r)}. \tag{4}$$

In our previous study [13], we tested $E_{ms}$ using simulated time series and SBF data. The results indicated that $E_{ms}$ yielded consistent values for various sampling rates, but $E_s$ cannot [13].

Another representative derivative of $E_s$ is the fuzzy entropy $E_f(m, n, \sigma, N)$ [17], which differs from $E_s$ in two aspects. First, when calculating the distance between two sequences $\mathbf{x}_m(i)$ and $\mathbf{x}_m(j)$, they are converted to have zero means. Second, the similarity between two sequences is quantified using an exponential function $\exp(-(d_{ij}^m)^n/\sigma)$, where $d_{ij}^m$ is the distance between them. It was reported that $E_f$ is more robust than $E_s$ when applying to short time series [17]. We tested the performance of $E_f$ and $E_s$ using sinusoidal signals with frequencies of 0.1, 0.3, and 1 Hz, respectively, and examined how they changed with increasing values of $r$ (or $\sigma$). The frequencies of the sinusoidal signals are roughly equal to the central frequencies of myogenic, respiratory, and cardiac components of BFO, respectively; their length is equal to that of SBF signals during the pressure loading period (3 min, 5760 points). The parameters $m = 2$, $n = 2$ were used. The later was selected using an approach recommended by the authors who proposed $E_f$ [17]. The results showed that $E_f$ yielded very small values for 0.3 Hz and especially 0.1 Hz sinusoidal signals (Figure 3B). This suggests that when applied to SBF signals, $E_f$ may be unable to reflect altered dynamics of the low-frequency components of BFO, e.g., metabolic (~0.01 Hz), neurogenic (~0.03 Hz), and myogenic (~0.1 Hz) components.

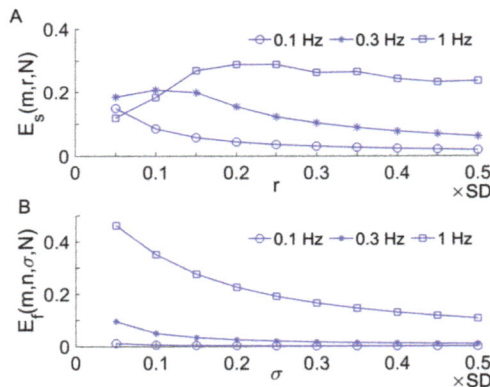

**Figure 3.** $E_s(m, r, N)$ (**A**) and $E_f(m, n, \sigma, N)$ (**B**) of 0.1, 0.3, and 1 Hz sinusoidal signals sampled at a rate of 32 Hz. The parameters $m = 2$, $n = 2$, and $N = 5760$ were used. The parameter $n = 2$ was selected using an approach recommended by the authors who proposed $E_f$ [17].

### 2.3. Multi-Scale Entropy

Despite $E_s$ being widely used to assess the complexity of time series, it is actually a measure of regularity, and this is the case for its various derivatives. Currently, complexity has not been well defined [18], and it is intuitively associated with "meaningful structural richness" [19]. A major problem of $E_s$ and its derivatives is that they yield the highest values for white noise, which is unpredictable but without structural complexity, and they may also yield higher values for physiological signals in health condition and lower values in pathological conditions [8,13]. In this context, several multiscale entropy (MSE) methods were introduced to quantify the regularity degree of a time series at multiple scales [20–22]. The first MSE method was proposed by Costa et al. [20], in which the original time series is divided into non-overlapping segments, and a new series is constructed using the average of each segment with its order being preserved. Then, $E_s$ is computed for each new series. This method, however, has been found to have several limitations. First, the procedure for constructing new series is similar to applying a low-pass filter to the original time series followed by a downsampling procedure. It has been found that the frequency response of the low-pass filter shows side lobes in the stop band, which lead to aliasing during the downsampling and thus produce artifacts [22]. Second, the SD of the new series likely decreases with increasing scales, whereas in the MSE algorithm, a constant tolerance (a constant proportion of the SD of the original time series) is used for all scales, resulting in decreasing entropy values with increasing scales. Finally, the length of the new series decreases rapidly with increasing scales, impeding reliable estimations of the entropy at large scales. The drawbacks of the MSE method and several improved algorithms have been discussed in Reference [23].

Recently, a technique called reshape scale (RS) method was proposed to construct new time series from the original one [21]. Its main procedures are as follows. For a time series $\{x(i), i = 1, \ldots, N\}$ and a scale factor $\tau$, a new time series is constructed as:

$$\mathbf{y}^{(\tau)} = \{\mathbf{b}_1, \mathbf{b}_2, \ldots, \mathbf{b}_\tau\}, \tag{5}$$

where $\mathbf{b}_i = \{x(i), x(i+\tau), \ldots, x(i+k\tau)\}$, $i = 1, 2, \ldots, \tau$, and $k$ is the maximal integer satisfying $i + k\tau \leq N$. Note that the length of $\mathbf{y}^{(\tau)}$ is also $N$ and when $\tau = 1$, it retrieves the original time series. Thus, a combination of the RS method and $E_s$ (or $E_f$) is a multiscale entropy method (denoted as RS-$E_s$ and RS-$E_f$, respectively). Figure 4 illustrates the relationships among the aforementioned entropy methods. Also, $E_{ms}(m, r, \tau, N)$ is a multiscale entropy method when $\tau$ takes multiple values. The RS method is essentially similar to the first procedure of the $E_{ms}(m, r, \tau, N)$ algorithm (Equation (1)). A main difference between them is that in the RS method the segment $\mathbf{b}_i$, $i = 1, 2, \ldots, \tau$, can be randomly appended to other segments.

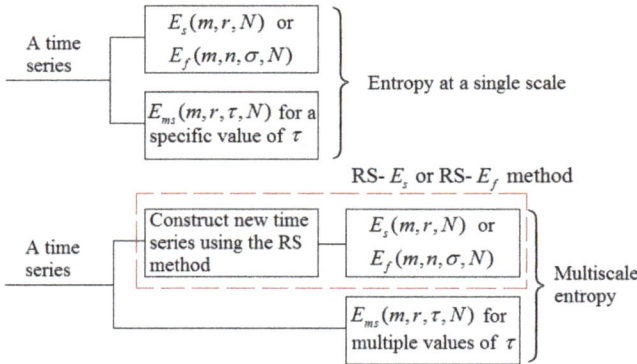

**Figure 4.** A flow diagram illustrating the relationships among the aforementioned entropy methods. RS, the reshape scale method [21].

We tested the performance of three MSE methods, i.e., RS-$E_s$, RS-$E_f$), and $E_{ms}$, using SBF signals. The following parameters were selected: $m=2$; for RS-$E_s$, $r$ was $0.2 \times$ SD of the constructed signal; for $E_{ms}$, $r$ was $0.2 \times$ SD of the original signal; for RS-$E_f$, by using an approach recommended by the authors who proposed $E_f$ [17], we selected $n = 2$ and $\sigma$ being $0.2 \times SD$ of the constructed signal. The results showed that $E_s$ was almost identical to $E_{ms}$ at each scale, both of which monotonically increased at the scales from 1 to around 10 and then reach a plateau. The values of $E_f$ showed a similar trend but were much smaller than $E_s$ and $E_{ms}$. Figure 5A shows an example of the testing results. Further, we performed the following tests. For a given SBF signal, we computed $E_s$, $E_f$, and $E_{ms}$ for 20 phase-randomized surrogate data sets at each scale. When computing $E_s$ and $E_f$, surrogate data were generated from the new signal constructed by using the RS method, whereas when computing $E_{ms}$, surrogate data were generated from the original signal. The results showed that each of RS-$E_s$, RS-$E_f$, and $E_{ms}$ for surrogate data showed similar trend compared to that for the real data (Figure 5A). However, RS-$E_f$ yielded smaller differences between surrogate data and real data compared to RS-$E_s$ and $E_{ms}$. A possible reason is that $E_f$ is insensitive to structural changes of low-frequency components (see Figure 3B) caused by the phase randomization procedure. These testing results suggest that RS-$E_f$ has no superiority over RS-$E_s$ or $E_{ms}$ for assessing the complexity of BFO.

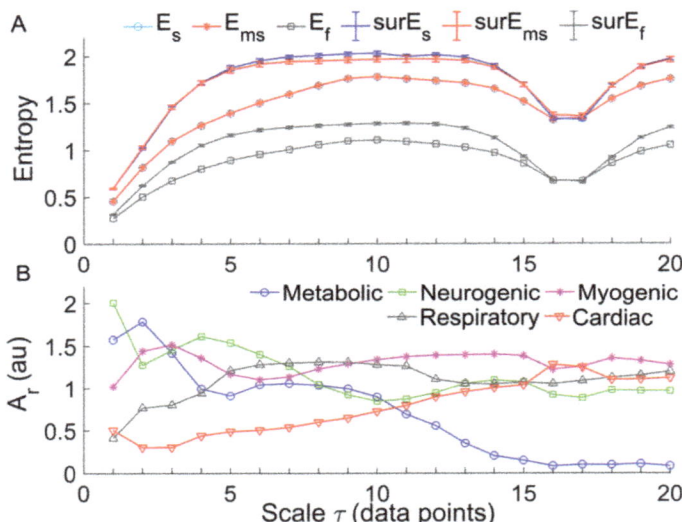

**Figure 5.** (**A**) Results of RS-$E_s$, RS-$E_f$, and $E_{ms}$ of a SBF signal from a diabetic subject (Figure 1A, 1–10 min). The following parameters were used: $m = 2$; $r = 0.2 \times SD$ (the generated signal at scale $\tau$) for $E_s(m, r, N)$; $r = 0.2 \times SD$ (the original signal) for $E_{ms}(m, r, \tau, N)$; $n = 2$; $\sigma = 0.2 \times SD$ (the generated signal at scale $\tau$). The results of surrogate tests are presented as means ± standard errors. *surE_s*, *surE_s*, and *surE_{ms}* refer to $E_s$, $E_f$, and $E_{ms}$ of phase-randomized surrogate data, respectively. For RS-$E_s$ and RS-$E_f$, 20 surrogate data sets were generated from the new signal at scale $\tau$; for $E_{ms}$, 20 surrogate data sets were also generated from the original signal for each scale $\tau$. (**B**) Relative wavelet amplitudes ($A_r$) of the metabolic, neurogenic, myogenic, respiratory, and cardiac frequencies of the generated signal at each scale $\tau$ using the RS method.

## 2.4. Multi-Scale Entropy of SBF Data

The above testing results (Figure 5A) indicate that $E_{ms}$ was almost identical to RS-$E_s$ and much larger than $E_f$ at each scale, while the transitions of $E_{ms}$ and RS-$E_s$ with increasing scales were similar to that of $E_f$. Therefore, we applied $E_{ms}$ to the SBF data collected from 18 diabetic patients and eight healthy controls. For the loading protocol, $E_{ms}$ was calculated for three segments of the SBF signal: baseline (1–10 min), loading period (11–13 min), and reactive hyperemia (a 5-min period following the peak hyperemia [8]); for the heating protocol, $E_{ms}$ was calculated for three segments of the SBF signal: baseline (1–10 min), P1 (a 5-min period following the beginning of the increase in SBF), and P2 (the last 10-min segment of the heating period). To eliminate the influences of possible ascending and/or descending trends as well as noise on $E_{ms}$, each data segment was filtered by decomposing it using the ensemble empirical mode decomposition method [24] and reconstructing a new signal from the intrinsic mode functions with frequencies between 0.0095 and 2 Hz [8]. Then $E_{ms}(m, r, \tau, N)$ was computed for the filtered data at the scales from $\tau = 1$ to 20. The parameters $m = 2$, $r = 0.2 \times$ SD were used.

## 2.5. Relative Wavelet Amplitude of BFO at Multiple Scales

To understand the underlying mechanisms responsible for the transition of $E_{ms}$, we applied wavelet analysis to the signals constructed from the original one using the RS method. For a constructed signal at scale $\tau$, $\mathbf{y}^{(\tau)} = \{y(i), i = 1, \ldots, N\}$ (Equation (5)), its continuous wavelet transform was defined as $w(s, t) = \int_{-\infty}^{+\infty} \psi_{s,t}(u) y(u) du$, where $\psi_{s,t} = \frac{1}{\sqrt{s}} \psi(\frac{u-t}{s})$, $\psi(u)$ is the mother wavelet function, $s$ is the scale corresponding to the central frequency of $\psi_{s,t}$, and $t$ is time. In this study, we used the Morlet wavelet $\psi(u) = \pi^{-1/4} e^{-i\omega_0 u} e^{-u^2/2}$ as the mother wavelet function, for which $s$ is the reciprocal of the central frequency of $\psi_{s,t}$ when $\omega_0 = 2\pi$. Then we calculated the average amplitudes of the wavelet transform over time and over the frequency interval of five frequency components: metabolic (0.0095–2 Hz), neurogenic (0.02–0.05 Hz), myogenic (0.05–0.15 Hz), respiratory (0.15–0.4 Hz), and cardiac (0.4–2 Hz) components. Finally, the averaged wavelet amplitudes of the five frequency components were normalized (divided) by that of the frequency interval 0.0095–2 Hz to yield relative wavelet amplitudes ($A_r$). Figure 3B shows the $A_r$ values of the five frequency components of the SBF signal for calculating $E_{ms}$. A prominent feature of the changes in $A_r$ was that $A_r$ of the metabolic and cardiac components show persistent decrease and increase with increasing scales, respectively.

## 2.6. Statistical Analysis

The differences in multiscale entropy $E_{ms}$ and relative wavelet amplitude $A_r$ between two groups were examined using Mann–Whitney U test; the within-group differences in these measures were examined using Wilcoxon signed-rank test. These tests were performed using SPSS 16 (SPSS, Chicago, IL, USA).

## 3. Results

A common feature of the multiscale entropy, $E_{ms}$, was that it rose with increasing scales at small scales and then reached a plateau (Figures 6 and 7). Applied pressure resulted in a significant increase in $E_{ms}$ at all scales in diabetics ($p < 0.01$, Figure 6C) but not in controls ($p > 0.05$, Figure 6A). During reactive hyperemia, $E_{ms}$ in diabetics showed little change compared to baseline (Figure 6C), whereas in controls it showed a significantly decrease at the scales $\tau = 3$ to 7 and $\tau = 10$ to 20 (Figure 6A). Compared to controls, $E_{ms}$ in diabetics was significantly lower at the scales $\tau = 1$ to 8 and $\tau = 14$ to 16 during baseline ($p < 0.05$, Figure 7A,B), but was significantly higher at the scales $\tau = 13$ to 18 during reactive hyperemia ($p < 0.05$, Figure 7E).

During thermally induced biphasic response, $E_{ms}$ in diabetics showed only small changes at all scales (Figure 6D), whereas in controls it significantly decreased at the scales $\tau = 1$ to 7 and $\tau = 16$ to 18 during P1 and at the scales $\tau = 1$ to 7 during P2 ($p < 0.05$, Figure 6B). Compared to controls, $E_{ms}$ in diabetics was significantly lower at the scales $\tau = 1$ to 4 and $\tau = 8$ to 12 during P1 ($p < 0.05$, Figure 7D) and at all scales during P2 ($p < 0.05$, Figure 7F).

Figure 8 shows the mean values of $A_r$ of the characteristic frequency components at multiple scales in two groups. A prominent feature was that $A_r$ of the metabolic component declined at small scales during baseline and the SBF responses except for thermally induced first peak (Figure 8A,B), while $A_r$ of cardiac component initially underwent a transient decrease followed by a sustained increase (Figure 8I,J). During thermally induced first peak, $A_r$ of metabolic component at small scales showed a pronounced increase followed by a substantial decrease (Figure 8B), while $A_r$ of neurogenic component exhibited similar changes but at larger scales (Figure 8D).

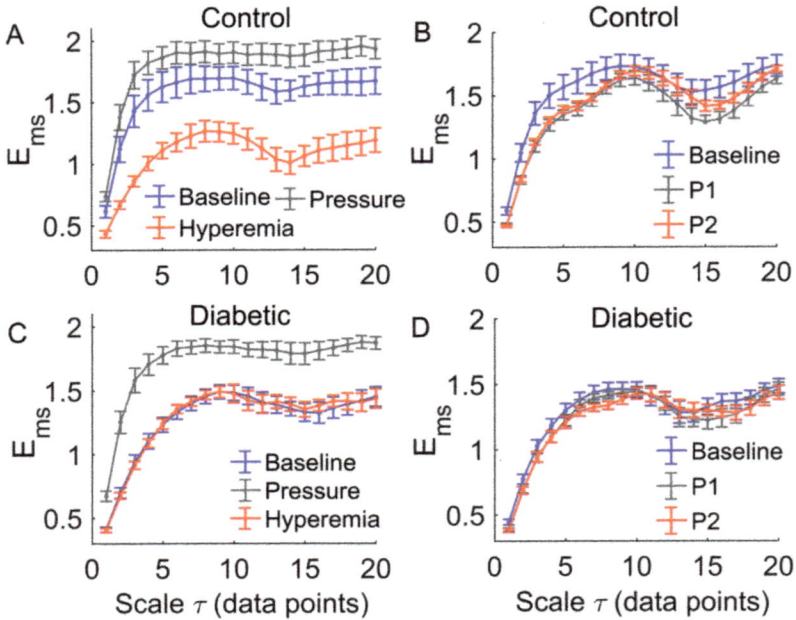

**Figure 6.** Within-group comparisons of $E_{ms}(m, r, \tau, N)$. Data are represented as means $\pm$ standard errors. (**A**) In controls, comparison of $E_{ms}$ between hyperemia period and baseline yielded $p < 0.05$ at $\tau = 1$ to 7 and $\tau = 16$ to 18. (**B**) In controls, comparison of $E_{ms}$ between P1 and baseline yielded $p < 0.05$ at $\tau = 1$ to 7 and $\tau = 16$ to 18, while comparison of $E_{ms}$ between P2 and baseline yielded $p < 0.05$ at $\tau = 1$ to 7. (**C**) In diabetics, comparison of $E_{ms}$ between the pressure period and baseline yielded $p < 0.01$ at all scales. (**D**) In diabetics, $E_{ms}(m, r, \tau, N)$ yielded similar values during three periods of heating protocol.

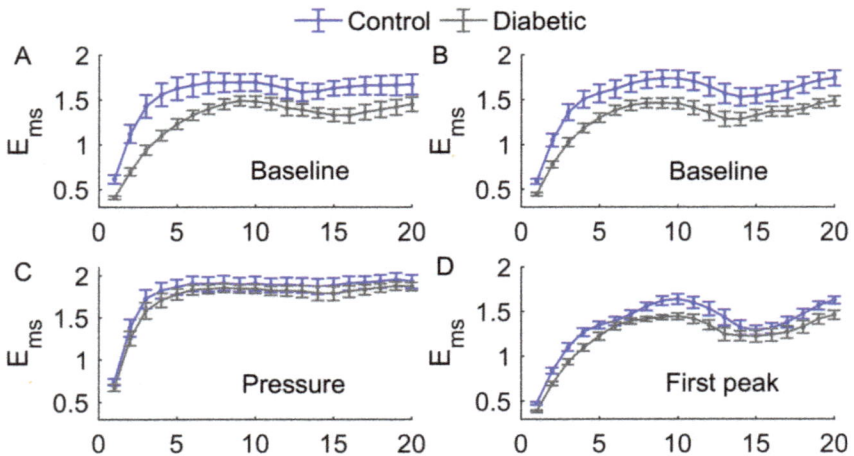

**Figure 7.** *Cont.*

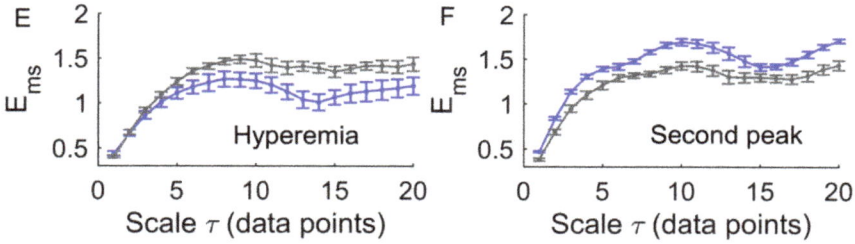

**Figure 7.** Comparisons of $E_{ms}(m, r, \tau, N)$ between two groups. Data are represented as means $\pm$ standard errors. (**A**) During the baseline period of the pressure loading protocol, $E_{ms}$ in diabetics was significantly lower ($p < 0.05$) at the scales $\tau = 1$ to 8 and $\tau = 14$ to 16. (**B**) During the baseline period of the local heating protocol, $E_{ms}$ in diabetics was significantly lower ($p < 0.05$) at all scales. (**C**) During the loading pressure period, there was no significant difference between two groups. (**E**) During hyperemia, $E_{ms}$ in diabetics was significantly higher at $\tau = 13$ to 18. (**D**) During P1, $E_{ms}$ in diabetics was significantly lower ($p < 0.05$) at $\tau = 1$ to 4 and $\tau = 8$ to 12. (**F**) During P2, $E_{ms}$ in diabetics was significantly lower ($p < 0.05$) at all scales.

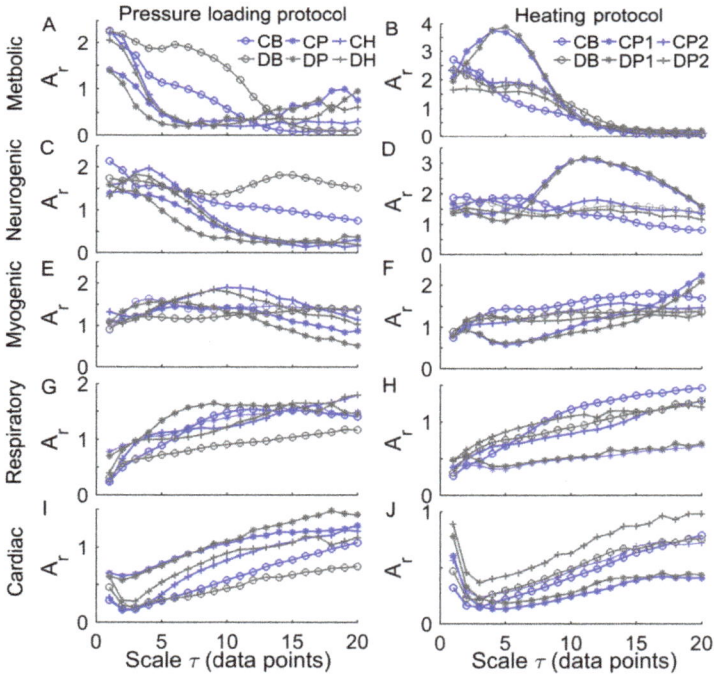

**Figure 8.** Relative wavelet amplitudes ($A_r$) of metabolic (**A,B**), neurogenic (**C,D**), myogenic (**E,F**), respiratory (**G,H**), and cardiac (**I,J**) components of blood flow oscillations (BFO) at multiple scales in response to loading pressure (left panels) and local heating (right panels). Data are represented as mean values in two groups. CB, CP, CH, CP1, and CP2 refer to baseline, loading pressure, hyperemia, first peak, and second peak periods in controls, respectively; DB, DP, DH, DP1, and DP2 refer to the same periods in diabetics.

## 4. Discussion

The main findings of this study are as follows. First, during reactive hyperemia and the biphasic response induced by local heating, $E_{ms}$ in diabetics showed only small changes compared to baseline but in controls it underwent significant changes (Figure 6). As a consequence, $E_{ms}$ in diabetics significantly differed from that in controls at multiple scales (Figure 7). Second, during baseline and the SBF responses except for the pressure loading period, $E_{ms}$ at small scales exhibited different transitions between two groups (Figure 7). These findings indicated that multiscale entropy could provide a more comprehensive assessment of SBF dynamics compared to entropy measures on a single scale. Since the SBF data were recorded from the first metatarsal head, one of the most common sites of DFU, our findings support the use of nonlinear measures of SBF responses induced by mechanical and thermal stresses to assess the risk of DFU.

In this study, we utilized a modified sample entropy algorithm [13] for computing multiscale entropy of BFO by varying the lag between neighboring data points of the sequences to be compared. We have demonstrated that this method yielded almost identical entropy values compared to the RS-$E_s$ method (Figure 5A), which is a combination of sample entropy [16], and a method for generating new signals from the original one, called reshape scale (RS) method [21]. We have also demonstrated that RS-$E_f$ has no superiority over $E_{ms}$ for assessing the complexity of BFO. Since computing $E_f$ is very time consuming, this measure may be more suitable for short series data, e.g., RR interval series.

A common feature of $E_{ms}$ of BFO was that it showed a rise with increasing scales at small scales, possibly including a rapid rise followed by a slow rise, and then reached a plateau (Figures 6 and 7). To get an insight of the underlying mechanisms for this phenomenon, we calculated the relative wavelet amplitudes ($A_r$) of the characteristic frequency components of BFO and examined how they changed with increasing scales. A prominent feature of changes in $A_r$ was that $A_r$ of metabolic component deceased with increasing scales during baseline and the SBF responses except for thermally induced first peak (Figure 8A,B), while $A_r$ of the cardiac component initially showed a transient decrease and then underwent a sustained increase (Figure 8I,J). Therefore, we speculate that $E_{ms}$ of BFO reflects the homogeneity degree of the combination of the characteristic frequency components. Thus an augmentation of any frequency component, e.g., metabolic or cardiac components, will contribute to lowering $E_{ms}$. In this sense, any entropy measures on a single scale, e.g., sample entropy, only reflect the homogeneity degree of a specific combination of the characteristic frequency components. For instance, sample entropy ($\tau = 1$) reflects the regularity degree of BFO where metabolic component and possibly cardiac component play a dominant role (Figure 8). In our previous study [8], because we used a fixed parameter $\tau = 12$, we observed a significant influence of cardiac component on $E_{ms}$ during the thermally induced second peak.

Our results showed that during baseline and the SBF responses except for the pressure loading period, $E_{ms}$ in diabetics significantly differed from that in controls at multiple scales (Figure 7). Also, the transition of $E_{ms}$ in diabetics was different from that in controls. For instance, during baseline, $E_{ms}$ in diabetics showed a fairly rapid increase at the scales from $\tau = 1$ to around 8 (Figure 7A,B), while in controls $E_{ms}$ showed a rapid increase at the scales from $\tau = 1$ to 3, followed by a slow rise at the scales from $\tau = 3$ to around 8 (Figure 7A,B). This distinct difference was partially due to different changes in $A_r$ of the metabolic component between two groups. As shown in Figure 8A,B, in diabetics $A_r$ of the metabolic component decayed slowly at small scales, whereas in controls it decayed rapidly.

The association between $E_{ms}$ of BFO and $A_r$ of the characteristic frequency components can also be observed during the SBF responses. However, it should be kept in mind that $E_{ms}$ is a global measure of the structural properties of BFO, and may be influenced by all frequency components and the interactions among them [8]. $E_{ms}$ could reveal some global features of BFO that cannot be distinctively reflect by $A_r$ and vice versa. For example, we observed significantly higher values of $E_{ms}$ in diabetics at the scales from $\tau = 6$ to 9 and from $\tau = 12$ to 18 during reactive hyperemia (Figure 7C), but $A_r$ of each frequency component in diabetics was similar to that in controls (Figure 8, left panels). In contrast, during the thermally induced first peak, $E_{ms}$ in diabetics showed only small changes compared to baseline (Figure 6D), but the $A_r$ of each frequency component showed pronounced changes (Figure 8, right panels). Therefore, $E_{ms}$ and wavelet analysis could be mutually complementary in assessing SBF responses.

There are several limitations of this study. The sample size was small, which might impede the power of the statistical analysis. We did not have an age, sex, or BMI matched control. Factors such as age, sex, and BMI may affect our results. Since ageing can lead to altered dynamics of skin BFO [12,13], we focused on examining whether age significantly contributed to the results observed in this study. For example, we observed a significant increase in $E_{ms}$ during the pressure loading period compared to baseline in diabetics (Figure 6C) but not in controls (Figure 6A) and a significant decrease in $E_{ms}$ during hyperemia compared to baseline in controls (Figure 6A), but not in diabetics (Figure 6C). We thus examined whether the changes in $E_{ms}$ (denoted as $\Delta E_{ms}$) were related to age. Figure 9 shows the results in the case of $\tau = 6$. On one hand, $\Delta E_{ms}$ was significantly different between two groups ($p < 0.05$ for pressure loading period and $p < 0.001$ for hyperemia). On the other hand, although $\Delta E_{ms}$ exhibited an increasing trend with age in both groups, the increasing rate (the slop of the fitting line) in diabetic group is much smaller than that in control group (Figure 9A,B). In particular, $\Delta E_{ms}$ in diabetic group during hyperemia was almost independent of age (Figure 9B). These results suggested that the significant difference in $\Delta E_{ms}$ between two groups was mainly attributed to impaired microvascular reactivity in the diabetic foot and that age has a marginal effect on BFO dynamics compared to diabetes. Future studies may follow up the development of DFU in a larger sample size with matched controls to validate our findings.

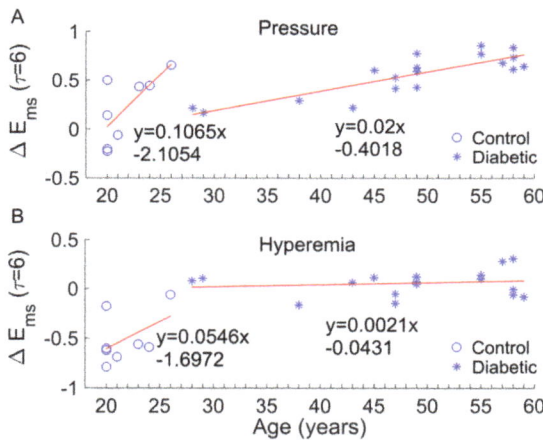

**Figure 9.** (**A**) Examination of the effect of age on changes in $E_{ms}$ during the pressure loading period compared to baseline (denoted as $\Delta E_{ms}$, scale factor $\tau = 6$). $\Delta E_{ms}$ was significantly larger in diabetics than in controls ($p < 0.05$). (**B**) Examination of the effect of age on $\Delta E_{ms}$ ($\tau = 6$) during hyperemia. The absolute value of $\Delta E_{ms}$ was significantly larger in controls ($p < 0.001$).

## 5. Conclusions

The present study indicated that during reactive hyperemia and the biphasic response induced by local heating, the regularity degree of SBF at the first metatarsal head of diabetics underwent only small changes compared to baseline, and significantly differed from that in healthy controls at multiple scales. On the other hand, the regularity degree of SBF in the diabetic foot displays distinctively different transitions compared to controls. This study suggests that multiscale entropy could provide a more comprehensive assessment of impaired microvascular reactivity in the diabetic foot compared to any entropy measures on a single scale and may be used to assess the risk for DFU.

**Acknowledgments:** This work was partially supported by Shaanxi Province Basic Research Program of Natural Science (2017JM8003).

**Author Contributions:** All authors conceived of the study. F.L. and Y.-K.J. designed and performed the experiments. F.L. and Y.-K.J. analyzed the data. F.L. and Y.-K.J. drafted the paper. All authors revised and approved the paper.

**Conflicts of Interest:** The authors declare no conflict of interest.

## References

1. Markakis, K.; Bowling, F.L.; Boulton, A.J. The diabetic foot in 2015: An overview. *Diabetes Metab. Res. Rev.* **2016**, *32* (Suppl. 1), 169–178. [CrossRef] [PubMed]

2. Van Netten, J.J.; Price, P.E.; Lavery, L.A.; Monteiro-Soares, M.; Rasmussen, A.; Jubiz, Y.; Bus, S.A. Prevention of foot ulcers in the at-risk patient with diabetes: A systematic review. *Diabetes Metab. Res. Rev.* **2016**, *32* (Suppl. 1), 84–98. [CrossRef] [PubMed]

3. Mills, J.L. Lower limb ischaemia in patients with diabetic foot ulcers and gangrene: Recognition, anatomic patterns and revascularization strategies. *Diabetes Metab. Res. Rev.* **2016**, *32* (Suppl. 1), 239–245. [CrossRef] [PubMed]

4. Holman, N.; Young, R.J.; Jeffcoate, W.J. Variation in the recorded incidence of amputation of the lower limb in England. *Diabetologia* **2012**, *55*, 1919–1925. [CrossRef] [PubMed]

5. Johannesson, A.; Larsson, G.U.; Ramstrand, N.; Turkiewicz, A.; Wiréhn, A.B.; Atroshi, I. Incidence of lower-limb amputation in the diabetic and nondiabetic general population: A 10-year population-based cohort study of initial unilateral and contralateral amputations and reamputations. *Diabetes Care* **2009**, *32*, 275–280. [CrossRef] [PubMed]

6. Burns, S.; Jan, Y.K. Diabetic Foot Ulceration and Amputation. In *Rehabilitation Medicine*; Kim, C.T., Ed.; InTech Publisher: Rijeka, Croatia, 2012; pp. 1–20.

7. Jan, Y.K.; Shen, S.; Foreman, R.D.; Ennis, W.J. Skin blood flow response to locally applied mechanical and thermal stresses in the diabetic foot. *Microvasc. Res.* **2013**, *89*, 40–46. [CrossRef] [PubMed]

8. Liao, F.; Jan, Y.K. Nonlinear dynamics of skin blood flow response to mechanical and thermal stresses in the plantar foot of diabetics with peripheral neuropathy. *Clin. Hemorheol. Microcirc.* **2017**, *66*, 197–210. [CrossRef] [PubMed]

9. Cracowski, J.L.; Minson, C.T.; Salvat-Melis, M.; Halliwill, J.R. Methodological issues in the assessment of skin microvascular endothelial function in humans. *Trends Pharmacol. Sci.* **2006**, *27*, 503–508. [CrossRef] [PubMed]

10. Stefanovska, A.; Bracic, M.; Kvernmo, H.D. Wavelet analysis of oscillations in the peripheral blood circulation measured by laser Doppler technique. *IEEE Trans. Biomed. Eng.* **1999**, *46*, 1230–1239. [CrossRef] [PubMed]

11. Kvandal, P.; Landsverk, S.A.; Bernjak, A.; Stefanovska, A.; Kvernmo, H.D.; Kirkebøen, K.A. Low-frequency oscillations of the laser Doppler perfusion signal in human skin. *Microvasc. Res.* **2006**, *72*, 120–127. [CrossRef] [PubMed]

12. Liao, F.; Garrison, D.W.; Jan, Y.K. Relationship between nonlinear properties of sacral skin blood flow oscillations and vasodilatory function in people at risk for pressure ulcers. *Microvasc. Res.* **2010**, *80*, 44–53. [CrossRef] [PubMed]

13. Liao, F.; Jan, Y.K. Using Modified Sample Entropy to Characterize Aging-Associated Microvascular Dysfunction. *Front. Physiol.* **2016**, *7*, 126. [CrossRef] [PubMed]

14. Parthimos, D.; Schmiedel, O.; Harvey, J.N.; Griffith, T.M. Deterministic nonlinear features of cutaneous perfusion are lost in diabetic subjects with neuropathy. *Microvasc. Res.* **2011**, *82*, 42–51. [CrossRef] [PubMed]

15. Liao, F.; Liau, B.Y.; Rice, I.M.; Elliott, J.; Brooks, I.; Jan, Y.K. Using local scale exponent to characterize heart rate variability in response to postural changes in people with spinal cord injury. *Front. Physiol.* **2015**, *6*, 142. [CrossRef] [PubMed]

16. Richman, J.S.; Moorman, J.R. Physiological time-series analysis using approximate entropy and sample entropy. *Am. J. Physiol. Heart Circ. Physiol.* **2000**, *278*, H2039–H2049. [CrossRef] [PubMed]

17. Chen, W.; Wang, Z.; Xie, H.; Yu, W. Characterization of surface EMG signal based on fuzzy entropy. *IEEE Trans. Neural Syst. Rehabil. Eng.* **2007**, *15*, 266–272. [CrossRef] [PubMed]

18. Von Tscharner, V.; Zandiyeh, P. Multi-scale transitions of fuzzy sample entropy of RR-intervals and their phase-randomized surrogates: A possibility to diagnose congestive heart failure. *Biomed. Signal Process. Control* **2017**, *31*, 350–356. [CrossRef]

19. Costa, M.; Goldberger, A.L.; Peng, C.K. Multiscale entropy analysis of biological signals. *Phys. Rev. E Stat. Nonlin. Soft Matter Phys.* **2005**, *71*, 021906. [CrossRef] [PubMed]

20. Costa, M.; Goldberger, A.L.; Peng, C.K. Multiscale entropy analysis of complex physiologic time series. *Phys. Rev. Lett.* **2002**, *89*, 068102. [CrossRef] [PubMed]

21. Zandiyeh, P.; von Tscharner, V. Reshape scale method: A novel multi scale entropic analysis approach. *Phys. Stat. Mech. Appl.* **2013**, *392*, 6265–6272. [CrossRef]

22. Valencia, J.F.; Porta, A.; Vallverdu, M.; Claria, F.; Baranowski, R.; Orlowska-Baranowska, E.; Caminal, P. Refined multiscale entropy: Application to 24-h Holter recordings of heart period variability in healthy and aortic stenosis subjects. *IEEE Trans. Biomed. Eng.* **2009**, *56*, 2202–2213. [CrossRef] [PubMed]

23. Humeau-Heurtier, A. The Multiscale Entropy Algorithm and Its Variants: A Review. *Entropy* **2015**, *17*, 3110–3123. [CrossRef]

24. Wu, Z.; Huang, N.E. Ensemble empirical mode decomposition: A noise-assisted data analysis method. *Adv. Adapt. Data Anal.* **2009**, *1*, 1–41. [CrossRef]

*entropy*

MDPI

*Article*

# Fuzzy Entropy Analysis of the Electroencephalogram in Patients with Alzheimer's Disease: Is the Method Superior to Sample Entropy?

**Samantha Simons [1], Pedro Espino [2] and Daniel Abásolo [1,***

[1]   Centre for Biomedical Engineering, Department of Mechanical Engineering Sciences, Faculty of Engineering and Physical Sciences, University of Surrey, Guildford GU2 7XH, UK; ssimons@fastmail.fm
[2]   Hospital Clínico Universitario de Valladolid, 47003 Valladolid, Spain; pedro@espino.es
*   Correspondence: d.abasolo@surrey.ac.uk; Tel.: +44-(0)1483-682971

Received: 29 November 2017; Accepted: 28 December 2017; Published: 3 January 2018

**Abstract:** Alzheimer's disease (AD) is the most prevalent form of dementia in the world, which is characterised by the loss of neurones and the build-up of plaques in the brain, causing progressive symptoms of memory loss and confusion. Although definite diagnosis is only possible by necropsy, differential diagnosis with other types of dementia is still needed. An electroencephalogram (EEG) is a cheap, portable, non-invasive method to record brain signals. Previous studies with non-linear signal processing methods have shown changes in the EEG due to AD, which is characterised reduced complexity and increased regularity. EEGs from 11 AD patients and 11 age-matched control subjects were analysed with Fuzzy Entropy (FuzzyEn), a non-linear method that was introduced as an improvement over the frequently used Approximate Entropy (ApEn) and Sample Entropy (SampEn) algorithms. AD patients had significantly lower FuzzyEn values than control subjects ($p < 0.01$) at electrodes T6, P3, P4, O1, and O2. Furthermore, when diagnostic accuracy was calculated using Receiver Operating Characteristic (ROC) curves, FuzzyEn outperformed both ApEn and SampEn, reaching a maximum accuracy of 86.36%. These results suggest that FuzzyEn could increase the insight into brain dysfunction in AD, providing potentially useful diagnostic information. However, results depend heavily on the input parameters that are used to compute FuzzyEn.

**Keywords:** Alzheimer's disease; electroencephalogram; non-linear analysis; complexity; irregularity; Fuzzy Entropy; Sample Entropy

---

## 1. Introduction

Alzheimer's disease (AD) is a form of dementia that is characterised by progressive impairments in cognition and memory [1]. The cause of AD is not known [2], and the course of the disease can last several years before death [1]. As AD is currently the most prevalent dementia worldwide [3,4] the impact of the disease is significant. Current clinical diagnosis is based on the National Institute of Neurological and Communicative Disorders and Stroke and Alzheimer's Disease and Related Disorders Association (NINCDS-ADRDA) criteria [5], and, although definite diagnosis is only possible by necropsy, a differential diagnosis with other types of dementia would be of great use.

The electroencephalogram (EEG), a recording of the electrical activity of the brain, shows great potential to characterise changes in brain activity as a result of AD. There are several reasons for this; the first being that AD is a cortical dementia [1] and, therefore, changes to the electrical activity of the brain resulting from AD could be registered on EEGs. Furthermore, the EEG can be recorded non-invasively, with portable equipment and at much lower cost than other imaging techniques that are used in AD diagnosis. Therefore, the application of signal processing algorithms to extract features from EEG signals may help in the characterisation of the changes that are associated with AD. In fact, several EEG features appear to be abnormal in AD patients, where a shift of the power spectrum to lower frequencies, a decrease of coherence among cortical areas, perturbed synchrony, and reduced complexity have been observed (for detailed reviews, please see [1,6]), although in the early stages of the disease, the EEG may show similar features to that of age-matched healthy controls [7]. In spite of these findings, there is room for the introduction of novel signal processing techniques for further study of the EEG. In particular, entropy algorithms quantifying irregularity in data could be useful to capture subtle changes in the EEG that might be caused by AD.

Different entropy algorithms have been introduced over the years to characterise the EEG, with greater entropy being associated with increased irregularity in the EEG. Embedding entropies are algorithms where entropy is used to provide information about how the EEG signal fluctuates with time by comparing the time series with a delayed version of itself [8]. The introduction of Approximate Entropy (ApEn) by Pincus [9] made the reliable characterisation of the entropy of short and noisy biomedical signals possible in ways that were, up until its introduction, not achievable. ApEn measures the regularity in data by examining time series for similar epochs and assigning a non-negative number to the sequence, with larger values corresponding to more complexity or irregularity in the data [9]. Given a time series with $N$ samples, a sample length $m$ and a tolerance window $r$, ApEn($m$, $r$, $N$) measures the logarithmic likelihood that samples of patterns that are close (within $r$) for $m$ contiguous observations remain close (within the same tolerance width $r$) on subsequent incremental comparisons [9]. The ApEn algorithm counts each sequence as matching itself to avoid the occurrence of $ln(0)$ in the calculations. The effect of self-matches provides a biased estimate of entropy, giving a false impression of determinism [10]. Furthermore, ApEn values depend heavily on the record length, with ApEn being lower than expected for short time series, and it also lacks consistency when different input parameter values are used to evaluate the same time series [11]. It was subsequently superseded by Sample Entropy (SampEn), as introduced by Richman and Moorman [11]. As it is also the case with ApEn, two input parameters, $m$ and $r$, must be specified to compute SampEn. SampEn($m$, $r$, $N$) is the negative logarithm of the conditional probability that two sequences similar for $m$ point vectors remain similar at the next point, where self-matches are not included in calculating the probability. SampEn would be lower for signals that show a higher degree of self-similarity, i.e., more regular. In addition to overcoming some of the limitations of ApEn, SampEn is easier to compute [11]. For both ApEn and SampEn, the recommended range of values for input parameters are $m$ = 1 or 2 and $r$ between 0.1 and 0.25 times the standard deviation of the input data [11].

Different algorithms that are attempting to improve SampEn have been suggested. Quadratic Sample Entropy (QSE) was introduced to reduce the influence of the arbitrary constants $m$ and $r$ on SampEn and to reduce the skewing of results when either the top or the bottom of the conditional probabilities was very small or very large [12]. Input variables are the same (i.e., a sample length $m$ and a tolerance window $r$) though the recommended values are different, with $r$ not being limited to the range suggested for ApEn and SampEn. Another attempt to improve SampEn was with the introduction of Fuzzy Entropy (FuzzyEn) [13], based on the concept of fuzzy sets to determine a fuzzy measurement of similarity of two vectors based on their shapes.

In this pilot study, FuzzyEn was used to characterise the broadband activity in the EEG of patients with AD. It was hypothesised that FuzzyEn would identify differences between the entropy of EEG signals from AD patients and age-matched control subjects, and that these differences could be used to help in the classification of EEG signals with respect to their class (AD patient or control subject). The quality of the classification would be evaluated using receiver operating characteristic (ROC) curves [14]. Furthermore, FuzzyEn results would be compared with SampEn to ascertain whether the claims that the former is a superior method to the latter hold true in the context of EEG analysis in AD.

The outline of the paper is as follows. Section 2 describes the EEG database that is used in this study and introduces FuzzyEn. Results obtained with all of the input parameters tested in this pilot study are presented in Section 3, whilst the discussion of the findings, focusing on a comparison of the results obtained with FuzzyEn and other related entropies, and the conclusions from this study follow in Section 4.

## 2. Materials and Methods

### 2.1. Subjects and EEG Recording

The database used in this study consisted of 11 patients with a diagnosis of AD (five men; six women; age: $72.5 \pm 8.3$ years, mean $\pm$ standard deviation (SD)), recruited from the Alzheimer's Patients' Relatives Association of Valladolid (AFAVA), Spain, and 11 age-matched controls (seven men; four women; age: $72.8 \pm 6.1$ years, mean $\pm$ SD). AD diagnosis was supported by clinical evaluation (clinical history, physical, and neurological examination) and Mini-Mental State Examination (MMSE), which is generally accepted as an effective way to evaluate cognitive function [15], was also performed. The average MMSE score for the AD patients was $13.1 \pm 5.9$ (mean $\pm$ SD), indicating moderate to severe dementia, and the score was 30 for all of the control subjects, indicating no mental impairment. All of the subjects and caregivers gave their informed consent for participation in the study.

EEG signals were recorded with the subjects in a relaxed state with eyes closed at the Hospital Clínico Universitario de Valladolid (Spain) using Oxford Instruments Profile Study Room 2.3.411 EEG equipment and the international 10–20 system with electrodes referenced to the linked ear lobes of each subject. More than 5 min of EEG data were recorded for each subject with a sampling rate of 256 Hz. Two AD patients were taking lorapezam at the time of recording the EEG, but no prominent rapid rhythms were observed in the visual examination of their EEGs. None of the other subjects who took part in the study were using medication that could be expected to influence the EEG.

Artefact-free sections of the EEG signals (split into 5-second epochs with no movement artefacts and no electroencephalographic signs of sleep) were selected by Dr Pedro Espino, the specialist physician that was overseeing the recording of the EEGs. On average, $30.0 \pm 12.5$ epochs (mean $\pm$ SD) were selected from each electrode for each subject. All of the epochs selected were filtered using a Hamming window FIR filter with order 426 and cut-off frequencies at 0.5 Hz and 40 Hz to remove residual noise (DC offset and mains hum) prior to the computation of FuzzyEn. Zero-phase filtering was used to make sure the use of a filter of such high order did not result in edge effects.

### 2.2. Fuzzy Entropy

Both ApEn and SampEn measure the similarity of the vectors being compared using a Heaviside function, which can be represented as:

$$\theta(z) = \begin{cases} 1, & if \ z \geq 0 \\ 0, & if \ z < 0 \end{cases} \tag{1}$$

This leads to a two-state binary classifier, where the vectors are either close or not. However, this might not be able to capture in the most appropriate way the boundaries between different classes, which in real biomedical data might be more ambiguous [13]. Therefore, FuzzyEn was introduced to overcome this limitation with a fuzzy function instead of the Heaviside function used to calculate the similarity degree between vectors.

Since its introduction, FuzzyEn has been used to characterise different types of biomedical signals, such as electromyograms [13,16–18], EEGs [19,20], gait [20], or heart rate variability [20,21]. Comparative studies with ApEn and SampEn suggest that FuzzyEn outperforms them [17,19]. Furthermore, recent evidence suggests that FuzzyEn is a robust entropy estimator when there are missing samples in the biomedical signals being analysed [20].

Given $N$ data points from a time series $\{x(n)\} = x(1), x(2), \ldots, x(N)$, FuzzyEn can be calculated using the following algorithm [13]:

1.  For $1 \leq i \leq N - m + 1$, form $m$-vectors $X_m(1) \ldots X_m(N - m + 1)$ defined as:

$$X_m(i) = [x(i), x(i+1), \ldots, x(i+m-1)] - x0(i) \tag{2}$$

These vectors represent $m$ consecutive $x$ values, commencing with the $i$th point, with the baseline $(x0(i) = \frac{1}{m} \sum_{j=0}^{m-1} x(i+j))$ removed.

2. Define the distance between vectors $X_m(i)$ and $X_m(j)$, $d_{ij,m}$, as the maximum absolute difference between their scalar components.

3. Given $n$ and $r$, calculate the similarity degree $D_{ij,m}$ of the vectors $X_m(i)$ and $X_m(j)$ with a fuzzy function:

$$D_{ij,m} = \mu(d_{ij,m}, r) = exp\left(\frac{-(d_{ij,m})^n}{r}\right) \tag{3}$$

4. Define the function $\phi_m$ as:

$$\phi_m(n,r) = \frac{1}{N-m} \sum_{i=1}^{N-m} \left(\frac{1}{N-m-1} \sum_{j=1, j\neq i}^{N-m} D_{ij,m}\right) \tag{4}$$

5. We increase the dimension to $m+1$, form vectors $X_{m+1}(i)$, and, subsequently, obtain the function $\phi_{m+1}$ repeating steps 2 to 4.

6. For time series with a finite number of samples $N$, FuzzyEn can be estimated with the following equation [13]:

$$FuzzyEn(m,n,r,N) = \ln\phi_m(n,r) - \ln\phi_{m+1}(n,r) \tag{5}$$

Given that FuzzyEn is based on the original SampEn algorithm, as introduced by Richman and Moorman in [11], it can be therefore computed as the negative logarithm of the conditional probability that two sequences similar for $m$ points—where similarity is measured using the fuzzy function introduced in Equation (3), instead of the Heaviside function used in the ApEn and SampEn algorithms—remain similar when the size of the vectors being considered is increased by one. The algorithm, as is also the case with SampEn, does not include self-matches when calculating the probability aforementioned. Thus, it does not show the bias that is associated with ApEn [11]. Furthermore, lower values of FuzzyEn indicate more self-similarity in the time series being characterised with this algorithm.

It is obvious that FuzzyEn values would depend on the values of the input parameters $m$, $n$, $r$, and $N$, and comparisons should only be attempted for fixed values of these parameters. $N$ is the length of the time series and is determined, in this particular study, by the sampling frequency of 256 Hz and the epoch length of 5 s. Parameter $m$ determines the length of the sequences to be compared, as in ApEn and SampEn. On the other hand, $r$ and $n$ determine the width and gradient of the fuzzy exponential function.

In principle, larger values of $m$ allow for a better reconstruction of the dynamics of the system being characterised. However, the accuracy and confidence of the entropy estimate improve with a greater number of matches of vectors of length $m$ and $m+1$. Therefore, it is usually recommended to choose small values of $m$ [11].

Figure 1 shows the changes in the shape of the fuzzy exponential function changes with $n$ and $r$. It has been recommended to use small integer values of $n$ [13] and set the tolerance width as $r$ times the standard deviation (SD) of the original data sequence [11]; the latter would give FuzzyEn scale invariance [9].

Based on these recommendations, in this pilot study, values of $m = 1$ and $m = 2$, $n = 1$, $n = 2$, and $n = 3$, and $r = 0.1$, $r = 0.15$, $r = 0.2$, and $r = 0.25$ times the SD of the original time series were used. This led to 24 variable combinations tested. FuzzyEn was therefore computed using 24 input parameter combinations for channels Fp1, Fp2, F3, F4, C3, C4, P3, P4, O1, O2, F7, F8, T3, T4, T5, and T6.

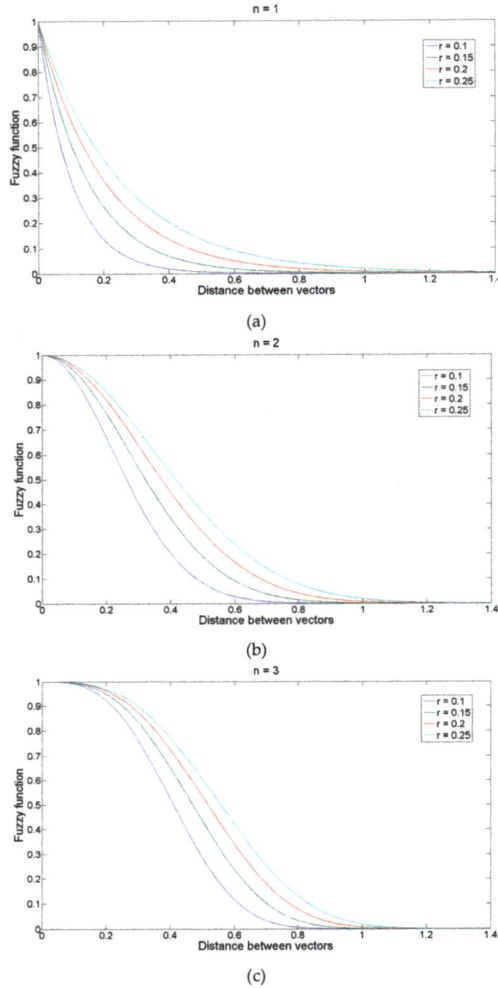

**Figure 1.** Fuzzy function for $n = 1$ (**a**), $n = 2$ (**b**), and $n = 3$ (**c**) and the different values of $r$ used in the study.

*2.3. Statistical Analysis*

The distribution of the FuzzyEn results was evaluated with the Lilliefors test. Depending on the results from the Lilliefors test, Student's $t$-test, or Kruskal-Wallis tests were used to evaluate the statistical significance of differences between groups of subjects at each electrode. In all of the above statistical analyses, statistical significance was at $p < 0.01$.

Results from the electrodes where statistically significant differences between AD patients and controls were found were then analysed with ROC curves, and sensitivity (true positive rate, i.e., percentage of AD patients correctly classified), specificity (true negative rate, i.e., proportion of control subjects correctly identified), accuracy (percentage of total subjects classified precisely), area under the curve, and the optimum threshold (FuzzyEn value that maximises diagnostic accuracy) were computed.

## 3. Results

FuzzyEn was computed for all the 24 input parameter values. Results were averaged based on all of the artefact-free five second epochs within the five-minute period of EEG recordings for the 22 subjects. For all the possible combinations of $m$, $n$, and $r$ values, and most electrodes, FuzzyEn was higher for the EEG of control subjects than that for AD patients. The tables in the Supplementary Materials section contain all of the results for the 24 combinations of input parameter values tested.

The results depended heavily on the choice of input parameters. For $n = 1$, the FuzzyEn values were found to follow a normal distribution. Therefore, Student's $t$ Test was used to evaluate the statistical significance of the findings. For all the values of $r$ and $m = 1$, FuzzyEn was significantly lower ($p < 0.01$) for AD patients at electrodes Fp1, T6, P3, and O2. With $m = 2$ and all the values of $r$, FuzzyEn was significantly lower ($p < 0.01$) for AD patients at electrodes T6, P3, P4, O1, and O2. These results suggest that AD is associated with a significant decrease of entropy—as estimated by FuzzyEn—in some, but not all, areas of the brain.

For $n = 2$ and $n = 3$ the results did not to follow a normal distribution and the Kruskal-Wallis test was used to evaluate the statistical significance of the findings. With $n = 2$, the number of electrodes where significant differences ($p < 0.01$) between both groups were found dropped significantly when compared to results obtained with $n = 1$. FuzzyEn was significantly lower in AD patients' EEGs at P3 (with $m = 1$ and all values of $r$, and $m = 2$ and $r = 0.15$, 0.2, and 0.25) and O2 (with $m = 1$ and $r = 0.1$, 0.15, and 0.2, and $m = 2$ and $r = 0.2$, and 0.25). With $n = 3$ and $m = 1$, FuzzyEn was only significantly lower ($p < 0.01$) in AD patients' EEGs at electrode O2 (for all combinations of $r$). With $n = 3$ and $m = 2$, the only electrode where FuzzyEn was significantly lower ($p < 0.01$) in AD patients was P3, and this only for $r = 0.2$ and $r = 0.25$. Furthermore, the dispersion of FuzzyEn values increased significantly for $n = 3$, suggesting a less reliable entropy estimate (see results in the Supplementary Materials section).

Figure 2 summarises the average FuzzyEn values for $n = 1$, $m = 2$, and $r = 0.25$ times the SD of the EEG time series, the combination of input parameters that highlights the biggest differences between both groups for all possible input parameter combinations. The decrease in entropy in AD patients is particularly evident for electrodes that are placed over the parietal, occipital, and temporal regions.

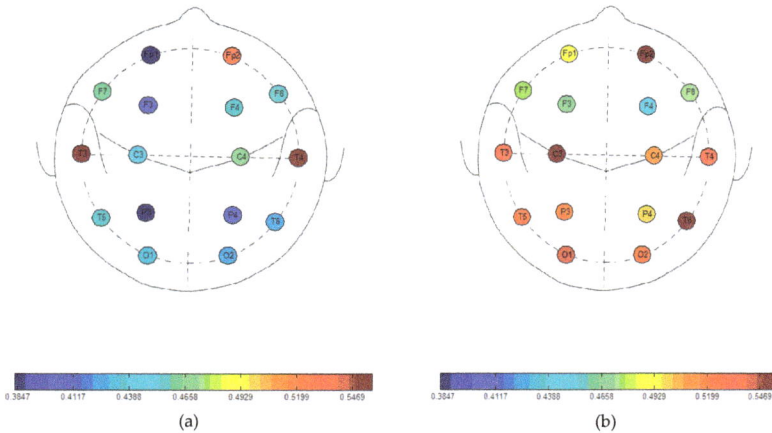

**Figure 2.** Average Fuzzy Entropy (FuzzyEn) values with $n = 1$, $m = 2$ and $r = 0.25$ times the Standard Deviation (SD) of the time series for Alzheimer's disease (AD) patients (**a**) and controls (**b**).

The possible usefulness of FuzzyEn in a diagnostic context was evaluated with ROC curves. The greatest accuracy, at 86.36%, was found when $n = 1$. This was the case in 9 of the 36 electrode and variable combinations where significant differences between the controls subjects and the AD patients had been found using that value of $n$. The largest area under the curve (0.9091) was found at electrode P3 when $n = 1$, $m = 2$, and $r = 0.1$ times the SD of the time series, closely followed by the area at P3, when $n = 1$, $m = 2$, and $r = 0.25$ times the SD of the time series with 0.9008. These did not correspond to the highest accuracy found, with 81.82% accuracy in both cases. Maximum sensitivity was 90.91%, whilst there were some combinations of electrode and input parameter values resulting in 100% specificity. The complete results for FuzzyEn with $n = 1$ are summarised in Table 1.

**Table 1.** Sensitivity, specificity, accuracy, and area under the ROC curve for FuzzyEn ($n = 1$) for all the electrodes and combinations of $m$ and $r$, for which statistically significant differences between AD patients and control subjects were found. The threshold is the FuzzyEn value that maximises accuracy.

| $m$ | $r$ | Electrode | Threshold | Sensitivity | Specificity | Accuracy | Area Under Curve |
|-----|-----|-----------|-----------|-------------|-------------|----------|------------------|
| 1 | 0.1 | Fp1 | 1.1169 | 63.64 | 81.82 | 72.73 | 0.7934 |
| | | T6 | 1.3916 | 81.82 | 81.82 | 81.82 | 0.8182 |
| | | P3 | 1.1755 | 81.82 | 81.82 | 81.82 | 0.8595 |
| | | O2 | 1.2918 | 81.82 | 90.91 | 86.36 | 0.8595 |
| 1 | 0.15 | Fp1 | 0.8015 | 63.64 | 81.82 | 72.73 | 0.7934 |
| | | T6 | 1.0217 | 81.82 | 81.82 | 81.82 | 0.8182 |
| | | P3 | 0.8516 | 81.82 | 90.91 | 86.36 | 0.8678 |
| | | O2 | 0.9393 | 81.82 | 90.91 | 86.36 | 0.8678 |
| 1 | 0.2 | Fp1 | 0.6248 | 63.64 | 81.82 | 72.73 | 0.7934 |
| | | T6 | 0.8040 | 81.82 | 81.82 | 81.82 | 0.8182 |
| | | P3 | 0.6669 | 81.82 | 90.91 | 86.36 | 0.8678 |
| | | O2 | 0.7362 | 81.82 | 81.82 | 81.82 | 0.8554 |
| 1 | 0.25 | Fp1 | 0.5105 | 63.64 | 81.82 | 72.73 | 0.7934 |
| | | T6 | 0.6662 | 81.82 | 81.82 | 81.82 | 0.8182 |
| | | P3 | 0.5473 | 81.82 | 90.91 | 86.36 | 0.8678 |
| | | O2 | 0.6054 | 81.82 | 81.82 | 81.82 | 0.8512 |
| 2 | 0.1 | T6 | 0.8755 | 81.82 | 81.82 | 81.82 | 0.8182 |
| | | P3 | 0.7847 | 81.82 | 81.82 | 81.82 | 0.9091 |
| | | P4 | 0.7380 | 72.73 | 81.82 | 77.27 | 0.8099 |
| | | O1 | 0.8414 | 81.82 | 72.73 | 77.27 | 0.8264 |
| | | O2 | 0.8197 | 90.91 | 81.82 | 86.36 | 0.8512 |
| 2 | 0.15 | T6 | 0.7127 | 81.82 | 81.82 | 81.82 | 0.8182 |
| | | P3 | 0.6295 | 81.82 | 81.82 | 81.82 | 0.8843 |
| | | P4 | 0.5885 | 63.64 | 90.91 | 77.27 | 0.8099 |
| | | O1 | 0.6879 | 81.82 | 72.73 | 77.27 | 0.8182 |
| | | O2 | 0.6617 | 90.91 | 81.82 | 86.36 | 0.8678 |
| 2 | 0.2 | T6 | 0.6018 | 81.82 | 81.82 | 81.82 | 0.8264 |
| | | P3 | 0.5301 | 81.82 | 81.82 | 81.82 | 0.8926 |
| | | P4 | 0.4895 | 63.64 | 100 | 81.82 | 0.8182 |
| | | O1 | 0.5743 | 81.82 | 72.73 | 77.27 | 0.8099 |
| | | O2 | 0.5523 | 90.91 | 81.82 | 86.36 | 0.8595 |
| 2 | 0.25 | T6 | 0.5206 | 81.82 | 81.82 | 81.82 | 0.8264 |
| | | P3 | 0.4564 | 81.82 | 81.82 | 81.82 | 0.9008 |
| | | P4 | 0.4182 | 63.64 | 100 | 81.82 | 0.8182 |
| | | O1 | 0.4926 | 81.82 | 72.73 | 77.27 | 0.8099 |
| | | O2 | 0.4727 | 90.91 | 81.82 | 86.36 | 0.8595 |

ROC results when $n = 2$ are summarised in Table 2. For this particular value of the fuzzy function, there were 12 combinations of electrode and input parameter values that showed a significant decrease of FuzzyEn in AD patients. The greatest accuracy was 81.82% and the largest area under the curve was 0.8843. Neither sensitivity nor specificity reached values that were greater than 81.82% in any case.

With $n = 3$, there were even fewer combinations (six in total) of electrode and input parameter values showing a significant decrease of FuzzyEn in AD. Accuracy reached a maximum value of 81.82%, whilst the largest area under the curve was 0.8678 (for an accuracy of 77.27%). As was also the case for $n = 2$, neither sensitivity, nor specificity, exceeded 81.82%. ROC results for these results are held in Table 3.

**Table 2.** Sensitivity, specificity, accuracy, and area under the ROC curve for FuzzyEn ($n = 2$) for all the electrodes and combinations of $m$ and $r$ for which statistically significant differences between AD patients and control subjects were found. The threshold is the FuzzyEn value that maximises accuracy.

| $m$ | $r$ | Electrode | Threshold | Sensitivity | Specificity | Accuracy | Area Under Curve |
|-----|-----|-----------|-----------|-------------|-------------|----------|------------------|
| 1 | 0.1 | P3 | 1.0782 | 81.82 | 81.82 | 81.82 | 0.8347 |
|   |     | O2 | 1.4320 | 72.73 | 81.82 | 77.27 | 0.8512 |
| 1 | 0.15 | P3 | 0.8648 | 81.82 | 81.82 | 81.82 | 0.8264 |
|   |     | O2 | 1.1963 | 72.73 | 81.82 | 77.27 | 0.8430 |
| 1 | 0.2 | P3 | 0.7279 | 81.82 | 81.82 | 81.82 | 0.8264 |
|   |     | O2 | 1.0326 | 72.73 | 81.82 | 77.27 | 0.8430 |
| 1 | 0.25 | P3 | 0.6377 | 81.82 | 81.82 | 81.82 | 0.8264 |
| 2 | 0.15 | P3 | 0.7627 | 81.82 | 81.82 | 81.82 | 0.8843 |
| 2 | 0.2 | P3 | 0.7231 | 81.82 | 81.82 | 81.82 | 0.8843 |
|   |     | O2 | 0.7832 | 72.73 | 81.82 | 77.27 | 0.8264 |
| 2 | 0.25 | P3 | 0.6875 | 81.82 | 81.82 | 81.82 | 0.8843 |
|   |     | O2 | 0.7493 | 72.73 | 81.82 | 77.27 | 0.8347 |

**Table 3.** Sensitivity, specificity, accuracy, and area under the receiver operating characteristic (ROC) curve for FuzzyEn ($n = 3$) for all the electrodes and combinations of $m$ and $r$ for which statistically significant differences between AD patients and control subjects were found. The threshold is the FuzzyEn value that maximises accuracy.

| $m$ | $r$ | Electrode | Threshold | Sensitivity | Specificity | Accuracy | Area Under Curve |
|-----|-----|-----------|-----------|-------------|-------------|----------|------------------|
| 1 | 0.1 | O2 | 1.4739 | 72.73 | 81.82 | 77.27 | 0.8430 |
| 1 | 0.15 | O2 | 1.3153 | 72.73 | 81.82 | 77.27 | 0.8554 |
| 1 | 0.2 | O2 | 1.2030 | 72.73 | 81.82 | 77.27 | 0.8678 |
| 1 | 0.25 | O2 | 1.1147 | 72.73 | 81.82 | 77.27 | 0.8512 |
| 2 | 0.2 | P3 | 0.7747 | 81.82 | 81.82 | 81.82 | 0.8306 |
| 2 | 0.25 | P3 | 0.7571 | 81.82 | 81.82 | 81.82 | 0.8347 |

## 4. Discussion and Conclusions

Resting state EEG activity of 11 AD patients and 11 control subjects was characterised with FuzzyEn in this pilot study. FuzzyEn was introduced to overcome some limitations of ApEn and SampEn [13], in particular, the fact that both algorithms use a Heaviside function to measure the similarity of the embedding vectors from the time series being compared [13].

FuzzyEn was lower in the EEG of AD patients for all possible combinations of $m$, $n$, and $r$ values and for most electrodes. The greatest number of electrodes (Fp1, T6, P3, P4, O1, O2) showing significant FuzzyEn differences between the EEG of AD patients and controls were seen when $n = 1$, i.e., the steepest gradient of the exponential function. Furthermore, the highest values of accuracy and area under the ROC curve were also obtained with this value of $n$. Our results suggest that brains affected by AD show a more regular electrophysiological behaviour in the parietal and occipital regions, something that is in agreement with previous studies [22–26].

Relevant findings in the changes in the EEG with AD using this same database include a significant reduction in complexity, as measured with the Lempel-Ziv algorithm, at electrodes T5, P3, P4, and O1, with classification accuracies between 72.73% and 81.82% [22]. Some of those electrodes showing reduced complexity (measured with Lempel-Ziv complexity) coincide with those where a significant decrease of irregularity in AD patients' EEGs has been highlighted by FuzzyEn. However, it is worth noting that FuzzyEn was able to find differences in a greater number of electrodes than Lempel-Ziv complexity. A significant loss of complexity in the EEG of AD patients at T5, T6, P3, P4, O1, and O2 was also found with a method based on auto-mutual information, with classification accuracies ranging from 81.82% to 90.91% [23], as well as with multiscale entropy (MSE), for which significant differences between the MSE of AD patients and controls were found at F3, F7, Fp1, Fp2, T5, T6, P3, P4, O1, and O2, with accuracies from 77.27% to 90.91% [24].

More relevant though are previous studies with this same EEG database using ApEn [23], SampEn [25], and QSE [26]. It was found that ApEn was significantly lower in AD patients than in controls at electrodes P3, P4, O1, and O2. However, classification accuracies that were obtained using ROC curves at these electrodes ranged from 72.72% to 77.27% (with the latter value found at P3, O1, and O2). The largest area under the curve was 0.8595 (at P3 and O1). Nevertheless, ApEn results should be interpreted with great care, as this is a biased entropy estimator, and, therefore, not as reliable as other algorithms [10,11]. SampEn values were also significantly lower for AD patients' EEGs than for age-matched controls' EEGs at O1, O2, P3, and P4. Moreover, with SampEn the classification accuracy obtained with ROC curves reached 77.27% at all of those electrodes, supporting the superior discriminating power of SampEn when compared to ApEn, which could arise from the fact of SampEn being an improvement over ApEn. Nevertheless, the largest area under the curve (0.8595 at O1) was similar to the one that is found with ApEn. These results are also supported by recent findings with QSE, with accuracies of 77.27% at P3, O1, and O2, for $m = 1$, $m = 2$, and a range of values of $r$ [26]. All of these results support that EEG activity of AD patients is significantly more regular (less complex) than in a normal brain in the parietal and occipital regions. Table 4 summarises the results obtained with all these methods.

It is worth noting that the detection of a significant decrease of entropy in the EEG of AD patients is heavily dependent on the input parameters that are used in the different entropy estimators. In the case of SampEn, the combination of input parameters that yielded the best results were $m = 1$ and $r = 0.25$ times the standard deviation of the time series [25], whilst for QSE, similar results were obtained with $m = 1$, $m = 2$, and a wide range of values of $r$ [26]. On the other hand, for FuzzyEn, the combination of input parameters that resulted in the greatest number of electrodes showing a statistically significant decrease of entropy in AD (five in total: T6, P3, P4, O1, O2) and the highest accuracies, was $n = 1$, $m = 2$, and $r = 0.25$ times the standard deviation of the time series. In fact, with $n = 1$, $m = 1$, and $r = 0.25$, FuzzyEn was significantly lower for AD patients at electrodes Fp1, T6, P3, and O2, but no significant differences were found at P4 and O1, unlike with SampEn or QSE. Therefore, recent claims of FuzzyEn being superior to ApEn and SampEn in the discrimination of EEG signals in AD patients [19] need to be evaluated with great care, as that might not necessarily be the case for all the combinations of input parameters. Furthermore, our results suggest that FuzzyEn becomes a much less reliable entropy estimator when $n = 3$ than when $n = 1$ or $n = 2$.

Our study is not the first time that the concept of FuzzyEn has been used to evaluate the complexity changes in the EEG in AD patients. Fuzzy versions of ApEn and SampEn (the latter corresponding to the algorithm used herein) were used to compute the complexity of the EEG in the delta, theta, alpha, and beta bands [19]. It was shown that the fuzzy entropies could distinguish EEGs of AD patients from those of controls in a better way than ApEn and SampEn, with a significant decrease in the alpha band, particularly at electrodes T3 and T4. A classification accuracy of 88.1% using fuzzy SampEn and a support vector machine classifier was reported. However, results cannot be compared directly to those that are presented above, as the analysis in [19] focused on different EEG frequency bands, with significant differences being found only in the alpha band. In our study, we have characterised the entropy of a much broader bandwidth of the EEG at rest, therefore limiting the impact of any technique used to extract the EEG activity in different frequency bands. Furthermore, recent evidence suggests that the presence of broadband activity of the EEG is required for a proper evaluation of complexity in the context of AD [27].

**Table 4.** Sensitivity, specificity, and accuracy for all the electrodes where significant differences between AD patients and control subjects were found with a selection of relevant non-linear methods previously used in the analysis of the same electroencephalogram (EEG) database.

| Method | Electrode | ROC Classification Results | | |
|---|---|---|---|---|
| | | Sensitivity | Specificity | Accuracy |
| LZC (3 symbol conversion) [22] | T5 | 72.73 | 72.73 | 72.73 |
| | P3 | 81.82 | 81.82 | 81.82 |
| | P4 | 72.73 | 90.91 | 81.82 |
| | O1 | 90.91 | 72.73 | 81.82 |
| Slope of MSE ($m = 1$, $r = 0.25$, 12 scales) for large time scales [24] | F3 | 81.82 | 81.82 | 81.82 |
| | F7 | 81.82 | 72.73 | 77.27 |
| | Fp1 | 90.91 | 90.91 | 90.91 |
| | Fp2 | 100 | 72.73 | 86.36 |
| | T5 | 90.91 | 81.82 | 86.36 |
| | T6 | 81.82 | 81.82 | 81.82 |
| | P3 | 81.82 | 90.91 | 86.36 |
| | P4 | 72.73 | 90.91 | 81.82 |
| | O1 | 81.82 | 90.91 | 86.36 |
| | O2 | 81.82 | 81.82 | 81.82 |
| ApEn ($m = 1$, $r = 0.25$) [23] | P3 | 72.73 | 81.82 | 77.27 |
| | P4 | 63.64 | 81.82 | 72.73 |
| | O1 | 81.82 | 72.73 | 77.27 |
| | O2 | 90.91 | 63.64 | 77.27 |
| AMI rate of decrease [23] | T5 | 90.91 | 72.73 | 81.82 |
| | T6 | 81.82 | 81.82 | 81.82 |
| | P3 | 100 | 81.82 | 90.91 |
| | P4 | 81.82 | 81.82 | 81.82 |
| | O1 | 81.82 | 81.82 | 81.82 |
| | O2 | 81.82 | 81.82 | 81.82 |
| SampEn ($m = 1$, $r = 0.25$) [25] | P3 | 72.73 | 81.82 | 77.27 |
| | P4 | 63.64 | 90.91 | 77.27 |
| | O1 | 81.82 | 72.73 | 77.27 |
| | O2 | 90.91 | 63.64 | 77.27 |
| * QSE ($m = 1$ and $m = 2$, different values of $r$) [26] | P3 | NR | NR | 77.27 |
| | P4 | NR | NR | 77.27 |
| | O1 | NR | NR | 77.27 |
| | O2 | NR | NR | 77.27 |

NR: not reported; * denotes the studies in which leave-one-subject-out cross-validation was used.

The reasons for the decrease of irregularity in the EEG of AD patients that are highlighted by FuzzyEn are not clear and might be a result of neuronal death, a consequence of neurotransmitter deficiency, and/or loss of connectivity of local neural networks as a result of nerve cell death [1]. These changes might be explained by the theory of AD being a disconnection syndrome [28]: the loss of connections between neurones in the cortex is a result from plaques and cell death [29], and this could lead to a much more regular EEG signal (recording of cortical brain activity).

Our pilot study has some limitations that should be mentioned. Although FuzzyEn is able to highlight subtle differences between EEG signals from AD patients and controls, the sample size used was small (11 AD patients and 11 control subjects). Therefore, the multiple comparisons might have resulted in an overestimation of the differences between the entropy of the EEG from AD patients and controls. Furthermore, the EEG changes that were detected by FuzzyEn might not be specific to AD. The detected increase of EEG regularity (or decrease of complexity) is also present in several physiological and pathological states, including, but not limited to, sleep [30], anaesthesia [31], the Creutzfeld-Jakob disease [32], vascular dementia [33], schizophrenia [34], or Parkinson's disease [35]. Thus, future studies on FuzzyEn of the EEG in patients suffering from other dementias or mild cognitive impairment need to be completed to ascertain the possible usefulness of this signal processing method in the diagnosis of AD.

Other potential future lines of research include the combination of FuzzyEn with MSE (used in [24] with SampEn as the entropy estimator) and the recently introduced refined composite MSE [36,37]. This could lead to further improvements in the characterisation of the EEG in AD. In fact, preliminary evidence suggests that refined multiscale FuzzyEn is able to detect differences due to AD in magnetoencephalograms [37]. Furthermore, a multivariate implementation of FuzzyEn could also be used in the analysis of the EEG in AD. This could potentially increase the discriminating power of the method, as shown with multivariate MSE with SampEn in [27]. However, it could also lead to the loss of the relevant spatial differences that are highlighted in this study (i.e., EEG changes in AD are not significant at all electrodes). Last, but not least, given that complementary information from EEG signals in AD can be highlighted by different methods, the combination of linear and non-linear signal processing algorithms could improve discrimination power. Among some of the entropy methods that could be tested, conditional entropy and corrected conditional entropy [38] and permutation entropy [39] show promise.

In summary, in spite of the aforementioned limitations, our findings with FuzzyEn suggest that this entropy estimator has potential to increase the insight into brain dysfunction in AD, as it detects subtle EEG differences between patients and controls with greater accuracy than SampEn or QSE. However, although our results generally support the notion of FuzzyEn outperforming these methods, as outlined in [17,19], one has to be very careful when comparing results, as that might be the case for certain combinations of input parameters, but not all.

**Supplementary Materials:** The following are available online at www.mdpi.com/1099-4300/20/1/21/s1, Table S1: FuzzyEn($n = 1$, $m = 1$, $r = 0.1$) results, Table S2: FuzzyEn($n = 1$, $m = 1$, $r = 0.15$) results, Table S3: FuzzyEn($n = 1$, $m = 1$, $r = 0.2$) results, Table S4: FuzzyEn($n = 1$, $m = 1$, $r = 0.25$) results, Table S5: FuzzyEn($n = 1$, $m = 2$, $r = 0.1$) results, Table S6: FuzzyEn($n = 1$, $m = 2$, $r = 0.15$) results, Table S7: FuzzyEn($n = 1$, $m = 2$, $r = 0.2$) results, Table S8: FuzzyEn($n = 1$, $m = 2$, $r = 0.25$) results, Table S9: FuzzyEn($n = 2$, $m = 1$, $r = 0.1$) results, Table S10: FuzzyEn($n = 2$, $m = 1$, $r = 0.15$) results, Table S11: FuzzyEn($n = 2$, $m = 1$, $r = 0.2$) results, Table S12: FuzzyEn($n = 2$, $m = 1$, $r = 0.25$) results, Table S13: FuzzyEn($n = 2$, $m = 2$, $r = 0.1$) results, Table S14: FuzzyEn($n = 2$, $m = 2$, $r = 0.15$) results, Table S15: FuzzyEn($n = 2$, $m = 2$, $r = 0.2$) results, Table S16: FuzzyEn($n = 2$, $m = 2$, $r = 0.25$) results, Table S17: FuzzyEn($n = 3$, $m = 1$, $r = 0.1$) results, Table S18: FuzzyEn($n = 3$, $m = 1$, $r = 0.15$) results, Table S19: FuzzyEn($n = 3$, $m = 1$, $r = 0.2$) results, Table S20: FuzzyEn($n = 3$, $m = 1$, $r = 0.25$) results, Table S21: FuzzyEn($n = 3$, $m = 2$, $r = 0.1$) results, Table S22: FuzzyEn($n = 3$, $m = 2$, $r = 0.15$) results, Table S23: FuzzyEn($n = 3$, $m = 2$, $r = 0.2$) results, Table S24: FuzzyEn($n = 3$, $m = 2$, $r = 0.25$) results.

**Acknowledgments:** We would like to thank the Department of Mechanical Engineering Sciences, University of Surrey, Postgraduate Scholarship and the IET Leslie H Paddle Postgraduate Scholarship 2013 for partially funding this work.

**Author Contributions:** Samantha Simons, Pedro Espino, and Daniel Abásolo conceived and designed the experiments. Pedro Espino recorded the EEG data. Samantha Simons performed the experiments and analysed the data. Samantha Simons, Pedro Espino and Daniel Abásolo contributed critically to revise the results and discuss them, and wrote the paper. All authors have read, revised and approved the final manuscript.

**Conflicts of Interest:** The authors declare no conflict of interest.

# References

1. Jeong, J. EEG dynamics in patients with Alzheimer's disease. *Clin. Neurophysiol.* **2004**, *115*, 1490–1505. [CrossRef] [PubMed]

2. Blennow, K.; de Leon, M.J.; Zetterberg, H. Alzheimer's disease. *Lancet* **2006**, *368*, 387–403. [CrossRef]

3. Bird, T.D. Alzheimer's disease and other primary dementias. In *Harrison's Principles of Internal Medicine*, 15th ed.; Braunwald, E., Fauci, A.S., Kasper, D.L., Hauser, S.L., Longo, D.L., Jameson, J.L., Eds.; McGraw-Hill: New York, NY, USA, 2001; pp. 2391–2399.

4. Kalaria, R.N.; Maestre, G.E.; Arizaga, R.; Friedland, R.P.; Galasko, D.; Hall, K.; Luchsinger, J.A.; Ogunniyi, A.; Perry, E.K.; Potocnik, F.; et al. Alzheimer's disease and vascular dementia in developing countries: Prevalence, management, and risk factors. *Lancet Neurol.* **2008**, *7*, 812–826. [CrossRef]

5. McKhann, G.; Drachman, D.; Folstein, M.; Katzman, R.; Price, D.; Stadlan, E.M. Clinical-diagnosis of Alzheimer's disease: Report of the NINCDS-ADRDA Work Group under the auspices of Department of

Health and Human Services Task Force on Alzheimer's Disease. *Neurology* **1984**, *34*, 939–944. [CrossRef] [PubMed]

6. Dauwels, J.; Vialatte, F.; Cichocki, A. Diagnosis of Alzheimer's disease from EEG signals: Where are we standing? *Curr. Alzheimer Res.* **2010**, *7*, 487–505. [CrossRef] [PubMed]

7. Markand, O.N. Organic brain syndromes and dementias. In *Current Practice of Clinical Electroencephalography*; Daly, D.D., Pedley, T.A., Eds.; Raven Press: New York, NY, USA, 1990; pp. 401–423.

8. Sleigh, J.W.; Steyn-Ross, D.A.; Grant, C.; Ludbrook, G. Cortical entropy changes with general anaesthesia: Theory and experiment. *Physiol. Meas.* **2004**, *25*, 921–934. [CrossRef] [PubMed]

9. Pincus, S.M. Approximate entropy as a measure of system complexity. *Proc. Natl. Acad. Sci. USA* **1991**, *88*, 2297–2301. [CrossRef] [PubMed]

10. Porta, A.; Gnecchi-Ruscone, T.; Tobaldini, E.; Guzzetti, S.; Furlan, R.; Montano, N. Progressive decrease of heart period variability entropy-based complexity during graded head-up tilt. *J. Appl. Physiol.* **2007**, *103*, 1143–1149. [CrossRef] [PubMed]

11. Richman, J.S.; Moorman, J.R. Physiological time-series analysis using approximate entropy and sample entropy. *Am. J. Physiol. (Heart Circ. Physiol.)* **2000**, *274*, 2039–2049. [CrossRef] [PubMed]

12. Lake, D.E. Improved entropy rate estimation in physiological data. In Proceedings of the 33rd Annual International Conference of the IEEE EMBS, Boston, MA, USA, 30 August–3 September 2011; pp. 1463–1466.

13. Chen, W.; Wang, Z.; Xie, H.; Yu, W. Characterization of surface EMG signal based on fuzzy entropy. *IEEE Trans. Neural Syst. Rehabil. Eng.* **2007**, *15*, 266–272. [CrossRef] [PubMed]

14. Fawcett, T. An introduction to ROC analysis. *Pattern Recognit. Lett.* **2006**, *27*, 861–874. [CrossRef]

15. Folstein, M.F.; Folstein, S.E.; McHugh, P.R. Mini-mental state. A practical method for grading the cognitive state of patients for the clinician. *J. Psychiatr. Res.* **1975**, *12*, 189–198. [CrossRef]

16. Chen, W.; Zhuang, J.; Yu, W.; Wang, Z. Measuring complexity using FuzzyEn, ApEn and SampEn. *Med. Eng. Phys.* **2009**, *31*, 61–68. [CrossRef] [PubMed]

17. Xie, H.B.; Chen, W.-T.; He, W.-X.; Liu, H. Complexity analysis of the biomedical signal using fuzzy entropy measurement. *Appl. Soft Comput.* **2011**, *11*, 2871–2879. [CrossRef]

18. Fu, A.; Wang, C.; Qi, H.; Li, F.; Wang, Z.; He, F.; Zhou, P.; Chen, S.; Ming, D. Electromyography-based analysis of human upper limbs during 45-day head-down bed-rest. *Acta Astronaut.* **2016**, *120*, 260–269. [CrossRef]

19. Cao, Y.; Cai, L.; Wang, J.; Wang, R.; Yu, H.; Cao, Y.; Liu, J. Characterization of complexity in the electroencephalogram activity of Alzheimer's disease based on fuzzy entropy. *Chaos* **2015**, *25*, 083136. [CrossRef] [PubMed]

20. Cirugeda-Roldan, E.; Cuesta-Frau, D.; Miro-Martinez, P.; Oltra-Crespo, S. Comparative study of entropy sensitivity to missing biosignal data. *Entropy* **2014**, *16*, 5901–5918. [CrossRef]

21. Liu, J.C.; Li, K.; Zhao, L.; Liu, F.; Zheng, D.; Liu, C.; Liu, S. Analysis of heart rate variability using fuzzy measure entropy. *Comput. Biol. Med.* **2013**, *43*, 100–108. [CrossRef] [PubMed]

22. Abásolo, D.; Hornero, R.; Gómez, C.; García, M.; López, M. Analysis of EEG background activity in Alzheimer's disease patients with Lempel-Ziv complexity and Central Tendency Measure. *Med. Eng. Phys.* **2006**, *28*, 315–322. [CrossRef] [PubMed]

23. Abásolo, D.; Escudero, J.; Hornero, R.; Gómez, C.; Espino, P. Approximate entropy and auto mutual information analysis of the electroencephalogram in Alzheimer's disease patients. *Med. Biol. Eng. Comput.* **2008**, *46*, 1019–1028. [CrossRef] [PubMed]

24. Escudero, J.; Abásolo, D.; Hornero, R.; Espino, P.; López, M. Analysis of electroencephalograms in Alzheimer's disease patients with multiscale entropy. *Physiol. Meas.* **2006**, *27*, 1091–1106. [CrossRef] [PubMed]

25. Abásolo, D.; Hornero, R.; Espino, P.; Álvarez, D.; Poza, J. Entropy analysis of the EEG background activity in Alzheimer's disease patients. *Physiol. Meas.* **2006**, *27*, 241–253. [CrossRef] [PubMed]

26. Simons, S.; Abásolo, D.; Escudero, J. Classification of Alzheimer's disease from Quadratic Sample Entropy of the EEG. *IET Healthc. Technol. Lett.* **2015**, *2*, 70–73. [CrossRef] [PubMed]

27. Azami, H.; Abásolo, D.; Simons, S.; Escudero, J. Univariate and multivariate generalized multiscale entropy to characterise EEG signals in Alzheimer's Disease. *Entropy* **2017**, *19*, 31. [CrossRef]

28. Delbeuck, X.; Van der Linden, M.; Collette, F. Alzheimer's disease as a disconnection syndrome? *Neuropsychol. Rev.* **2003**, *13*, 79–92. [CrossRef] [PubMed]

29. Morrison, J.H.; Scherr, S.; Lewis, D.A.; Campbell, M.J.; Bloom, F.E.; Rogers, L.; Benoit, R. The laminar and regional distribution of neocortical somatostatin and neuritic plaques: Implications for Alzheimer's disease as a global neocortical disconnection syndrome. In *The Biological Substrates of Alzheimer's Disease*; Scheibel, A.B., Wechsler, A.F., Eds.; Academic Press: Orlando, FA, USA, 1986; pp. 115–131.

30. Burioka, N.; Cornélissen, G.; Halberg, F.; Kaplan, D.T.; Suyama, H.; Sako, T.; Shimizu, E. Approximate entropy of human respiratory movement during eye-closed waking and different sleep stages. *Chest* **2003**, *123*, 80–86. [CrossRef] [PubMed]

31. Zhang, X.S.; Roy, R.J. Derived fuzzy knowledge model for estimating the depth of anesthesia. *IEEE Trans. Biomed. Eng.* **2001**, *48*, 312–323. [CrossRef] [PubMed]

32. Babloyantz, A.; Destexhe, A. The Creutzfeldt-Jakob disease in the hierarchy of chaotic attractors. In *From Chemical to Biological Organization*; Markus, M., Müller, S.C., Nicolis, G., Eds.; Springer: Berlin, Germany, 1988; pp. 307–316.

33. Jeong, J.; Kim, S.J.; Han, S.H. Non-linear dynamical analysis of the EEG in Alzheimer's disease with optimal embedding dimension. *Electroenceph. Clin. Neurophysiol.* **1998**, *106*, 220–228. [CrossRef]

34. Röschke, J.; Fell, J.; Beckmann, P. Non-linear analysis of sleep EEG data in schizophrenia: Calculation of the principal Lyapunov exponent. *Psychiatr. Res.* **1995**, *56*, 257–269. [CrossRef]

35. Stam, C.J.; Jelles, B.; Achtereekte, H.A.M.; Rombouts, S.A.R.B.; Slaets, J.P.J.; Keunen, R.W.M. Investigation of EEG nonlinearity in dementia and Parkinson's disease. *Electroenceph. Clin. Neurophysiol.* **1995**, *95*, 309–317. [CrossRef]

36. Azami, H.; Rostaghi, M.; Abásolo, D.; Escudero, J. Refined composite multiscale dispersion entropy and its application to biomedical signals. *IEEE Trans. Biomed. Eng.* **2017**, *64*, 2972–2979. [CrossRef] [PubMed]

37. Azami, H.; Fernández, A.; Escudero, J. Refined multiscale fuzzy entropy based on standard deviation for biomedical signal analysis. *Med. Biol. Eng. Comput.* **2017**, *55*, 2037–2052. [CrossRef] [PubMed]

38. Porta, A.; De Maria, B.; Bari, V.; Marchi, A.; Faes, L. Are nonlinear model-free conditional entropy approaches for the assessment of cardiac control complexity superior to the linear model-based one? *IEEE Trans. Biomed. Eng.* **2017**, *64*, 1287–1296. [CrossRef] [PubMed]

39. Bandt, C.; Pompe, B. Permutation entropy: A natural complexity measure for time series. *Phys. Rev. Lett.* **2002**, *88*, 174102. [CrossRef] [PubMed]

*entropy*

MDPI

*Article*

# Automated Multiclass Classification of Spontaneous EEG Activity in Alzheimer's Disease and Mild Cognitive Impairment

Saúl J. Ruiz-Gómez [1] , Carlos Gómez [1,*], Jesús Poza [1,2,3], Gonzalo C. Gutiérrez-Tobal [1], Miguel A. Tola-Arribas [4], Mónica Cano [5] and Roberto Hornero [1,2,3]

[1] Biomedical Engineering Group, E.T.S.I. de Telecomunicación, Universidad de Valladolid, 47011 Valladolid, Spain; saul.ruiz@gib.tel.uva.es (S.J.R.-G.); jesus.poza@tel.uva.es (J.P.); gonzalo.gutierrez@gib.tel.uva.es (G.C.G.-T.); roberto.hornero@tel.uva.es (R.H.)
[2] Instituto de Investigación en Matemáticas (IMUVA), Universidad de Valladolid, 47011 Valladolid, Spain
[3] Instituto de Neurociencias de Castilla y León (INCYL), Universidad de Salamanca, 37007 Salamanca, Spain
[4] Servicio de Neurología, Hospital Universitario Río Hortega, 47012 Valladolid, Spain; mtola.nrl@gmail.com
[5] Servicio de Neurofisiología Clínica, Hospital Universitario Río Hortega, 47012 Valladolid, Spain; mcanopo@saludcastillayleon.es
* Correspondence: carlos.gomez@tel.uva.es; Tel.: +34-983-423-981

Received: 15 December 2017; Accepted: 5 January 2018; Published: 9 January 2018

**Abstract:** The discrimination of early Alzheimer's disease (AD) and its prodromal form (i.e., mild cognitive impairment, MCI) from cognitively healthy control (HC) subjects is crucial since the treatment is more effective in the first stages of the dementia. The aim of our study is to evaluate the usefulness of a methodology based on electroencephalography (EEG) to detect AD and MCI. EEG rhythms were recorded from 37 AD patients, 37 MCI subjects and 37 HC subjects. Artifact-free trials were analyzed by means of several spectral and nonlinear features: relative power in the conventional frequency bands, median frequency, individual alpha frequency, spectral entropy, Lempel–Ziv complexity, central tendency measure, sample entropy, fuzzy entropy, and auto-mutual information. Relevance and redundancy analyses were also conducted through the fast correlation-based filter (FCBF) to derive an optimal set of them. The selected features were used to train three different models aimed at classifying the trials: linear discriminant analysis (LDA), quadratic discriminant analysis (QDA) and multi-layer perceptron artificial neural network (MLP). Afterwards, each subject was automatically allocated in a particular group by applying a trial-based majority vote procedure. After feature extraction, the FCBF method selected the optimal set of features: individual alpha frequency, relative power at delta frequency band, and sample entropy. Using the aforementioned set of features, MLP showed the highest diagnostic performance in determining whether a subject is not healthy (sensitivity of 82.35% and positive predictive value of 84.85% for HC vs. all classification task) and whether a subject does not suffer from AD (specificity of 79.41% and negative predictive value of 84.38% for AD vs. all comparison). Our findings suggest that our methodology can help physicians to discriminate AD, MCI and HC.

**Keywords:** Alzheimer's disease; mild cognitive impairment; electroencephalography (EEG); spectral analysis; nonlinear analysis; multiclass classification approach

---

## 1. Introduction

Dementia due to Alzheimer's disease (AD) is a progressive neurodegenerative disorder associated with cognitive, behavioral and functional alterations. AD prevalence increases exponentially with age, from 1% in people between 60 and 64 years up to 38% in people over 85 years [1]. Since AD is increasingly being recognized as a modern epidemic, growing efforts have been devoted to exploring its underlying brain dynamics. Despite the considerable progress made to understand AD pathophysiology, a better characterization of its early stages is still required [1]. Mild cognitive impairment (MCI) subjects exhibit a memory impairment beyond what would be expected for their age, but do not fully accomplish the criteria for dementia diagnosis [2]. In this regard, further research is essential to identify incipient AD, since subjects with MCI have high risk of developing it [3]. Recent studies estimated that the conversion rate from MCI to AD is approximately 15% per year [4], whereas this rate is only 1–2% from global population [1]. Despite the fact that current pharmacological treatments and non-pharmacological therapies are not able to heal AD or MCI, an early diagnosis is still crucial since these are more effective in the first stages of the dementia [5].

Several neuroimaging techniques have been used during the last decades with the aim of distinguishing AD and MCI patients from cognitively healthy control (HC) subjects: functional magnetic resonance imaging (fMRI), positron emission tomography (PET), magnetic resonance spectroscopy, electroencephalography (EEG), and magnetoencephalography (MEG), among others [6]. PET and fMRI show a good structural accuracy, but both offer a limited temporal resolution. By contrast, EEG and MEG are non-invasive techniques with high temporal resolution, allowing for studying the dynamical processes involved in the regulation of complex functional brain systems [7]. Particularly, EEG is widely used due to its portability, low cost, and availability. Moreover, EEG has already shown its usefulness to characterize brain dynamics in AD and MCI [7–14].

The abnormalities that AD and MCI elicit in EEG activity have been traditionally analyzed using simple signal processing methods, such as spectral techniques [13,14]. Spectral analyses seem to discriminate AD and MCI patients from HC subjects through a power increase in low frequency bands, as well as a decrease in higher frequencies [13,14]. Since the mid 1990s, nonlinear analysis techniques have also been widely used in order to provide complementary information to spectral measures [10]. Previous studies suggested a more regular EEG activity for AD and MCI patients when compared to HC subjects [11,14]. Other authors reported a decrease of variability and complexity as the disease worsens [7–9,12]. However, almost all these studies only applied one or a few methods to partially characterize the brain dynamics in AD and MCI.

The main objective of this study is to evaluate the diagnostic usefulness of an EEG-based methodology by means of different multiclass classifiers: logistic discriminant analysis (LDA), quadratic discriminant analysis (QDA) and multi-layer perceptron neural network (MLP). We hypothesize that the combination of spectral measures and nonlinear methods can be useful to help in AD and MCI diagnosis. For this reason, our proposed methodology is based on both frequency (spectral features) and time domain (nonlinear features) analyses applied to EEG recordings. However, this exhaustive characterization of EEG may lead to obtaining redundant features sharing similar information. In order to avoid this issue, an automatic feature selection stage based on the fast correlation-based filter (FCBF) is followed [15]. Finally, a classification approach is also conducted. Previous studies performed a binary classification approach facing AD vs. HC, MCI vs. HC and AD vs. MCI [16–20]. Only McBride et al. reported a three-way classification, but via binary classifiers [21]. Additionally, their approach was validated through a leave-one-out cross-validation procedure, leading to multiple models. By contrast, our proposal focuses on building a single multiclass model to determine the group for each subject. This is an essential feature for a simplified screening protocol in the future. Afterwards, the group for each subject was settled with a trial-based majority vote procedure, as proposed in previous studies involving early AD recognition [22].

## 2. Materials and Methods

### 2.1. Subjects

EEG data were recorded from 111 subjects: 37 AD patients, 37 MCI patients, and 37 elderly HC subjects. Patients with dementia or MCI due to AD were diagnosed according to the clinical National Institute on Aging and Alzheimer's Association (NIA-AA) criteria, whereas HC were elderly subjects without a cognitive impairment and with no history of neurological or psychiatric disorder [23]. Inclusion and exclusion criteria for each group can be found in our previous study [20].

All participants and patients' caregivers were informed about the research background and the study protocol. Moreover, all of them gave their written informed consent to be included in the study. The Ethics Committee at the Río Hortega University Hospital (Valladolid, Spain) endorsed the study protocol, according to The Code of Ethics of the World Medical Association (Declaration of Helsinki).

## 2.2. EEG Recording

Five minutes of spontaneous EEG activity were recorded using a 19-channel EEG system (XLTEK®, Natus Medical, Pleasanton, CA, USA). Specifically, EEG activity was acquired from Fp1, Fp2, Fz, F3, F4, F7, F8, Cz, C3, C4, T3, T4, T5, T6, Pz, P3, P4, O1, and O2, at a sampling frequency of 200 Hz. Subjects were asked to stay in a relaxed state, awake, and with closed eyes during EEG acquisition. During the recording procedure, EEG traces were visually monitored in real time, and muscle activity was identified to avoid high-frequency noise. Additionally, independent component analysis (ICA) was performed to minimize the presence of oculographic, cardiographic, and myographic artifacts [7]. Afterwards, EEG signals were digitally filtered using a finite impulse response filter designed with a Hamming window between 1 and 70 Hz and a notch filter to remove the power line frequency interference (50 Hz, Butterworth filter). Finally, an experienced technician selected artifact-free epochs of 5-s by visual inspection.

We randomly divided our EEG database into training and test sets. The training set was formed by: 20 AD patients (45.85 ± 8.36 trials per subject, mean ± standard deviation, SD), 20 MCI subjects (46.85 ± 10.68 trials per subject) and 20 HC subjects (45.60 ± 7.93 trials per subject). The recordings not selected for the training set were assigned to the test set: 17 AD patients (44.53 ± 10.10 trials per subject), 17 MCI subjects (49.82 ± 8.29 trials per subject) and 17 HC subjects (44.24 ± 7.81 trials per subject). No statistically significant differences were found in age ($p$-value > 0.05, Kruskal–Wallis test) and gender ($p$-value > 0.05, chi-squared test) among AD, MCI, and HC groups. Table 1 shows relevant socio-demographic and clinical data for each group.

**Table 1.** Social-demographic and clinical data for each group.

|  | Training Set | | | Test Set | | |
|---|---|---|---|---|---|---|
|  | HC | MCI | AD | HC | MCI | AD |
| Number of subjects | 20 | 20 | 20 | 17 | 17 | 17 |
| Number of trials | 912 | 937 | 917 | 752 | 847 | 757 |
| Age (years) | 75.6 | 77.9 | 80.7 | 76.4 | 75.3 | 82.4 |
| (median [IQR]) | [74.1, 77.6] | [67.9, 79.8] | [74.7, 83.3] | [73.6, 78.9] | [69.8, 82.0] | [77.7, 83.9] |
| Gender (Male:Female) | 8:12 | 8:12 | 5:15 | 4:13 | 8:9 | 7:10 |
| MMSE [1] | 29 | 27.5 | 21 | 29 | 27 | 22 |
| (median [IQR]) | [28, 30] | [26.5, 29] | [18.5, 22.5] | [28, 30] | [27, 28] | [20, 24] |
| B-ADL [2] | 1.1 | 2.9 | 5.8 | 1.2 | 2.8 | 6.4 |
| (median [IQR]) | [1.0, 1.2] | [2.4, 3.3] | [5.1, 7.2] | [1.0, 1.3] | [2.3, 2.5] | [5.0, 4.3] |
| Education level (A:B) [3] | 5:15 | 11:9 | 8:12 | 5:12 | 12:5 | 10:7 |

[1] MMSE: Mini Mental State Examination; [2] B-ADL: Bayer-Activities of Daily Living; [3] A: primary education or below, B: secondary education or above.

## 2.3. Methods

The methodology followed in this study is represented in Figure 1. After EEG-signal recording and data pre-processing, both spectral and nonlinear features were computed. Then, FCBF was applied to the training set to automatically select an optimum set of features. Finally, three different multiclass classification approaches (LDA, QDA, and MLP) were adopted to settle the group for each trial and subject.

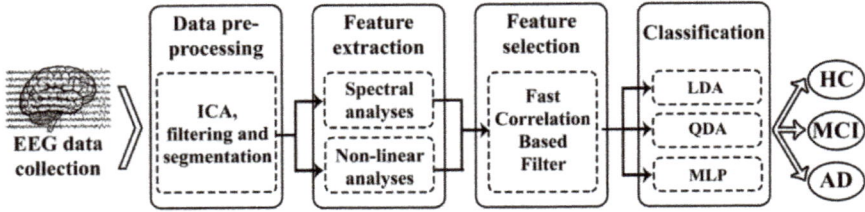

**Figure 1.** Block diagram of the steps followed in the EEG analysis: data collection, pre-processing, feature extraction, feature selection and classification.

### 2.3.1. Feature Extraction

#### Spectral Analysis

A typical approach to characterize electromagnetic brain recordings is based on the analysis of their spectral content [24–26]. Spectral parameters are based on the normalized power spectral density in the frequency band of interest ($PSD_n$). In this request, the following spectral parameters have been calculated from the $PSD_n$: relative power (*RP*), median frequency (*MF*), individual alpha frequency (*IAF*), and spectral entropy (*SE*).

- *RP* represents the relative contribution of different frequency components to the global power spectrum. *RP* is more appropriate than absolute power to analyze EEG data, as *RP* provides independent thresholds from the measurement equipment and lower inter-subject variability [27]. *RP* is obtained by summing the contribution of the desired spectral components:

$$RP(f_1, f_2) = \sum_{f_1}^{f_2} PSD_n(f),$$ 
(1)

where $f_1$ and $f_2$ are the low and the high cut-off frequencies of each band, respectively.
In this study, *RP* was calculated in the conventional EEG frequency bands: delta ($\delta$, 1–4 Hz), theta ($\theta$, 4–8 Hz), alpha ($\alpha$, 8–13 Hz), beta-1 ($\beta_1$, 13–19 Hz), beta-2 ($\beta_2$, 19–30 Hz) and gamma ($\gamma$, 30–70 Hz).

- *MF* offers an alternative way to quantify the spectral changes of the EEG, and it is a simple index that summarizes the whole spectral content of the $PSD_n$. *MF* is defined as the frequency that comprises 50% of the $PSD_n$ power:

$$\sum_{1Hz}^{MF} PSD_n(f) = 0.5 \sum_{1Hz}^{70Hz} PSD_n(f).$$ 
(2)

Previous studies suggested that *MF* provides a better performance for the characterization of brain activity than mean frequency, whose original definition is based on the computation of the spectral centroid [28].

- *IAF* evaluates the frequency at which the maximum alpha power is reached. Alpha oscillations are dominant in the EEG of resting normal subjects, with the exception of irregular activity in the delta band and lower frequencies. This issue involves that the *PSD* displays a peak around the alpha band. The *IAF* estimation in the present work is based on the calculation of the *MF* in the extended alpha band (4–15 Hz), as previous EEG studies on AD recommended [29]. This is shown in the following equation:

$$\sum_{1Hz}^{IAF} PSD_n(f) = 0.5 \sum_{4Hz}^{15Hz} PSD_n(f).$$ 
(3)

- *SE* estimates the signal irregularity in terms of the flatness of the power spectrum [30]. On the one hand, a uniform power spectrum with a broad spectral content (e.g., a highly irregular signal like white noise) provides a high entropy value. On the other hand, a narrow power spectrum with only a few spectral components (e.g., a highly predictable signal like a sum of sinusoids) yields a low *SE* value. The equation for calculating *SE* would be:

$$SE = - \sum_{1\mathrm{Hz}}^{70\mathrm{Hz}} PSD_n(f) \cdot \log[PSD_n(f)]. \tag{4}$$

Nonlinear Analysis

Alterations caused by AD and MCI also modify complexity, variability and the irregularity of the EEG activity [9,12,31–34]. Hence, to complement the spectral analysis, five global nonlinear methods were also calculated: Lempel–Ziv complexity (*LZC*), central tendency measure (*CTM*), sample entropy (*SampEn*), fuzzy entropy (*FuzzyEn*), and auto-mutual information (*AMI*).

- *LZC* estimates the complexity of a finite sequence of symbols. *LZC* analysis is based on a coarse-graining of measurements. Therefore, the EEG signal must be previously transformed into a finite symbol string. In this study, we used the simplest possible way: a binary sequence conversion (zeros and ones). By comparison with a threshold $T_d$, the original signal samples are converted into a 0–1 sequence $P = s(1), s(2), \ldots, s(N)$ with $s(i)$ defined by:

$$s(i) = \begin{cases} 0 \; if \; x(i) < T_d \\ 1 \; if \; x(i) \geq T_d \end{cases}. \tag{5}$$

The threshold $T_d$ is estimated as the median value of the signals amplitude in each channel because it is more robust to outliers. The string $P$ is then scanned from left to right and a complexity counter $c(N)$ is increased by one every time a new subsequence of consecutive characters is encountered in the scanning process. In order to obtain a complexity measure that is independent of the sequence length, $c(N)$ should be normalized. For a binary conversion, the upper bound of $c(N)$ is given by $b(N) = N/\log_2(N)$ and $c(N)$ can be normalized via $b(N)$:

$$LZC = \frac{c(N)}{b(N)}. \tag{6}$$

*LZC* values are normalized between 0 and 1, with higher *LZC* values for more complex time series. The detailed algorithm for *LZC* measure can be found in [35].
- *CTM* quantifies the variability of a given time series on the basis of its first-order differences. For *CTM* calculation, scatter plots of first differences of the data are drawn. The value of *CTM* is computed as the proportion of points in the plot that fall within a radius $\rho$, which must be specified [36]. For a time series with $N$ samples, $N - 2$ would be the total number of points in the scatter plot that can be plotted by representing $x(n+2) - x(n+1)$ versus $x(n+1) - x(n)$. Subsequently, the *CTM* of the time series can be computed as:

$$CTM = \frac{\sum_{i=1}^{N-2} \delta(d_i)}{N-2}, \tag{7}$$

where

$$\delta(d_i) = \begin{cases} 1 \; if \; \left[ (x(i+2) - x(i+1))^2 + (x(i+1) - x(i))^2 \right]^{\frac{1}{2}} < \rho \\ 0 \; otherwise \end{cases}. \tag{8}$$

Thus, *CTM* ranges between 0 and 1, with higher values corresponding to points more concentrated around the center of the plot (i.e., corresponding to less degree of variability).

- *SampEn* is an embedding entropy used to quantify the irregularity. It can be applied to short and relatively noisy time series [37]. To compute *SampEn*, two input parameters should be specified: a run length $m$ and a tolerance window $r$. *SampEn* is the negative natural logarithm of the conditional probability that two sequences similar for $m$ points remain similar at the next point, within a tolerance $r$, excluding self-matches [37]. Thus, *SampEn* assigns a nonnegative number to a time series, with larger values corresponding to greater signal irregularity. For a time series of $N$ points, $X(n) = \{x(1), x(2), \ldots, x(N)\}$, the $k = 1, \ldots, N - m + 1$ vectors of length $m$ are formed as $X_m(k) = \{x(k+i), i = 0, \ldots, m-1\}$. The distances among vectors are calculated as the maximum absolute distance between their corresponding scalar elements. $B_i$ is the number of vectors that satisfy the condition that their distance is less than $r$. The counting number of different vectors is calculated and normalized as [37]:

$$B^m(r) = \frac{1}{N-m} \sum_{i=1}^{N-m} \frac{B_i}{N-m-1}. \tag{9}$$

Repeating the process for vectors of length $m + 1$, $B^{m+1}(r)$ can be obtained and *SampEn* can be defined as:

$$SampEn(m,r) = -ln\left[\frac{B^{m+1}(r)}{B^m(r)}\right]. \tag{10}$$

- *FuzzyEn* provides information about how a signal fluctuates with time by comparing the time series with a delayed version of itself [38]. As *SampEn*, higher *FuzzyEn* values are associated with more irregular time series. To compute *FuzzyEn*, three parameters must be fixed. The first parameter, $m$, is the length of the vectors to be compared, like in *SampEn*. The other ones, $r$ and $n$, are the width and the gradient of the boundary of the exponential function, respectively [38]. Given a time series $X(n) = \{x(1), x(2), \ldots, x(N)\}$, the *FuzzyEn* algorithm reads as follows:

1.  Compose $N - m + 1$ vectors of length $m$ such that:

$$X_i^m = \{x(i), x(i+1), \ldots, x(i+m-1)\} - x_0(i), \tag{11}$$

    where $x_0(i)$ is given by:

$$x_0(i) = \frac{1}{m} \sum_{j=0}^{m-1} x(i+j). \tag{12}$$

2.  Compute the distance, $d_{ij}^m$, between each two vectors, $X_i^m$ and $X_j^m$, as the maximum absolute difference of their corresponding scalar components. Given $n$ and $r$, calculate the similarity degree, $D_{ij}^m$, between $X_i^m$ and $X_j^m$ through a fuzzy function $\mu(d_{ij}^m, n, r)$:

$$D_{ij}^m(n,r) = \mu\left(d_{ij}^m, n, r\right) = \exp\left[-\frac{(d_{ij}^m)^n}{r}\right]. \tag{13}$$

3.  Define the function $\phi^m$ as:

$$\phi^m(n,r) = \frac{1}{N-m} \sum_{i=1}^{N-m} \left(\frac{1}{N-m+1} \sum_{j=1,\, j\neq i}^{N-m} D_{ij}^m\right). \tag{14}$$

4.  Increase the dimension to $m + 1$, form the vector $X_i^{m+1}$ and the function $\phi^{m+1}$. Finally, $FuzzyEn(m, n, r)$ is defined as the negative natural logarithm of the deviation of $\phi^m$ from $\phi^{m+1}$:

$$FuzzyEn(m,\ n,\ r) = \ln[\phi^m(n,r)] - \ln\left[\phi^{m+1}(n,r)\right]. \tag{15}$$

- *AMI* is the particularization of mutual information applied to time-delayed versions of the same sequence. Mutual information is a metric derived from Shannon's information theory to estimate the information gain from observations of one random event on another [31]. *AMI* estimates, on average, the degree to which a time-delayed version of a signal can be predicted from the original one. Thus, more predictable time series, and accordingly more regular, lead to higher *AMI* values. The *AMI* between $X(n)$ and $X(n+k)$ is [31]:

$$AMI = \sum_{X(n),\ X(n+k)} P_{XX\tau}[X(n),\ X(n+k)] \log_2\left\{\frac{P_{XXk}[X(n),X(n+k)]}{P_{Xk}[X(n)]\ P_{Xk}[X(n+k)]}\right\}, \tag{16}$$

where $P_{Xk}[X(n)]$ is the probability density for the measurement $X(n)$, while $P_{XXk}[X(n),\ X(n+k)]$ is the joint probability density for the measurements of $X(n)$ and $X(n+k)$. In this study, the *AMI* was estimated over a time delay from 0 to 0.5 s and was then normalized, so that $AMI(k = 0) = 1$.

### 2.3.2. Feature Selection: Fast-Correlation-Based Filter

The aforementioned characterization of the EEG may lead to the extraction of several features that provide similar information about the brain dynamics in AD, MCI, and HC. Consequently, a feature selection stage was also included. In our study, FCBF was used to discard those redundant features that share more information with the other ones than with the variable that defines the group membership. FCBF is based on symmetrical uncertainty (*SU*), which is a normalized quantification of the information gain between each feature and the group membership variables [15]. It consists of two steps: relevance and redundancy analyses of the features.

- In the first step, a relevance analysis of the features is done. Thus, *SU* between each feature $X_i$ and the group membership $Y$ is computed as follows:

$$SU(X_i, Y) = 2\left[\frac{H(X_i) - H(X_i|Y)}{H(X_i) + H(Y)}\right], \quad i = 1, 2, \ldots, I, \tag{17}$$

where $H(\cdot)$ is the well-known Shannon's entropy, $H(X_i|Y)$ is the Shannon's entropy of $X_i$ conditioned on $Y$, and $I$ is the number of features extracted (in our study, $I = 14$ features). *SU* is normalized to the range [0, 1], with a value of $SU = 1$, indicating that, when knowing one feature, it is possible to completely predict the other, and a value of $SU = 0$ indicates that the two variables are independent. Then, a ranking of features is done based on their relevance since the higher the value of *SU* is, the more relevant the feature is.

- The second step is a redundancy analysis used to discard redundant features. *SU* between each pair of features $SU(X_i, X_j)$ is sequentially estimated beginning from the first-ranked ones. If $X_i$ shares more information with $X_j$ than with the corresponding group $Y$, $SU(X_i, X_j) \geq SU(X_i, Y)$ (with $X_i$ being more highly ranked than $X_j$), the feature $j$ is discarded due to redundancy and it is not considered in subsequent comparisons. The optimal features are those not discarded when the algorithm ends.

### 2.3.3. Classification Approach

The described AD-MCI-HC diagnosis problem corresponds to a pattern classification task. Specifically, it can be modeled as a three-class classification problem. Bayesian decision theory establishes the rule to make such a decision to minimize the probability of misclassification [39]. We have implemented LDA, QDA, and MLP models to ensure that our conclusions take into account a variety of classification methodologies. In this study, we classify trials using each trained model, and, then, every subject is classified by means of a majority vote of all its trials [22].

Linear and Quadratic Discriminant Analysis (LDA and QDA)

LDA takes an input vector and assigns it to one out of the $K$ classes using linear hyperplanes as decision surfaces [40]. This classifier assumes that different classes generate data based on different Gaussian distributions, whose parameters are estimated with the fitting function during the training. In order to predict the classes of new data, the trained model finds the class with the smallest misclassification cost assuming that the covariance matrices of each class are identical (homoscedasticity) [40].

QDA is a classification approach closely related to LDA. However, there is no assumption that the covariance of all classes are identical among them and it establishes a quadratic decision boundary between classes in the feature space [40].

Multi-Layer Perceptron Artificial Neural Network (MLP)

MLP is an artificial neural network that maps an input vector onto a set of output variables using a nonlinear function controlled by a vector of adjustable parameters. The use of neural networks for classification issues has some advantages. First, no prior assumptions about the distribution of the data are required, since neural network algorithms adjust themselves to the environment by means of the training or learning process. Thus, complex relationships can be modeled by these algorithms [41].

An MLP consists of three or more layers (an input and an output layer with one or more hidden layers) of neurons, with each layer fully connected to the next one. In our study, we have evaluated MLP networks with a single hidden layer of neurons, since networks with this architecture are capable of universal approximation [42]. MLP utilizes backpropagation in conjunction with an optimization method, such as gradient descent, with the aim of finding appropriate weights to connect neurons each other. Backpropagation is based on the definition of a suitable error function, which is minimized by updating the weights in the network [39].

In order to predict the classes for new data, the trained MLP model provides the posterior probability of belonging to each class. A three-class classification problem involves the use of three output neurons, one neuron per group. In our study, the number of neurons in the hidden layer ($n_h$) and a regularization parameter ($u$) were optimized by cross-validation leaving all trials of a subject out in every iteration in the training set. This procedure was carried out 30 times to minimize the effect of network random initialization and then the results were averaged [43]. NETLAB toolbox was used to implement the neural network classifier [44].

*2.4. Statistical Analysis*

The three-class diagnostic ability of the models was assessed in terms of accuracy (*Acc*, overall percentage of subjects rightly classified) and Cohen's kappa (*k*). *k* measures the agreement between predicted and observed classes, avoiding the part of agreement by chance [45]. On the other hand, the performance of the models for HC vs. all and AD vs. all comparison was described by sensitivity (*Se*, percentage of positive subjects appropriately classified), specificity (*Sp*, percentage of negative subjects correctly classified), *Acc*, positive predictive value (*PPV*, proportion of positive estimations of the models that are true positive results) and negative predictive value (*NPV*, proportion of negative estimations of the models that are true negative results).

**3. Results**

According to the proposed methods, we calculated 14 features from each EEG channel. Nine spectral features: $RP(d)$ (where $RP(d)$ represents de $RP$ value for the $d$ band), $RP(q)$, $RP(a)$, $RP(b_1)$,

$RP(\beta_2)$, $RP(g)$, *MF*, *IAF*, and *SE*, and five derived from the nonlinear methods: *LZC*, *CTM*, *SampEn*, *FuzzyEn*, and *AMI*. The results were obtained based on all the artifact-free trials within the five-minute period of recording. Results from all EEG channels were averaged in order to achieve one value per trial for each method.

## 3.1. Training Set

In order to select the optimal value of the different input parameters of each feature, only a training set was used. The optimal value for $r$ (*CTM*) was obtained by evaluating the range $r \in [0.01, 0.5]$ (step = 0.005). Values of $r < 0.01$ were not considered, since they led to a *CTM* value close to 0 for every subject, whereas values of $r > 0.5$ were also discarded since they led to *CTM* values equal to 1 regardless the group. For both *SampEn* and *FuzzyEn*, $m$ and $r$ optimal values were obtained by evaluating all the combinations for $m = 1, 2$ and $r \in (0.1 \cdot SD, 0.25 \cdot SD)$ (step = 0.05), where SD is the standard deviation of the time series [38,46]. In the case of *FuzzyEn*, values of $n = 1, 2, 3$ were also evaluated to obtain its optimal value [38]. We chose those configurations ($r = 0.075$ for *CTM*; $m = 1$ and $r = 0.1 \cdot SD$ for *SampEn*; and $m = 1$, $r = 0.1 \cdot SD$, and $n = 3$ for *FuzzyEn*) for which the corresponding *CTM*, *SampEn*, and *FuzzyEn* values showed the lowest *p*-value (Kruskal–Wallis test) among the three groups. Table 2 summarizes the averaged results for each group, taking into account only the training set. After feature extraction, FCBF was applied to the training set. The final FCBF optimal set was composed of three features: two spectral measures (*IAF* and *RP(d)*) and a nonlinear one (*SampEn*).

**Table 2.** Averaged results (median (interquartile range)) for each group and for each feature taking into account only the training set.

| Features | HC | MCI | AD |
|---|---|---|---|
| $RP(\delta)$ | 0.227 [0.179, 0.277] | 0.164 [0.102, 0.221] | 0.158 [0.103, 0.229] |
| $RP(\theta)$ | 0.111 [0.083, 0.131] | 0.122 [0.087, 0.155] | 0.143 [0.103, 0.188] |
| $RP(\alpha)$ | 0.243 [0.174, 0.291] | 0.317 [0.224, 0.544] | 0.279 [0.192, 0.447] |
| $RP(\beta_1)$ | 0.128 [0.101, 0.155] | 0.101 [0.081, 0.160] | 0.101 [0.073, 0.141] |
| $RP(\beta_2)$ | 0.111 [0.084, 0.138] | 0.105 [0.048, 0.135] | 0.091 [0.060, 0.119] |
| $RP(\gamma)$ | 0.097 [0.074, 0.168] | 0.087 [0.037, 0.145] | 0.089 [0.047, 0.141] |
| MF | 10.584 [9.690, 11.900] | 10.467 [8.639, 12.285] | 9.971 [9.030, 10.997] |
| IAF | 9.502 [8.751, 9.996] | 9.404 [8.519, 9.972] | 8.811 [8.510, 9.474] |
| SE | 0.813 [0.760, 0.822] | 0.796 [0.695, 0.816] | 0.782 [0.733, 0.809] |
| LZC | 0.684 [0.6331, 0.7360] | 0.667 [0.551, 0.731] | 0.663 [0.589, 0.713] |
| CTM | 0.101 [0.076, 0.129] | 0.111 [0.086, 0.165] | 0.116 [0.077, 0.183] |
| SampEn | 1.366 [1.288, 1.540] | 1.312 [1.103, 1.489] | 1.274 [1.034, 1.489] |
| FuzzyEn | 0.532 [0.466, 0.624] | 0.514 [0.395, 0.618] | 0.508 [0.427, 0.584] |
| AMI | −0.149 [−0.184, −0.130] | −0.149 [−0.175, −0.124] | −0.145 [−0.164, −0.128] |

The MLP model was obtained according to the optimal values for $n_h$ and $u$. Both were optimized by cross-validation, leaving all trials for each subject out in every iteration. For each value of $u$ between 0 and 100 (step = 5), we varied the number of neurons in the hidden layer from 1 to 20 (step = 1) in order to compute the $k$ value. This procedure was carried out 30 times to minimize the effect of network random initialization. Then, the $k$ values were averaged [43]. The optimal values (highest $k$ for trials) were $u = 45$ and 11 neurons in the hidden layer, as Figure 2 shows. On the other hand, since LDA and QDA models have no tuning parameters to be optimized, these were trained using all trials in the training set.

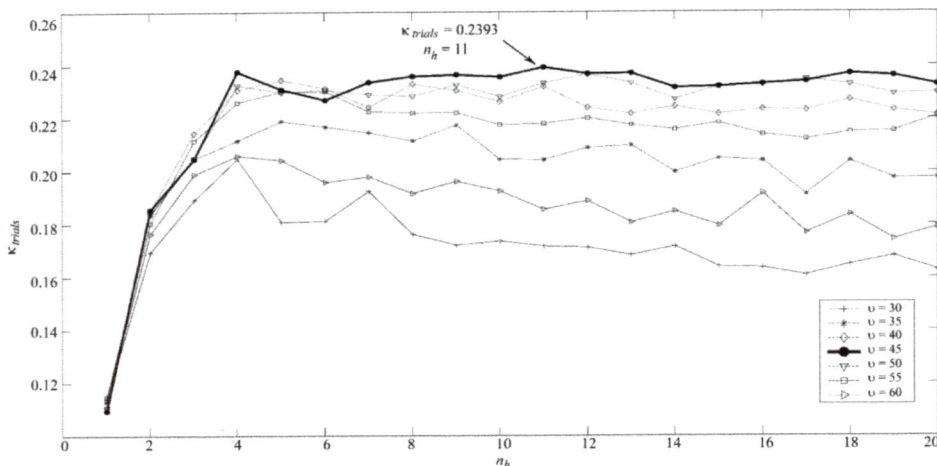

**Figure 2.** Optimal regularization parameter ($v$) and number of neurons in the hidden layer ($n_h$) for MLP.

### 3.2. Test Set

Once the models were trained, their diagnostic ability was only evaluated using the test set. The overall accuracy of the models in the three-class classification task was 58.82% with LDA, 60.78% with QDA, and 62.75% with MLP. Additionally, we obtained $k$ values of 0.3824 with LDA, 0.4118 with QDA and 0.4412 with MLP. These results show that MLP outperformed the discriminant analyses classifiers.

Table 3 displays the confusion matrices of each model, i.e., the model class estimation for each subject versus their actual group. As expected, the three models had higher difficulties when classifying MCI trials and subjects, as this is an intermediate state between HC and AD.

**Table 3.** Confusion matrices of each model: trials and subjects' classification in the test set.

| | LDA | | | QDA | | | MLP | | |
|---|---|---|---|---|---|---|---|---|---|
| Actual ↓\Estimated → | HC | MCI | AD | HC | MCI | AD | HC | MCI | AD |
| HC | 11 | 4 | 2 | 13 | 3 | 1 | 12 | 3 | 2 |
| MCI | 4 | 7 | 6 | 4 | 7 | 6 | 4 | 8 | 5 |
| AD | 2 | 3 | 12 | 3 | 3 | 11 | 2 | 3 | 12 |

Table 4 shows *Se, Sp, Acc, PPV* and *NPV* for each method for HC vs. all and AD vs. all, derived from confusion matrices. MLP showed the highest diagnostic performance when determining whether a subject is not healthy (HC vs. all classification tasks: *Se* = 82.35% and *PPV* = 84.85%). Furthermore, the network showed the highest diagnostic capability when determining whether a subject does not suffer from AD (AD vs. all comparison: *Sp* = 79.41% and *NPV* = 84.38%). LDA and QDA showed similar tendencies although reaching lower diagnostic performance than MLP, as Table 4 shows.

**Table 4.** Diagnostic performance for HC vs. all and AD vs. all, derived from confusion matrices.

|  | HC vs. All | | | AD vs. All | | |
|---|---|---|---|---|---|---|
|  | **LDA** | **QDA** | **MLP** | **LDA** | **QDA** | **MLP** |
| *Se* (%) | 82.35 | 79.41 | 82.35 | 70.59 | 64.71 | 70.59 |
| *Sp* (%) | 64.71 | 76.47 | 70.59 | 76.47 | 79.41 | 79.41 |
| *Acc* (%) | 76.47 | 78.43 | 78.43 | 74.51 | 74.51 | 76.47 |
| *PPV* (%) | 82.35 | 87.10 | 84.85 | 60.00 | 61.11 | 63.16 |
| *NPV* (%) | 64.71 | 65.00 | 66.67 | 83.87 | 81.82 | 84.38 |

## 4. Discussion

### 4.1. Spectral and Nonlinear Characterization of AD and MCI

Our spectral results suggested that AD and MCI elicit a slowing of spontaneous EEG activity. Further inspection of *RP* values revealed that AD patients reached higher *RP* values in low frequency bands (*q*) and lower *RP* values in high frequency bands ($b_1$, $b_2$ and *g*) than HC subjects. For the MCI group, a slight slowing of neural oscillations was found in comparison with HC. This increase of slow rhythms in spontaneous EEG activity was also observed by means of *MF* and *IAF*. Both spectral parameters were lower for AD patients than for MCI and HC subjects. These findings confirm the trend reported in previous studies: AD and MCI are accompanied by a progressive slow-down of EEG [24,25]. Finally, our *SE* results showed changes in the frequency distribution of the power spectrum. However, the physiological explanations for all of these alterations are not clear. The most extended hypothesis is that a significant cerebral cholinergic deficit underlies cognitive symptoms, as memory loss. A loss of cholinergic innervation of the neocortex might play a critical role in the EEG slowing associated with AD [24]. Analogously, the slowing of neural oscillations in AD could also be due to the loss of neurotransmitter acetylcholine, since the cholinergic system modulates spontaneous cortical activity at low frequencies [26].

Regarding the nonlinear parameters that quantify the complexity and irregularity of EEG recordings, our findings showed lower *LZC*, *SampEn*, *FuzzyEn* and higher *AMI* values for AD patients than for HC subjects. For these measures, MCI subjects showed intermediate values between AD and HC. Previous EEG studies also reported a loss of complexity and irregularity associated with early AD and MCI by means of nonlinear measures [9,12,31–34]. Additionally, *CTM* values were higher in AD patients and lower in HC subjects. This result suggests a decrease on variability in AD, as Abásolo et al. previously reported [12]. Taking into account the different nature of the nonlinear parameters, our results showed that the brain activity from AD patients is less complex, more regular and less variable than in MCI and HC subjects. These changes can be associated with both loss of information content and alterations in information processing at the cerebral cortex [47]. The decrease of EEG complexity can also be due to the loss of neurons or synapses, since they are associated with the complex dynamical processing within the brain neural networks [33].

### 4.2. Towards a Screening Protocol of AD

Previous studies explored several EEG features for AD and MCI discrimination from HC, focusing on binary discrimination problems (AD vs. HC, MCI vs. HC and AD vs. MCI) [16–20]. To the best of our knowledge, only one study performed a three-way classification, although via binary classifiers [21]. McBride et al. reached an accuracy value of 85.42% when comparing HC vs. all and 83.33% for AD vs. all (eyes closed resting condition) [21]. Although their results are slightly higher than ours (78.43% and 76.47% for both comparisons, respectively), several advantages of our methodology should be noticed. Firstly, their database was composed by only 47 subjects, in contrast to the 111 subjects recruited for our study. This data limitation also led the authors to validate its proposal through a leave-one-out cross-validation procedure instead of using a hold-out approach (training and test sets). As they obtained a different model for each iteration, the inclusion of new subjects would imply changes in every iteration of cross-validation. However, once our model is trained, the subsequent runtime to apply new data is trivial. It allows us to classify new data just feeding the trained model with the standardized version of them, simplifying the screening protocol.

In contrast to the above-mentioned studies, our MLP single model can be used not only for the three-class classification task but also in binary assessments of healthy vs. cognitively impaired subjects. As derived from Tables 3 and 4, it has shown the ability to detect whether a subject suffers from AD or MCI in 28 out of the 34 non-healthy subjects (82.53% *Se*)—with a positive post-test probability of 84.85% (28 subjects rightly classified out of 33 subjects predicted as AD or MCI)—and only predicting two out of 17 AD patients as HC. In addition, the same model also showed the ability to discard AD in 27 out of the 34 subjects not suffering from it (79.41% *Sp*), including 15 out of the 17 HC (88.24%). These results highlight the clinical usefulness of our proposal, which might be expressed as a screening strategy similar to:

1. If the MLP model predicts AD, recommend beginning a treatment since most probably (89.47%, 17 out of 19 subjects) the patient suffers from AD or MCI.
2. If the MLP model predicts HC, do not treat the patient, since most probably (88.89%, 16 out of 18 subjects) he/she does not suffer from AD; consider a regular evaluation of the subject in the persistence of symptoms in order to minimize the number of AD and MCI missed subjects.
3. If the MLP predicts MCI, conduct a regular evaluation of the patient since doubts arise about the cognitive status of the subject.

### 4.3. Limitations and Future Research Lines

Despite the fact that we showed the usefulness of our proposal, some limitations need to be addressed. Although we used a large data sample to train and validate the models (5122 trials), they were obtained from 111 subjects. Hence, analyzing more recordings from different subjects would enhance the generalization ability of our results. Moreover, taking into account the MCI heterogeneity, it would be useful to characterize different subtypes and conduct a longitudinal analysis to characterize subjects with stable MCI and those who progress to AD. Finally, only three classification approaches (LDA, QDA, and MLP) have been used in this study. In future research works, the usefulness of other advanced classification methods, such as spiking neural networks and support vector machines, should be evaluated.

### 5. Conclusions

To sum up, our results show that both AD and MCI elicit changes in the EEG background activity: a slowing of EEG rhythms, alterations in the frequency distribution of the power spectrum, a complexity loss, a regularity increase and a variability decrease. Our proposal has shown that spectral and nonlinear features allows us to characterize the brain abnormalities associated with AD and MCI. In addition, we have shown the high diagnostic ability of different three-class models trained with this EEG information, particularly when predicting AD and HC status. These results highlight the usefulness of our proposal in order to help physicians classify AD, MCI and HC from EEG data.

**Acknowledgments:** This research was supported by the "Ministerio de Economía y Competitividad" and "European Regional Development Fund" (FEDER) under project TEC2014-53196-R, by "European Commission" and FEDER under project "Análisis y correlación entre el genoma completo y la actividad cerebral para la ayuda en el diagnóstico de la enfermedad de Alzheimer" ("Cooperation Programme Interreg V-A Spain-Portugal, POCTEP 2014–2020"), and by "Consejería de Educación de la Junta de Castilla y León" and FEDER under project VA037U16. Saúl J. Ruiz-Gómez has a predoctoral scholarship from the "Junta de Castilla y León" and European Social Fund.

**Author Contributions:** Saúl J. Ruiz-Gómez processed the signals, analyzed the data, and wrote the manuscript. Carlos Gómez and Roberto Hornero designed the study and interpreted the results. Jesús Poza and Gonzalo C. Gutiérrez-Tobal interpreted the results. Miguel A. Tola-Arribas and Mónica Cano took part in the diagnosis of subjects and the collection of data. All authors have read and approved the final manuscript.

**Conflicts of Interest:** The authors declare no conflict of interest.

## References

1. Alzheimer's Association. 2017 Alzheimer's disease facts and figures. *Alzheimer's Dement.* **2017**, *13*, 325–373. [CrossRef]
2. Petersen, R.C. Alzheimer's disease: Progress in prediction. *Lancet Neurol.* **2010**, *9*, 4–5. [CrossRef]
3. Mufson, E.J.; Binder, L.; Counts, S.E.; DeKosky, S.T.; DeToledo-Morrell, L.; Ginsberg, S.D.; Ikonomovic, M.D.; Perez, S.E.; Scheff, S.W. Mild cognitive impairment: Pathology and mechanisms. *Acta Neuropathol.* **2012**, *123*, 13–30. [CrossRef] [PubMed]
4. Davatzikos, C.; Bhatt, P.; Shaw, L.M.; Batmanghelich, K.N.; Trojanowski, J.Q. Prediction of MCI to AD conversion, via MRI, CSF biomarkers, and pattern classification. *Neurobiol. Aging* **2011**, *32*. [CrossRef] [PubMed]
5. Lin, P.-J.; Neumann, P.J. The economics of mild cognitive impairment. *Alzheimers Dement.* **2013**, *9*, 58–62. [CrossRef] [PubMed]
6. Ewers, M.; Sperling, R.A.; Klunk, W.E.; Weiner, M.W.; Hampel, H. Neuroimaging markers for the prediction and early diagnosis of Alzheimer's disease dementia. *Trends Neurosci.* **2011**, *34*, 430–442. [CrossRef] [PubMed]
7. Poza, J.; Gómez, C.; García, M.; Corralejo, R.; Fernández, A.; Hornero, R. Analysis of neural dynamics in mild cognitive impairment and Alzheimer's disease using wavelet turbulence. *J. Neural Eng.* **2014**, *11*, 26010. [CrossRef] [PubMed]
8. Fernández, A.; Hornero, R.; Gómez, C.; Turrero, A.; Gil-Gregorio, P.; Matías-Santos, J.; Ortiz, T. Complexity analysis of spontaneous brain activity in Alzheimer disease and mild cognitive impairment: An MEG study. *Alzheimer Dis. Assoc. Disord.* **2010**, *24*, 182–189. [CrossRef] [PubMed]
9. Hornero, R.; Abasolo, D.; Escudero, J.; Gomez, C. Nonlinear analysis of electroencephalogram and magnetoencephalogram recordings in patients with Alzheimer's disease. *Philos. Trans. R. Soc. A Math. Phys. Eng. Sci.* **2009**, *367*, 317–336. [CrossRef] [PubMed]
10. Stam, C.J. Nonlinear dynamical analysis of EEG and MEG: Review of an emerging field. *Clin. Neurophysiol.* **2005**, *116*, 2266–2301. [CrossRef] [PubMed]
11. Woon, W.L.; Cichocki, A.; Vialatte, F.; Musha, T. Techniques for early detection of Alzheimer's disease using spontaneous EEG recordings. *Physiol. Meas.* **2007**, *28*, 335–347. [CrossRef] [PubMed]
12. Abásolo, D.; Hornero, R.; Gómez, C.; García, M.; López, M. Analysis of EEG background activity in Alzheimer's disease patients with Lempel–Ziv complexity and central tendency measure. *Med. Eng. Phys.* **2006**, *28*, 315–322. [CrossRef] [PubMed]
13. Gasser, U.S.; Rousson, V.; Hentschel, F.; Sattel, H.; Gasser, T. Alzheimer disease versus mixed dementias: An EEG perspective. *Clin. Neurophysiol.* **2008**, *119*, 2255–2259. [CrossRef] [PubMed]
14. Baker, M.; Akrofi, K.; Schiffer, R.; Boyle, M.W.O. EEG Patterns in Mild Cognitive Impairment (MCI) Patients. *Open Neuroimag. J.* **2008**, *2*, 52–55. [CrossRef] [PubMed]
15. Yu, L.; Liu, H. Efficient Feature Selection via Analysis of Relevance and Redundancy. *J. Mach. Learn. Res.* **2004**, *5*, 1205–1224. [CrossRef]
16. Bertè, F.; Lamponi, G.; Calabrò, R.S.; Bramanti, P. Elman neural network for the early identification of cognitive impairment in Alzheimer's disease. *Funct. Neurol.* **2014**, *29*, 57–65. [PubMed]
17. Buscema, M.; Vernieri, F.; Massini, G.; Scrascia, F.; Breda, M.; Rossini, P.M.; Grossi, E. An improved I-FAST system for the diagnosis of Alzheimer's disease from unprocessed electroencephalograms by using robust invariant features. *Artif. Intell. Med.* **2015**, *64*, 59–74. [CrossRef] [PubMed]
18. Huang, C.; Wahlund, L.-O.; Dierks, T.; Julin, P.; Winblad, B.; Jelic, V. Discrimination of Alzheimer's disease and mild cognitive impairment by equivalent EEG sources: A cross-sectional and longitudinal study. *Clin. Neurophysiol.* **2000**, *111*, 1961–1967. [CrossRef]
19. Iqbal, K.; Alonso, A.D.C.; Chen, S.; Chohan, M.O.; El-Akkad, E.; Gong, C.-X.; Khatoon, S.; Li, B.; Liu, F.; Rahman, A.; et al. Tau pathology in Alzheimer disease and other tauopathies. *Biochim. Biophys. Acta Mol. Basis Dis.* **2005**, *1739*, 198–210. [CrossRef] [PubMed]
20. Poza, J.; Gómez, C.; García, M.; Tola-Arribas, M.A.; Carreres, A.; Cano, M.; Hornero, R. Spatio-Temporal Fluctuations of Neural Dynamics in Mild Cognitive Impairment and Alzheimer's Disease. *Curr. Alzheimer Res.* **2017**, *14*, 924–936. [CrossRef] [PubMed]

21. McBride, J.C.; Zhao, X.; Munro, N.B.; Smith, C.D.; Jicha, G.A.; Hively, L.; Broster, L.S.; Schmitt, F.A.; Kryscio, R.J.; Jiang, Y. Spectral and complexity analysis of scalp EEG characteristics for mild cognitive impairment and early Alzheimer's disease. *Comput. Methods Programs Biomed.* **2014**, *114*, 153–163. [CrossRef] [PubMed]

22. Petrosian, A.A.; Prokhorov, D.V.; Lajara-Nanson, W.; Schiffer, R.B. Recurrent neural network-based approach for early recognition of Alzheimer's disease in EEG. *Clin. Neurophysiol.* **2001**, *112*, 1378–1387. [CrossRef]

23. Albert, M.S.; DeKosky, S.T.; Dickson, D.; Dubois, B.; Feldman, H.H.; Fox, N.C.; Gamst, A.; Holtzman, D.M.; Jagust, W.J.; Petersen, R.C.; et al. The diagnosis of mild cognitive impairment due to Alzheimer's disease: Recommendations from the National Institute on Aging-Alzheimer's Association workgroups on diagnostic guidelines for Alzheimer's disease. *Alzheimer's Dement.* **2011**, *7*, 270–279. [CrossRef] [PubMed]

24. Jeong, J. EEG dynamics in patients with Alzheimer's disease. *Clin. Neurophysiol.* **2004**, *115*, 1490–1505. [CrossRef] [PubMed]

25. Dauwels, J.; Vialatte, F.-B.; Cichocki, A. Diagnosis of alzheimers disease from eeg signals: Where are we standing? *Curr. Alzheimer Res.* **2010**, *7*, 1–43. [CrossRef]

26. Osipova, D.; Ahveninen, J.; Kaakkola, S.; Jääskeläinen, I.P.; Huttunen, J.; Pekkonen, E. Effects of scopolamine on MEG spectral power and coherence in elderly subjects. *Clin. Neurophysiol.* **2003**, *114*, 1902–1907. [CrossRef]

27. Rodriguez, G.; Copello, F.; Vitali, P.; Perego, G.; Nobili, F. EEG spectral profile to stage Alzheimer's disease. *Clin. Neurophysiol.* **1999**, *110*, 1831–1837. [CrossRef]

28. Poza, J.; Hornero, R.; Abásolo, D.; Fernández, A.; García, M. Extraction of spectral based measures from MEG background oscillations in Alzheimer's disease. *Med. Eng. Phys.* **2007**, *29*, 1073–1083. [CrossRef] [PubMed]

29. Moretti, D.V.; Babiloni, C.; Binetti, G.; Cassetta, E.; Dal Forno, G.; Ferreric, F.; Ferri, R.; Lanuzza, B.; Miniussi, C.; Nobili, F.; et al. Individual analysis of EEG frequency and band power in mild Alzheimer's disease. *Clin. Neurophysiol.* **2004**, *115*, 299–308. [CrossRef]

30. Powell, G.E.; Percival, I.C. A spectral entropy method for distinguishing regular and irregular motion of Hamiltonian systems. *J. Phys. A Math. Gen.* **1979**, *12*, 2053–2071. [CrossRef]

31. Abásolo, D.; Escudero, J.; Hornero, R.; Gómez, C.; Espino, P. Approximate entropy and auto mutual information analysis of the electroencephalogram in Alzheimer's disease patients. *Med. Biol. Eng. Comput.* **2008**, *46*, 1019–1028. [CrossRef] [PubMed]

32. Gómez, C.; Hornero, R.; Abásolo, D.; Fernández, A.; Escudero, J. Analysis of the magnetoencephalogram background activity in Alzheimer's disease patients with auto-mutual information. *Comput. Methods Programs Biomed.* **2007**, *87*, 239–247. [CrossRef] [PubMed]

33. Jeong, J.; Gore, J.C.; Peterson, B.S. Mutual information analysis of the EEG in patients with Alzheimer's disease. *Clin. Neurophysiol.* **2001**, *112*, 827–835. [CrossRef]

34. Cao, Y.; Cai, L.; Wang, J.; Wang, R.; Yu, H.; Cao, Y.; Liu, J. Characterization of complexity in the electroencephalograph activity of Alzheimer's disease based on fuzzy entropy. *Chaos* **2015**, *25*, 83116. [CrossRef] [PubMed]

35. Lempel, A.; Ziv, J. On the complexity of finite sequences. *IEEE Trans. Inf. Theory* **1976**, *22*, 75–81. [CrossRef]

36. Cohen, M.E.; Hudson, D.L.; Deedwania, P.C. Applying continuous chaotic modeling to cardiac signal analysis. *IEEE Eng. Med. Biol. Mag.* **1996**, *15*, 97–102. [CrossRef]

37. Ben-Mizrachi, A.; Procaccia, I.; Grassberger, P. Characterization of experimental (noisy) strange attractors. *Phys. Rev. A* **1984**, *29*, 975–977. [CrossRef]

38. Monge, J.; Gómez, C.; Poza, J.; Fernández, A.; Quintero, J.; Hornero, R. MEG analysis of neural dynamics in attention-deficit/hyperactivity disorder with fuzzy entropy. *Med. Eng. Phys.* **2015**, *37*, 416–423. [CrossRef] [PubMed]

39. Bishop, C.M. *Neural Networks for Pattern Recognition*; Oxford University Press: Oxford, UK, 1995; ISBN 9780198538646.

40. Bishop, C.M. Pattern Recognition and Machine Learning. *J. Electron. Imaging* **2007**, *16*, 49901. [CrossRef]

41. Zhang, G.P. Neural networks for classification: A survey. *IEEE Trans. Syst. Man Cybern. Part C Appl. Rev.* **2000**, *30*, 451–462. [CrossRef]

42. Hornik, K. Approximation capabilities of multilayer feedforward networks. *Neural Netw.* **1991**, *4*, 251–257. [CrossRef]

*Entropy* **2018**, *20*, 35

43. Gutiérrez-Tobal, G.C.; Álvarez, D.; Marcos, J.V.; Del Campo, F.; Hornero, R. Pattern recognition in airflow recordings to assist in the sleep apnoea-hypopnoea syndrome diagnosis. *Med. Biol. Eng. Comput.* **2013**, *51*, 1367–1380. [CrossRef] [PubMed]

44. Nabney, I.T. *NETLAB: Algorithms for Pattern Recognition*; Springer Science & Business Media: New York, NY, USA, 2002.

45. Witten, I.H.; Frank, E.; Hall, M.A. *Data Mining: Practical Machine Learning Tools and Techniques*; Morgan Kaufmann Publishers: Burlington, ON, Canada, 2011; ISBN 0080890369.

46. Richman, J.S.; Moorman, J.R. Physiological time-series analysis using approximate entropy and sample entropy. *Am. J. Physiol. Heart Circ. Physiol.* **2000**, *278*, 2039–2049. [CrossRef] [PubMed]

47. Baraniuk, R.G.; Flandrin, P.; Janssen, A.J.E.M.; Michel, O.J.J. Measuring time-frequency information content using the Reényi entropies. *IEEE Trans. Inf. Theory* **2001**, *47*, 1391–1409. [CrossRef]

entropy

MDPI

*Article*

# Automated Diagnosis of Myocardial Infarction ECG Signals Using Sample Entropy in Flexible Analytic Wavelet Transform Framework

Mohit Kumar [1], Ram Bilas Pachori [1,*] and U. Rajendra Acharya [2,3,4,*]

1   Discipline of Electrical Engineering, Indian Institute of Technology Indore, Indore 453552, India;
    phd1401202005@iiti.ac.in
2   Department of Electronics and Computer Engineering, Ngee Ann Polytechnic, Singapore 599489, Singapore
3   Department of Biomedical Engineering, School of Science and Technology, SUSS University,
    Singapore 599491, Singapore
4   Department of Biomedical Engineering, Faculty of Engineering, University of Malaya,
    Kuala Lumpur 50603, Malaysia
*   Correspondence: pachori@iiti.ac.in (R.B.P.); aru@np.edu.sg (U.R.A.)

Received: 28 July 2017; Accepted: 8 September 2017; Published: 13 September 2017

**Abstract:** Myocardial infarction (MI) is a silent condition that irreversibly damages the heart muscles. It expands rapidly and, if not treated timely, continues to damage the heart muscles. An electrocardiogram (ECG) is generally used by the clinicians to diagnose the MI patients. Manual identification of the changes introduced by MI is a time-consuming and tedious task, and there is also a possibility of misinterpretation of the changes in the ECG. Therefore, a method for automatic diagnosis of MI using ECG beat with flexible analytic wavelet transform (FAWT) method is proposed in this work. First, the segmentation of ECG signals into beats is performed. Then, FAWT is applied to each ECG beat, which decomposes them into subband signals. Sample entropy (SEnt) is computed from these subband signals and fed to the random forest (RF), J48 decision tree, back propagation neural network (BPNN), and least-squares support vector machine (LS-SVM) classifiers to choose the highest performing one. We have achieved highest classification accuracy of 99.31% using LS-SVM classifier. We have also incorporated Wilcoxon and Bhattacharya ranking methods and observed no improvement in the performance. The proposed automated method can be installed in the intensive care units (ICUs) of hospitals to aid the clinicians in confirming their diagnosis.

**Keywords:** Myocardial infarction (MI); electrocardiogram (ECG) beats; flexible analytic wavelet transform (FAWT); sample entropy; classification

## 1. Introduction

Myocardial infarction (MI) is a condition that indicates the injury of a heart cell due to the lack of oxygenated blood in the cardiac arteries [1]. The main cause of MI is coronary heart disease (CHD), which is responsible for nearly one-third of all deaths in the age group of above 35 years [2,3]. MI is silent in nature and may lead to fast and non-recoverable damage to the muscles of heart [3]. If MI is not controlled timely, then myocardial structure and functions of the left ventricle (LV) continue to be damaged further. For the diagnosis of MI, the electrocardiogram (ECG) is used due to its low operating cost and non-invasive nature [4]. Vital information related to the functioning of the heart can be assessed by analyzing the ECG signals. Moreover, the MI results in the ST deviations and T wave abnormalities in the ECG signal [4]. Manual identification of the changes in the ECG signals is a difficult task. Only 82% ST-segment elevation in MI subjects may be recognised by the experienced cardiologists [5]. Therefore, an automated identification system for MI patients is needed to facilitate the clinicians in their accurate diagnosis.The classification of ECG signals and the extracted beats from ECG signals have been studied in the literature for diagnosis of heart disorders [6–8].

In literature, various studies are performed for the detection of MI patients. In [9], a total integral of one ECG cycle and T-wave integral are suggested as features for the detection of MI. Time-domain features computed from 12-lead ECG signals are explored with fuzzy multi-layer perception (FMLP) network to classify the MI ECG signals [10]. In [11], a new multiple instance learning based approach is proposed for the detection of MI. The Hermite basis functions are used to decompose the multilead ECG signals and the obtained coefficients are found effective for the detection of acute MI [12]. The phase space fractal dimension features and the artificial neural network classifier are explored to detect the MI [13]. In [4], authors have applied neuro-fuzzy approach for the diagnosis of MI patients using multilead ECG signals. A hybrid approach based on hidden Markov models (HMMs) and Gaussian mixture models (GMMs) is proposed to distinguish the MI and normal ECG signals in [14].

In [15], characterization of the QRS complex of normal and MI subjects is performed using discrete wavelet transform (DWT). Three different wavelets are used to decompose the ECG signals up to the fourth level of decomposition. The Daubechies wavelet performed best among the three chosen wavelets. The DWT technique is incorporated to extract the QRS complex of ECG signals, and it is found that identification of the MI subject is possible by detecting the QRS complex [16]. In [17], the phase of the complex wavelet coefficients obtained from the dual tree complex wavelet transform (DTCWT) of 12-lead ECG signals is computed. Then, multiscale phase alteration values are used as features to identify the normal, MI, and other abnormal ECG signals. In [18], the ECG signals of normal, MI, and coronary artery disease (CAD) are applied to DWT, empirical mode decomposition (EMD), and discrete cosine transform (DCT) techniques. The authors in this study achieved the best performance when features obtained using DCT technique are subjected to the k-nearest neighbour classifier (k-NN) classifier. Contourlet transform (CT) and shearlet transform (ST) based technique is proposed to distinguish normal, MI, CAD, and congestive heart failure (CHF) subjects using ECG beats in [19]. The performance of the CT based technique is found to be better in comparison to the ST based method.

Our aim is to develop a method for automated diagnosis of MI patients in this work. We have analyzed normal and ECG beats using sample entropy (SEnt) in flexible analytic wavelet transform (FAWT) [20,21] framework. First, preprocessing is performed to remove the baseline wandering and other noise present in the ECG signals. Then, ECG signals are segmented into the beats. Furthermore, these beats are decomposed up to the 24th level of decomposition using FAWT. Sample entropy (SEnt) is computed from each subband signal, which is reconstructed from the corresponding coefficients of the FAWT based decomposition. The computed features are subjected to the random forest (RF) [22], J48 decision tree [23,24], back propagation neural network (BPNN) [25], and least-squares support vector machines (LS-SVM) [26] classifiers for separating the ECG beats of MI and normal classes. The steps performed in the present work are shown in Figure 1.

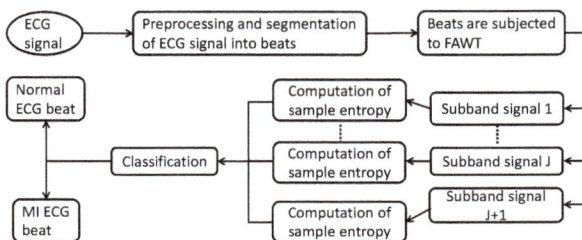

**Figure 1.** The proposed method to diagnose the myocardial infarction (MI) patients.

The organization of the remaining sections of the paper is as follows: the dataset used, preprocessing and segmentation of the ECG signals into beats, FAWT, SEnt, and classification methods are provided in Section 2. The obtained results in this work are given in Section 3 and discussed in Section 4. Finally, Section 5 presents the conclusions of the work.

## 2. Methodology

### 2.1. Dataset Studied in This Work

The dataset, containing normal and MI ECG signals, has been obtained from Physikalisch Technische Bundesanstalt (PTB) diagnostic ECG database from the Physiobank [27,28]. Each signal was acquired at the sampling rate of 1000 Hz. The dataset contains normal ECG recordings of 52 subjects and MI ECG recordings of 148 subjects. The ECG signals obtained from the lead-2 have been used in present work.

### 2.2. Preprocessing and Segmentation of ECG Signals

We have used Daubechies 6 (db6) wavelet basis function to eliminate baseline wander and noise present in the ECG signals [29]. After preprocessing, each ECG signal is segmented into beats based on R-peak detection. The Pan–Tompkins algorithm is applied to identify the R-peaks [30]. The 250 samples from the left and 400 samples from the right of the R-point are considered as one ECG beat [3]. Thus, each ECG beat contains 651 samples. Finally, we have 40,182 MI ECG beats and 10,546 normal ECG beats.

### 2.3. Computation of Features in FAWT Framework

In this work, we have computed SEnt in FAWT domain to classify MI and normal ECG beats. The brief explanation of FAWT method is given below.

- FAWT

The FAWT is a rational-dilation wavelet transform, which allows one to easily adjust the dilation factor, Quality (Q)-factor, and redundancy (R). The FAWT employs the fractional sampling rate in high pass and low pass channels. Moreover, it provides analytic bases by separating positive and negative frequencies in high-pass channels [20]. Employing fractional sampling rate and analytic bases in FAWT provides shift-invariance, tunable oscillatory bases, and flexible time-frequency covering [21]. These properties make this transform suitable for analysing the transient and oscillatory components of the signals. Q-factor controls the frequency resolution of FAWT. The high Q-factor provides finer filter banks for analysing the signals in the frequency domain. For fixed dilation and Q-factors, the redundancy controls the position of the wavelet.

The mathematical expressions of the filter banks for FAWT are given in Table 1. This table also provides expression for perfect reconstruction condition for FAWT.

In Table 1, $A$ and $B$ are used to adjust the sampling rate of the low-pass channel, and $C$ and $D$ are used to adjust the sampling rate of high-pass channels. The $\omega_p$ and $\omega_s$ are the cutoff frequencies of the pass-band and stop-band for the low pass filter, respectively. The $\omega_0$ and $\omega_3$ are the stop-band cutoff frequencies for the high pass filter, and $\omega_1$ and $\omega_2$ are the pass-band cutoff frequencies.

The relation between $\beta$ and the Q-factor is as follows [20]:

$$Q = \frac{2 - \beta}{\beta}. \tag{1}$$

Redundancy is defined as follows [20]:

$$R = \left(\frac{C}{D}\right) \frac{1}{1 - (A/B)}. \tag{2}$$

We have used $A = 5$, $B = 6$, $C = 1$, $D = 2$ and $\beta = (0.8 \times C)/D$ [21,31,32] in the present work. Level of decomposition is kept at $J = 24$. The selection procedure of $J$ is given in Section 3. FAWT has been utilized for detecting the CAD in [31,32], in order to diagnose CHF in [33], to identify electroencephalogram (EEG) signals of focal and non-focal classes in [34], and for the faults identification in rotating machinery [21]. We have used the Matlab toolbox (İ Bayram, İstanbul Technical University, İstanbul, Turkey) available for FAWT implementation at [35].

**Table 1.** Mathematical expressions for filters and perfect reconstruction condition of flexible analytic wavelet transform (FAWT).

| Filters and Parameters Used in FAWT | Mathematical Expressions |
|---|---|
| Low pass filter [20] | $H(w) = \begin{cases} (AB)^{1/2}, & \|w\| < w_p \\ (AB)^{1/2}\theta\left(\frac{w - w_p}{w_s - w_p}\right), & w_p \leq w \leq w_s \\ (AB)^{1/2}\theta\left(\frac{\pi - w + w_p}{w_s - w_p}\right), & -w_s \leq w \leq -w_p \\ 0, & \|w\| \geq w_s \end{cases}$ where, $w_p = \frac{(1 - \beta)\pi}{A} + \frac{\epsilon}{A}, w_s = \frac{\pi}{B}$ |
| High pass filter [20] | $G(w) = \begin{cases} (2CD)^{1/2}\theta\left(\frac{\pi - w - w_0}{w_1 - w_0}\right), & w_0 \leq w < w_1 \\ (2CD)^{1/2}, & w_1 \leq w < w_2 \\ (2CD)^{1/2}\theta\left(\frac{w - w_2}{w_3 - w_2}\right), & w_2 \leq w \leq w_3 \\ 0, & w \in [0, w_0) \cup (w_3, 2\pi) \end{cases}$ where, $w_0 = \frac{(1 - \beta)\pi + \epsilon}{C}, w_1 = \frac{A\pi}{BC}, w_2 = \frac{\pi - \epsilon}{C},$ $w_3 = \frac{\pi + \epsilon}{C}, \epsilon \leq \frac{A - B + \beta B}{A + B}\pi.$ |
| Condition for perfect reconstruction [20] | $\|\theta(\pi - w)\|^2 + \|\theta(w)\|^2 = 1$ $\left(1 - \frac{A}{B}\right) \leq \beta \leq \left(\frac{C}{D}\right)$ where, $\theta(w) = \frac{[2 - \cos(w)]^{1/2}[1 + \cos(w)]}{2}$ for $w \in [0, \pi]$ |

## 2.4. Sample Entropy

SEnt [36] measures the complexity of the time series. It improves the performance by excluding the bias due to the self matches counted in the computation of approximate entropy. Higher values of SEnt indicate more complexity of the signal; on the other hand, the lower value of SEnt shows less complexity of the signal.

Let us consider a time-series $(y_1, y_2, \ldots\ldots, y_P)$ of length $P$ for which the SEnt can be computed as [37]:

$$SEnt(m, r, P) = -\ln\left(\frac{I^{m+1}(r)}{I^m(r)}\right), \tag{3}$$

where $I^m(r)$ is defined as follows [37]:

$$I^m(r) = \frac{1}{(P - m\tau)} \sum_{j=1}^{P - m\tau} C_j^m(r) \tag{4}$$

and

$$C_j^m(r) = \frac{S_j^r}{P - (m+1)\tau},$$ (5)

where $S_j^r$ is the total count for which $L\,[Y(j), Y(k)] \le r$ without considering the self-matches.

The parameter $L\,[Y(j), Y(k)]$ is the distance between $Y(j)$ and $Y(k)$ vectors. $Y(j)$ and $Y(k)$ can be given as [37]:

$$Y(j) = \{y_j, y_{j+\tau}, \ldots\ldots, y_{j+(m-1)\tau}\},$$

$$Y(k) = \{y_k, y_{k+\tau}, \ldots\ldots, y_{k+(m-1)\tau}\},$$

where $j$ and $k$ vary from 1 to $P - m\tau$ and $k \ne j$.

In this work, we have experimentally chosen threshold $(r) = 0.35$, delay $(\tau) = 1$, and embedding dimension $(m) = 5$. Parameter selection procedure has been explained in the Section 3.

### 2.5. Studied Classification Techniques

We have used RF, J48 decision tree, BPNN, and LS-SVM in this work to perform the classification of normal and MI ECG beats based on the extracted features. In this work, we have used Waikato environment for knowledge analysis (WEKA) toolbox (version 3.7.13, The University of Waikato, Hamilton, New Zeland) for the implementation of RF, J48 decision tree, and BPNN classifiers [38]. We have used default parameters provided in WEKA toolbox for RF, J48 decision tree, and BPNN classifiers. Recently, RF, J48 decision tree, and BPNN classifiers have been used to analyze the sleep stages from EEG signals in [39].

Mathematical expression for decision making function of LS-SVM is given as follows [26]:

$$I = \text{sign}\left[\sum_{z=1}^{Z} \alpha_z w_z E(y, y_z) + b\right].$$ (6)

In the above expression, $E(y, y_z)$, $\alpha_z$, $y_z$, $b$, $w_z$, and $Z$ represent a kernel function, Lagrangian multiplier, the $z$-th input vector of $D$-dimensions, bias term, target vector, and total data points, respectively.

Kernel functions are used with LS-SVM to map the input space to the higher dimension space and the two classes can be separated using an optimal hyperplane [26]. In this work, linear, polynomial, radial basis function (RBF), and Morlet wavelet kernels are employed with LS-SVM to perform the classification. The mathematical expressions of these kernels are provided in Table 2. In Table 2, $x$ represents the order of the polynomial kernel, $\sigma$ determines the width of RBF kernel, and $q$ and $D$ represent the scale factor of the Morlet wavelet kernel and the dimension of the feature set, respectively. LS-SVM is widely used in various biomedical signals classifications [40–44].

In the present work, we have evaluated the classification performance in terms of accuracy, specificity, and sensitivity [45].

Table 2. Different kernel functions and their mathematical expressions.

| Kernel Functions | Mathematical Expressions |
|---|---|
| Linear [26] | $E(y, y_z) = y_z^T y$ |
| Polynomial [26] | $E(y, y_z) = (y_z^T y + 1)^x$ |
| Radial basis function (RBF) [46] | $E(y, y_z) = e^{\frac{-\|y - y_z\|^2}{2\sigma^2}}$ |
| Morlet wavelet [47,48] | $E(y, y_z) = \prod_{n=1}^{D} \cos\left[k_0 \frac{y^n - y_z^n}{q}\right] e^{\frac{-\|y^n - y_z^n\|^2}{2q^2}}$ |

## 3. Results

First, we have segmented the MI and normal ECG signals into the beats. Each ECG beat of both classes is decomposed into different subband signals using FAWT. The sample entropies are computed from these different subband signals. We start performing the experiments with $J = 5$ and initial parameters for SEnt $m = 2$, $\tau = 1$, and $r = 0.15$ are choosen [37]. Typical subband signals extracted from the decomposition of normal and MI ECG beats at the 5th level of decomposition using FAWT are shown in Figure 2a,b, respectively. In Figure 2, subband signals $SB_1$ to $SB_5$ are reconstructed from the detail coefficients from level 1 to level 5, and $SB_6$ is reconstructed from the approximate coefficients at level 5.

We fed the features to the RF classifier for selecting the suitable parameters. Variation of classification accuracies for various values for $m$ and $\tau$ is provided in Table 3. It can be inferred from the table that accuracy of classification is highest for $m = 5$ and $\tau = 1$. Classification accuracy for various values of $r$ is shown in Figure 3. We have achieved the maximum classification accuracy for $r = 0.35$. Hence, we have used the parameters $m = 5$, $\tau = 1$, and $r = 0.35$ to compute the sample entropies in this work. Moreover, we have increased the decomposition level to $J = 6$, and observed that the classification accuracy is increased to 91.95%. Hence, we further increased the decomposition level up to the maximum possible decomposition level using FAWT with parameters values $A = 5$, $B = 6$, $C = 1$, $D = 2$ and $\beta = (0.8 \times C)/D$, which is $J = 24$ [35]. The plot of classification accuracy versus decomposition levels is shown in Figure 4. We can observe that accuracy is increasing with increase in the decomposition level. The highest classification accuracy of 97.10% is achieved with RF classifier at $J = 24$. We have employed a 10-fold cross-validation procedure for the training and testing of the classifier [49]. The classification accuracy achieved using J48 decision tree and BPNN classifiers are 93.97% and 92.85%, respectively.

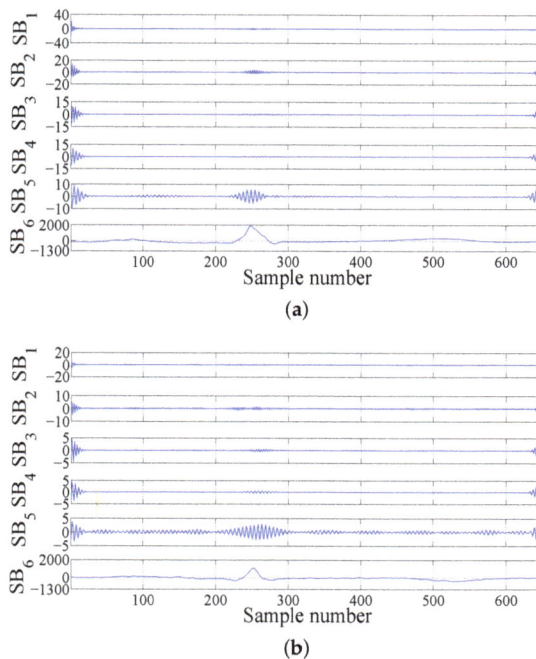

**Figure 2.** Plot of decomposed subband signals: (**a**) normal electrocardiogram (ECG) beat, (**b**) MI ECG beat.

**Table 3.** Classification accuracies computed using random forest (RF) classifier for different values of $m$ and $\tau$ using Sample entropy (SEnt) with $r = 0.15$.

| $m \longrightarrow$ | 2 | 3 | 4 | 5 |
|---|---|---|---|---|
| $\tau\downarrow$ | | | | |
| 1 | 87.716% | 89.353% | 89.353% | 89.629% |
| 2 | 88.92% | 89.128% | 89.32% | 89.075% |
| 3 | 89.126% | 88.84% | 88.739% | 88.84% |

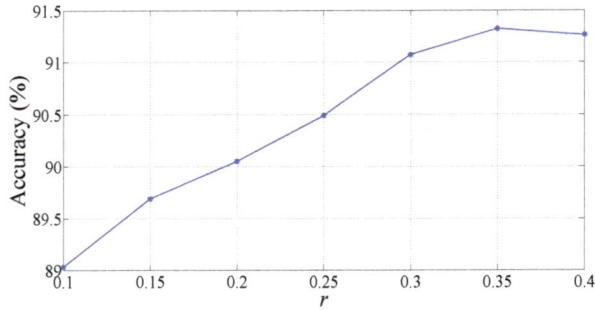

**Figure 3.** Plot of accuracy (%) versus $r$ of SEnt with RF classifier.

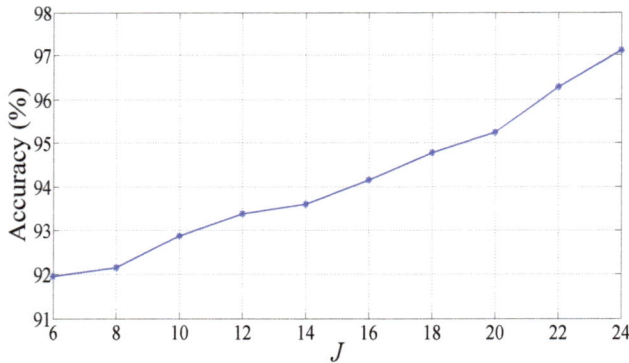

**Figure 4.** Plot of accuracy (%) versus level of decomposition with RF classifier.

Furthermore, we have also tested the features with LS-SVM classifier with different kernel functions, namely, polynomial, linear, RBF, and Morlet wavelet kernels at $J = 24$. Variation of classification accuracy (%) with the RBF kernel parameter $\sigma$ is shown in Figure 5. We can observe from Figure 5 that classification accuracy of LS-SVM is at a maximum for RBF kernel parameter $\sigma = 2.2$. Plot of changes in the value of accuracy (%) for variation in the parameter $q$ of Morlet wavelet kernel is shown in Figure 6. LS-SVM showed maximum accuracy with Morlet wavelet kernel at $q = 11$. The performance of LS-SVM, using four different kernels used in our work, is summarized in Table 4. LS-SVM yielded the highest classification performance with RBF kernel, and achieved an accuracy, specificity, and sensitivity of 99.31%, 98.12%, and 99.62%, respectively.

We have also employed Wilcoxon and Bhattacharya ranking methods for improving the performance of the proposed system [50,51]. The plots of the classification accuracy (%) for various ranked features are shown in Figures 7 and 8 for RBF and Morlet wavelet kernels, respectively.

It can be noted that the ranking methods are not able to improve the classification performance. The discrimination ability of the features is determined by computing the *p*-values using the Kruskal–Wallis (KW) test [52]. Recently, the KW test has been explored to test the statistical significance of the features in various biomedical signal analysis applications [53–55]. The *p*-values are found significantly low ($p < 0.0001$) for all the features (SEnt computed from 25 subband signals), which indicate good discrimination ability of all the computed features. Mean and standard deviation values for features are provided in Table 5. In Table 5, SEnt refers to the sample entropy and the subscript refers to the corresponding subband signal from which SEnt is computed.

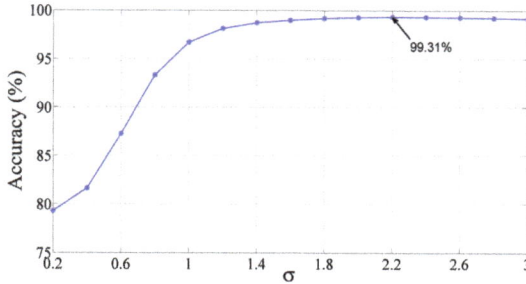

**Figure 5.** Plot of accuracies versus $\sigma$ of RBF kernel.

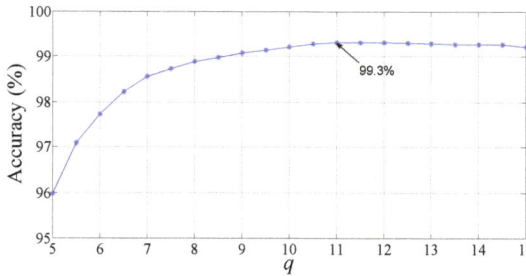

**Figure 6.** Plot of accuracies versus $q$ of Morlet wavelet kernel.

**Table 4.** Classification accuracy (%) of least-squares support vector machine (LS-SVM) for different kernel functions.

| Kernel Function | Parameters | Accuracy (%) | Sensitivity (%) | Specificity (%) |
|---|---|---|---|---|
| Linear | | 83.32 | 81.83 | 89.02 |
| Polynomial | $x = 2$ | 96.30 | 96.01 | 97.43 |
| | $x = 3$ | 96.74 | 96.44 | 97.92 |
| RBF | $\sigma = 2.2$ | 99.31 | 99.62 | 98.12 |
| Morlet wavelet | $q = 11, k_0 = 0.25$ | 99.30 | 99.64 | 97.92 |

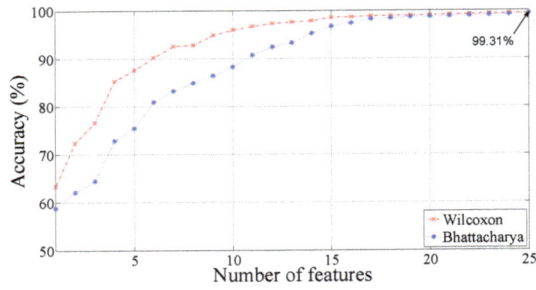

**Figure 7.** Plot of accuracy (%) versus number of features using LS-SVM with RBF kernel.

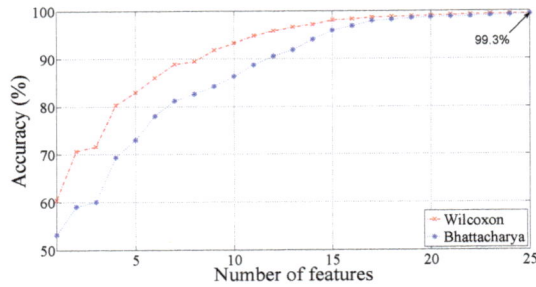

**Figure 8.** Plot of accuracy (%) versus number of features using LS-SVM with Morlet wavelet kernel.

**Table 5.** Mean ($M$), standard deviation ($SD$) for normal and MI classes.

| Feature | Normal Class ($M \pm SD$) | MI Class ($M \pm SD$) |
|---------|---------------------------|------------------------|
| $SEnt_{SB_1}$ | $0.0111 \pm 0.0248$ | $0.0448 \pm 0.0651$ |
| $SEnt_{SB_2}$ | $0.0250 \pm 0.0385$ | $0.0742 \pm 0.0664$ |
| $SEnt_{SB_3}$ | $0.0030 \pm 0.0039$ | $0.0071 \pm 0.0112$ |
| $SEnt_{SB_4}$ | $0.0032 \pm 0.0026$ | $0.0058 \pm 0.0070$ |
| $SEnt_{SB_5}$ | $0.0282 \pm 0.0298$ | $0.0660 \pm 0.0491$ |
| $SEnt_{SB_6}$ | $0.0625 \pm 0.0431$ | $0.0971 \pm 0.0483$ |
| $SEnt_{SB_7}$ | $0.0727 \pm 0.0390$ | $0.0973 \pm 0.0420$ |
| $SEnt_{SB_8}$ | $0.0696 \pm 0.0388$ | $0.0965 \pm 0.0413$ |
| $SEnt_{SB_9}$ | $0.0501 \pm 0.0324$ | $0.0722 \pm 0.0338$ |
| $SEnt_{SB_{10}}$ | $0.0493 \pm 0.0246$ | $0.0596 \pm 0.0257$ |
| $SEnt_{SB_{11}}$ | $0.0569 \pm 0.0251$ | $0.0680 \pm 0.0244$ |
| $SEnt_{SB_{12}}$ | $0.0674 \pm 0.0305$ | $0.0902 \pm 0.0205$ |
| $SEnt_{SB_{13}}$ | $0.0627 \pm 0.0354$ | $0.0928 \pm 0.0288$ |
| $SEnt_{SB_{14}}$ | $0.0599 \pm 0.0340$ | $0.0754 \pm 0.0374$ |
| $SEnt_{SB_{15}}$ | $0.0501 \pm 0.0305$ | $0.0663 \pm 0.0380$ |
| $SEnt_{SB_{16}}$ | $0.0480 \pm 0.0221$ | $0.0597 \pm 0.0329$ |
| $SEnt_{SB_{17}}$ | $0.0521 \pm 0.0162$ | $0.0607 \pm 0.0247$ |
| $SEnt_{SB_{18}}$ | $0.0894 \pm 0.0151$ | $0.0978 \pm 0.0227$ |
| $SEnt_{SB_{19}}$ | $0.1437 \pm 0.0129$ | $0.1442 \pm 0.0157$ |
| $SEnt_{SB_{20}}$ | $0.1491 \pm 0.0056$ | $0.1515 \pm 0.0070$ |
| $SEnt_{SB_{21}}$ | $0.1501 \pm 0.0066$ | $0.1475 \pm 0.0087$ |
| $SEnt_{SB_{22}}$ | $0.1230 \pm 0.0100$ | $0.1197 \pm 0.0104$ |
| $SEnt_{SB_{23}}$ | $0.0904 \pm 0.0030$ | $0.0911 \pm 0.0038$ |
| $SEnt_{SB_{24}}$ | $0.0665 \pm 0.0010$ | $0.0663 \pm 0.0013$ |
| $SEnt_{SB_{25}}$ | $0.0420 \pm 0.0087$ | $0.0363 \pm 0.0107$ |

Furthermore, we test the classification performance with balanced dataset (10,546 beats of each class) with the same parameter values of SEnt and FAWT. Classification accuracy for this case using LS-SVM classifier is presented in Table 6.

**Table 6.** Classification accuracy (%) of LS-SVM with different kernel functions for balanced dataset.

| Kernel Function | Parameters | Accuracy (%) | Sensitivity (%) | Specificity (%) |
|---|---|---|---|---|
| Linear | | 85.74 | 84.64 | 86.83 |
| Polynomial | $x = 2$ | 94.06 | 92.61 | 95.52 |
| | $x = 3$ | 96.88 | 95.98 | 97.77 |
| RBF | $\sigma = 2.2$ | 98.27 | 99.13 | 97.40 |
| Morlet wavelet | $q = 11, k_0 = 0.25$ | 98.19 | 99.20 | 97.17 |

## 4. Discussions

In the present work, the ECG beats are decomposed into the 24th level of FAWT for obtaining subband signals. Furthermore, SEnt is computed from each of the subband signal. We can observed from Table 5 that SEnt computed from the lower frequency subband signals (SB21, SB22, SB24, and SB25) showed higher values for normal ECG beats in comparison to the MI ECG beats. Therefore, lower frequency subband signals show higher complexity for normal ECG beats than MI ECG beats. However, lower values of SEnt are observed for higher frequency subband signals extracted from normal ECG beats. Hence, complexity of higher frequency subband signals is lower for normal ECG beats. Finally, our method achieved 99.31% accuracy using LS-SVM classifier with RBF kernel.

Summary of the comparison of the present work with the other existing work is provided in Table 7. In [56], and time-domain features are computed from 12-lead ECG signals. The computed features are fed to the BPNN classifier, which yielded sensitivity of 97.5%. In [57], the time-domain method is used for extracting the features from the ECG signals to diagnose the MI patients. The authors have used 12-lead ECG signals of 20 normal and 20 MI subjects. They achieved a sensitivity of 85% to detect the MI subjects. In [58], the authors proposed a method based on the spectral differences of cross wavelet transform (XWT) of the ECG signals. Furthermore, they proposed threshold based classifier and achieved 97.6% classification accuracy. In [5], an algorithm based on the parametrization of ECG signal is developed. In this algorithm, a 20th order polynomial is fitted with the ECG signal. Their method showed 94.4% classification accuracy with J48 decision tree model for the diagnosis of MI. The approach presented in [59] utilized the evaluation of multiscale energy and eigenspace (MEES) features. The suggested method used support vector machine (SVM) classifier with RBF kernel and achieved 96.15% classification accuracy. In [3], ECG beats are decomposed up to the 4th level of decomposition using DWT. From the DWT coefficients, 12 nonlinear parameters are extracted. The authors achieved 98.8% accuracy using a k-NN classifier. They also performed statistical tests for determining the significance levels of the studied features. A method to automatically detect the MI using ECG signals is also proposed in [60]. The achieved accuracies were 93.53% and 95.22% using convolutional neural network (CNN) algorithms for the ECG beats with noise and without noise removal, respectively.

We have achieved highest accuracy in comparison to the existing methods that are mentioned in Table 7. Moreover, the methods suggested in [5,56–59] used ECG recordings of the multiple leads. However, our method uses only lead-2 ECG recordings, which makes our method less complex than multiple leads methods. The method suggested in [3] also requires ECG records of one lead (lead-11) only. However, the method in [3] achieved 98.8% classification accuracy with 47 features. In comparison to the method in [3], our method has achieved 99.31% accuracy with 25 features. Our method showed better results than the method in [3] with a lesser number of features. The study proposed in [60] also used lead-2 ECG signals and achieved 95.22% accuracy with an 11-layer deep neural network. This method is more complex than our method and also time-consuming.

**Table 7.** Summary of automated diagnosis of MI using ECG.

| Author | Year | Dataset | Analyzing Method | Number of Leads | Classification Method Used | 10-Fold Cross Validation | Classification Performance (%) |
|---|---|---|---|---|---|---|---|
| Arif et al. [56] | 2010 | PTB diagnostic ECG dtabase | Time-domain method | 12-lead | BPNN | No | Sensitivity = 97.5 |
| Al-Kindi et al. [57] | 2011 | PTB diagnostic ECG dtabase | Time-domain method | 12-lead | - | No | Sensitivity = 85 |
| Banerjee et al. [58] | 2014 | PTB diagnostic ECG dtabase | XWT based method | 3-lead | Threshold based classifier | No | Accuracy = 97.6 |
| Liu et al. [5] | 2015 | PTB diagnostic ECG dtabase | ECG polynomial fitting | 12-lead | J48 decision tree | No | Accuracy = 94.4 |
| Sharma et al. [59] | 2015 | PTB diagnostic ECG dtabase | MEES based method | 12-lead | SVM with RBF kernel | No | Accuracy = 96.15 |
| Acharya et al. [3] | 2016 | PTB diagnostic ECG dtabase | DWT, Nonlinear features | One lead (lead-11) | k-NN | Yes | Accuracy = 98.8 |
| Acharya et al. [60] | 2017 | PTB diagnostic ECG dtabase | No feature extraction and selection | One lead (lead-2) | CNN | Yes | Accuracy = 95.22 |
| Present method | | PTB diagnostic ECG dtabase | FAWT and SEnt | One lead (lead-2) | LS-SVM | Yes | Accuracy = 99.31 |

## 5. Conclusions

In this work, normal and MI ECG beats are analyzed using SEnt in FAWT framework. We have achieved the highest classification performance using lead-2 ECG signals as compared to the reported works. We have identified the suitable parameters to compute the SEnt in FAWT domain for the detection of MI subjects accurately. Parameters for the computation of SEnt and the decomposition level in FAWT domain are selected on the basis of classification accuracy computed using an RF classifier. Achieved classification accuracies with RF, J48 decision tree, BPNN, and LS-SVM classifiers are 97.10%, 93.97%, 92.85%, and 99.31%, respectively, using the entire dataset. Our method achieved classification accuracy of 98.27% with LS-SVM using balanced data set. Therefore, we can conclude that our methodology has performed well for the detection of MI patients using both balanced and unbalanced (entire) datasets. Our automated system can be used to assist cardiologists to cross check their diagnosis. It can be extended to diagnose the severity of MI. Along with the echocardiography, it can be used to localize the MI.

**Author Contributions:** Ram Bilas Pachori and U. Rajendra Acharya designed the research problem. Mohit Kumar carried out the research work and wrote the manuscript. All authors edited the manuscript. All authors have read and approved the final manuscript.

**Conflicts of Interest:** The authors declare no conflict of interest.

## References

1. Thygesen, K.; Alpert, J.S.; Jaffe, A.S.; Simoons, M.L.; Chaitman, B.R.; White, H.D. Third universal definition of myocardial infarction. *Eur. Heart J.* **2012**, *33*, 2551–2567.
2. Sanchis-Gomar, F.; Perez-Quilis, C.; Leischik, R.; Lucia, A. Epidemiology of coronary heart disease and acute coronary syndrome. *Ann. Transl. Med.* **2016**, *4*, 256.
3. Acharya, U.R.; Fujita, H.; Sudarshan, V.K.; Oh, S.L.; Adam, M.; Koh, J.E.W.; Tan, J.H.; Ghista, D.N.; Martis, R.J.; Chua, C.K.; et al. Automated detection and localization of myocardial infarction using electrocardiogram: A comparative study of different leads. *Knowl. Based Syst.* **2016**, *99*, 146–156.
4. Lu, H.L.; Ong, K.; Chia, P. An automated ECG classification system based on a neuro-fuzzy system. *Comput. Cardiol.* **2000**, *27*, 387–390.
5. Liu, B.; Liu, J.; Wang, G.; Huang, K.; Li, F.; Zheng, Y.; Luo, Y.; Zhou, F. A novel electrocardiogram parameterization algorithm and its application in myocardial infarction detection. *Comput. Biol. Med.* **2015**, *61*, 178–184.
6. Crippa, P.; Curzi, A.; Falaschetti, L.; Turchetti, C. Multi-class ECG beat classification based on a Gaussian mixture model of Karhunen-Loève transform. *Int. J. Simul. Syst. Sci. Technol.* **2015**, *16*, doi:10.5013/IJSSST.a.16.

7.  Biagetti, G.; Crippa, P.; Curzi, A.; Orcioni, S.; Turchetti, C. A Multi-class ECG beat classifier based on the truncated KLT representation. In Proceedings of the European Modelling Symposium, Pisa, Italy, 21–23 October 2014; pp. 93–98.

8.  Li, H.; Liang, H.; Miao, C.; Cao, L.; Feng, X.; Tang, C.; Li, E. Novel ECG signal classification based on KICA nonlinear feature extraction. *Circuits Syst. Signal Process.* **2016**, *35*, 1187–1197.

9.  Safdarian, N.; Dabanloo, N.J.; Attarodi, G. A new pattern recognition method for detection and localization of myocardial infarction using T-wave integral and total integral as extracted features from one cycle of ECG signal. *J. Biomed. Sci. Eng.* **2014**, *7*, 818–824.

10.  Bozzola, P.; Bortolan, G.; Combi, C.; Pinciroli, F.; Brohet, C. A hybrid neuro-fuzzy system for ECG classification of myocardial infarction. In Proceedings of the Computers in Cardiology, Indianapolis, IN, USA, 8–11 September 1996; pp. 241–244.

11.  Sun, L.; Lu, Y.; Yang, K.; Li, S. ECG analysis using multiple instance learning for myocardial infarction detection. *IEEE Trans. Biomed. Eng.* **2012**, *59*, 3348–3356.

12.  Haraldsson, H.; Edenbrandt, L.; Ohlsson, M. Detecting acute myocardial infarction in the 12-lead ECG using Hermite expansions and neural networks. *Artif. Intell. Med.* **2004**, *32*, 127–136.

13.  Lahiri, T.; Kumar, U.; Mishra, H.; Sarkar, S.; Roy, A.D. Analysis of ECG signal by chaos principle to help automatic diagnosis of myocardial infarction. *J. Sci. Ind. Res.* **2009**, *68*, 866–870.

14.  Chang, P.C.; Hsieh, J.C.; Lin, J.J.; Chou, Y.H.; Liu, C.H. A hybrid system with hidden Markov models and Gaussian mixture models for myocardial infarction classification with 12-lead ECGs. In Proceedings of the 11th IEEE International Conference on High Performance Computing and Communications, Seoul, Korea, 25–27 June 2009; pp. 110–116.

15.  McDarby, G.; Celler, B.G.; Lovell, N.H. Characterising the discrete wavelet transform of an ECG signal with simple parameters for use in automated diagnosis. In Proceedings of the 2nd International Conference on Bioelectromagnetism, Melbourne, Australia, 15–18 February 1998; pp. 31–32.

16.  Banerjee, S.; Mitra, M. ECG feature extraction and classification of anteroseptal myocardial infarction and normal subjects using discrete wavelet transform. In Proceedings of the International Conference on Systems in Medicine and Biology, Kharagpur, India, 16–18 December 2010; pp. 55–60.

17.  Tripathy, R.K.; Dandapat, S. Detection of cardiac abnormalities from multilead ECG using multiscale phase alternation features. *J. Med. Syst.* **2016**, *40*, 143.

18.  Acharya, U.R.; Fujita, H.; Adam, M.; Lih, O.S.; Sudarshan, V.K.; Hong, T.J.; Koh, J.E.W.; Hagiwara, Y.; Chua, C.K.; Poo, C.K.; et al. Automated characterization and classification of coronary artery disease and myocardial infarction by decomposition of ECG signals: A comparative study. *Inf. Sci.* **2017**, *377*, 17–29.

19.  Acharya, U.R.; Fujita, H.; Sudarshan, V.K.; Oh, S.L.; Adam, M.; Tan, J.H.; Koo, J.H.; Jain, A.; Lim, C.M.; Chua, K.C. Automated characterization of coronary artery disease, myocardial infarction, and congestive heart failure using contourlet and shearlet transforms of electrocardiogram signal. *Knowl. Based Syst.* **2017**, *132*, 156–166.

20.  Bayram, İ. An analytic wavelet transform with a flexible time-frequency covering. *IEEE Trans. Signal Process.* **2013**, *61*, 1131–1142.

21.  Zhang, C.; Li, B.; Chen, B.; Cao, H.; Zi, Y.; He, Z. Weak fault signature extraction of rotating machinery using flexible analytic wavelet transform. *Mech. Syst. Signal Process.* **2015**, *64–65*, 162–187.

22.  Breiman, L. Random forests. *Mach. Learn.* **2001**, *45*, 5–32.

23.  Quinlan, J.R. Induction of decision trees. *Mach. Learn.* **1986**, *1*, 81–106.

24.  Quinlan, J.R. *C4.5: Programs for Machine Learning*; Morgan Kaufmann: San Francisco, CA, USA, 1993.

25.  Jang, J.S.R.; Sun, C.T.; Mizutani, E. *Neuro-Fuzzy and Soft Computing: A Computational Approach to Learning and Machine Intelligence*; Pearson: London, UK, 1997.

26.  Suykens, J.A.K.; Vandewalle, J. Least squares support vector machine classifiers. *Neural Process. Lett.* **1999**, *9*, 293–300.

27.  Bousseljot, R.; Kreiseler, D.; Schnabel, A. Nutzung der EKG-Signaldatenbank CARDIODAT der PTB über das Internet. *Biomed. Tech. Biomed. Eng.* **1995**, *40*, 317–318.

28.  Goldberger, A.L.; Amaral, L.A.; Glass, L.; Hausdorff, J.M.; Ivanov, P.C.; Mark, R.G.; Mietus, J.E.; Moody, G.B.; Peng, C.K.; Stanley, H.E. Physiobank, physiotoolkit, and physionet: Components of a new research resource for complex physiologic signals. *Circulation* **2000**, *101*, e215–e220.

29. Martis, R.J.; Acharya, U.R.; Min, L.C. ECG beat classification using PCA, LDA, ICA and discrete wavelet transform. *Biomed. Signal Process. Control* **2013**, *8*, 437–448.

30. Pan, J.; Tompkins, W.J. A real-time QRS detection algorithm. *IEEE Trans. Biomed. Eng.* **1985**, *32*, 230–236.

31. Kumar, M.; Pachori, R.B.; Acharya, U.R. An efficient automated technique for CAD diagnosis using flexible analytic wavelet transform and entropy features extracted from HRV signals. *Expert Syst. Appl.* **2016**, *63*, 165–172.

32. Kumar, M.; Pachori, R.B.; Acharya, U.R. Characterization of coronary artery disease using flexible analytic wavelet transform applied on ECG signals. *Biomed. Signal Process. Control* **2017**, *31*, 301–308.

33. Kumar, M.; Pachori, R.B.; Acharya, U.R. Use of accumulated entropies for automated detection of congestive heart failure in flexible analytic wavelet transform framework based on short-term HRV signals. *Entropy* **2017**, *19*, 92.

34. Gupta, V.; Priya, T.; Yadav, A.K.; Pachori, R.B.; Acharya, U.R. Automated detection of focal EEG signals using features extracted from flexible analytic wavelet transform. *Pattern Recogn. Lett.* **2017**, *94*, 180–188.

35. Bayram, İ. An Analytic Wavelet Transform with a Flexible Time-Frequency Covering. Available online: http://web.itu.edu.tr/ibayram/AnDWT/ (accessed on 11 June 2017).

36. Richman, J.S.; Moorman, J.R. Physiological time–series analysis using approximate entropy and sample entropy. *Am. J. Physiol. Heart Circ. Physiol.* **2000**, *278*, H2039–H2049.

37. Sokunbi, M.O.; Fung, W.; Sawlani, V.; Choppin, S.; Linden, D.E.J.; Thome, J. Resting state fMRI entropy probes complexity of brain activity in adults with ADHD. *Psychiatry Res. Neuroimaging* **2013**, *214*, 341–348.

38. Hall, M.; Frank, E.; Holmes, G.; Pfahringer, B.; Reutemann, P.; Witten, I.H. The WEKA data mining software: An update. *SIGKDD Explor.* **2009**, *11*, 10–18.

39. Sharma, R.; Pachori, R.B.; Upadhyay, A. Automatic sleep stages classification based on iterative filtering of electroencephalogram signals. *Neural Comput. Appl.* **2017**, 1–20, doi:10.1007/s00521-017-2919-6.

40. Sharma, R.; Pachori, R.B.; Acharya, U.R. An integrated index for the identification of focal electroencephalogram signals using discrete wavelet transform and entropy measures. *Entropy* **2015**, *17*, 5218–5240.

41. Pachori, R.B.; Kumar, M.; Avinash, P.; Shashank, K.; Acharya, U.R. An improved online paradigm for screening of diabetic patients using RR-interval signals. *J. Mech. Med. Biol.* **2016**, *16*, 1640003.

42. Sharma, R.; Pachori, R.B.; Acharya, U.R. Application of entropy measures on intrinsic mode functions for the automated identification of focal electroencephalogram signals. *Entropy* **2015**, *17*, 669–691.

43. Sharma, R.; Pachori, R.B. Classification of epileptic seizures in EEG signals based on phase space representation of intrinsic mode functions. *Expert Syst. Appl.* **2015**, *42*, 1106–1117.

44. Sharma, R.; Kumar, M.; Pachori, R.B.; Acharya, U.R. Decision support system for focal EEG signals using tunable-Q wavelet transform. *J. Comput. Sci.* **2017**, *20*, 52–60.

45. Azar, A.T.; El-Said, S.A. Performance analysis of support vector machines classifiers in breast cancer mammography recognition. *Neural Comput. Appl.* **2014**, *24*, 1163–1177.

46. Khandoker, A.H.; Lai, D.T.H.; Begg, R.K.; Palaniswami, M. Wavelet-based feature extraction for support vector machines for screening balance impairments in the elderly. *IEEE Trans. Neural Syst. Rehabil. Eng.* **2007**, *15*, 587–597.

47. Zavar, M.; Rahati, S.; Akbarzadeh-T, M.R.; Ghasemifard, H. Evolutionary model selection in a wavelet-based support vector machine for automated seizure detection. *Expert Syst. Appl.* **2011**, *38*, 10751–10758.

48. Bajaj, V.; Pachori, R.B. Classification of seizure and nonseizure EEG signals using empirical mode decomposition. *IEEE Trans. Inf. Technol. Biomed.* **2012**, *16*, 1135–1142.

49. Kohavi, R. A study of cross-validation and bootstrap for accuracy estimation and model selection. In Proceedings of the 14th International Joint Conference on Artificial Intelligence, Montreal, QC, Canada, 20–25 August 1995; pp. 1137–1143.

50. Derryberry, D.R.; Schou, S.B.; Conover, W.J. Teaching rank-based tests by emphasizing structural similarities to corresponding parametric tests. *J. Stat. Educ.* **2010**, *18*, 1–19.

51. Theodoridis, S.; Koutroumbas, K. Feature Selection. In *Pattern Recognition*, 2nd ed.; Academic Press: San Diego, CA, USA, 2003; pp. 163–205.

52. McKight, P.E.; Najab, J. Kruskal-Wallis Test. *Corsini Encyclopedia of Psychology*; John Wiley and Sons, Inc.: Hoboken, NJ, USA, 2010.

53. Pachori, R.B. Discrimination between ictal and seizure-free EEG signals using empirical mode decomposition. *Res. Lett. Signal Process.* **2008**, doi:10.1155/2008/293056.

54. Pachori, R.B.; Avinash, P.; Shashank, K.; Sharma, R.; Acharya, U.R. Application of empirical mode decomposition for analysis of normal and diabetic RR-interval signals. *Expert Syst. Appl.* **2015**, *42*, 4567–4581.
55. Sood, S.; Kumar, M.; Pachori, R.B.; Acharya, U.R. Application of empirical mode decomposition-based features for analysis of normal and CAD heart rate signals. *J. Mech. Med. Biol.* **2016**, *16*, 1640002.
56. Arif, M.; Malagore, I.A.; Afsar, F.A. Automatic detection and localization of myocardial infarction using back propagation neural networks. In Proceedings of the 4th International Conference on Bioinformatics and Biomedical Engineering, Chengdu, China, 18–20 June 2010; pp. 1–4.
57. Al-Kindi, S.G.; Ali, F.; Farghaly, A.; Nathani, M.; Tafreshi, R. Towards real-time detection of myocardial infarction by digital analysis of electrocardiograms. In Proceedings of the 1st Middle East Conference on Biomedical Engineering, Sharjah, UAE, 21–24 February 2011; pp. 454–457.
58. Banerjee, S.; Mitra, M. Application of cross wavelet transform for ECG pattern analysis and classification. *IEEE Trans. Instrum. Meas.* **2014**, *63*, 326–333.
59. Sharma, L.N.; Tripathy, R.K.; Dandapat, S. Multiscale energy and eigenspace approach to detection and localization of myocardial infarction. *IEEE Trans. Biomed. Eng.* **2015**, *62*, 1827–1837.
60. Acharya, U.R.; Fujita, H.; Oh, S.L.; Hagiwara, Y.; Tan, J.H.; Adam, M. Application of deep convolutional neural network for automated detection of myocardial infarction using ECG signals. *Inf. Sci.* **2017**, *415*, 190–198.

*entropy*

MDPI

Article

# Changes in the Complexity of Heart Rate Variability with Exercise Training Measured by Multiscale Entropy-Based Measurements

Frederico Sassoli Fazan [1], Fernanda Brognara [1], Rubens Fazan Junior [1], Luiz Otavio Murta Junior [2] and Luiz Eduardo Virgilio Silva [1,3,*]

[1] Department of Physiology, School of Medicine of Ribeirão Preto, University of São Paulo, Ribeirão Preto, SP 14049-900, Brazil; frederico.fazan@usp.br (F.S.F.); fernanda.brognara@usp.br (F.B.); rfazan@usp.br (R.F.J.)
[2] Department of Computing and Mathematics, School of Philosophy, Sciences and Languages of Ribeirão Preto, University of São Paulo, Ribeirão Preto, SP 14040-901, Brazil; murta@usp.br
[3] Department of Computer Science, Institute of Mathematics and Computer Sciences, University of São Paulo, São Carlos, SP 13566-590, Brazil
* Correspondence: luizeduardo@usp.br

Received: 12 December 2017; Accepted: 8 January 2018; Published: 17 January 2018

**Abstract:** Quantifying complexity from heart rate variability (HRV) series is a challenging task, and multiscale entropy (MSE), along with its variants, has been demonstrated to be one of the most robust approaches to achieve this goal. Although physical training is known to be beneficial, there is little information about the long-term complexity changes induced by the physical conditioning. The present study aimed to quantify the changes in physiological complexity elicited by physical training through multiscale entropy-based complexity measurements. Rats were subject to a protocol of medium intensity training ($n = 13$) or a sedentary protocol ($n = 12$). One-hour HRV series were obtained from all conscious rats five days after the experimental protocol. We estimated MSE, multiscale dispersion entropy (MDE) and multiscale SDiff$_q$ from HRV series. Multiscale SDiff$_q$ is a recent approach that accounts for entropy differences between a given time series and its shuffled dynamics. From SDiff$_q$, three attributes ($q$-attributes) were derived, namely SDiff$_{q_{max}}$, $q_{max}$ and $q_{zero}$. MSE, MDE and multiscale $q$-attributes presented similar profiles, except for SDiff$_{q_{max}}$. $q_{max}$ showed significant differences between trained and sedentary groups on Time Scales 6 to 20. Results suggest that physical training increases the system complexity and that multiscale $q$-attributes provide valuable information about the physiological complexity.

**Keywords:** sample entropy; dispersion entropy; multiscale entropy; complexity; heart rate variability; rat; exercise; physical training; conditioning

---

## 1. Introduction

The study of system complexity is very challenging and has attracted much attention in the past few years [1–3]. Physiological complexity reflects the interoperability and correct functioning of regulatory processes as a whole, so the higher the complexity, the higher the system ability to adapt to different situations in daily life [4].

Heart rate variability (HRV) series, derived from the electrocardiogram (ECG) or arterial pressure signals, is one of the most important sources of information about system physiological status. Heart rate is actively controlled by the autonomic nervous system and can respond to many situations when the organism is challenged. A number of studies demonstrated that many indices extracted from HRV are powerful risk predictors of morbidity and death, for cardiac and non-cardiac diseases [5–7].

One of the most substantial challenges in the quantification of complexity from HRV time series is the difficulty in finding out a single measurement capable of doing this task consistently.

In other words, most of the complexity measurements are capable of extracting some properties that regard complexity itself, but none of them are enough to characterize all the complex traits of a system. Mono- and multi-fractal measurements [8,9], irreversibility estimations [10,11], symbolic methods [12,13], network analysis [14,15], as well as entropy-based approaches [16,17] have been proposed to infer the system complexity.

Multiscale entropy (MSE) is an important example of an approach that has been shown to be quite robust and consistent to characterize the system complexity from HRV time series. Like many other approaches, it has some limitations depending on the situation, and improvements or refinements have been proposed since MSE has emerged [18]. For example, the entropy estimator used in MSE (sample entropy) can be replaced by other estimators, such as permutation entropy [19], fuzzy entropy [20], distribution entropy [21], dispersion entropy [22], Rényi entropy [23] and bubble entropy [24], among others. Some other entropy-based proposals, such as entropy of entropy [25] and multiscale SDiff$_q$ (a measure of entropic differences) [26], are markedly different from the MSE original framework, although notably inspired by MSE.

Mild intensity aerobic exercise has been shown to improve several systemic functions and prepare the organism for sudden changes in the body. Experimental models using physical training have demonstrated that gaining physical conditioning, before an induced pathology, can reduce the disturbances caused by the disease [27,28]. In other words, physical conditioning seems to increase the system physiological complexity level. However, controversial findings have been reported about complexity and exercise, and scarce studies applied multiscale complexity approaches to identify how the aerobic training can increase the complexity in healthy subjects [29–33].

In the present study, we applied MSE and two other complexity measurements derived from MSE, namely multiscale dispersion entropy and multiscale SDiff$_q$, to quantify the increase of complexity with physical training in experimental models of healthy rats. Results show that all measurements point to the same direction, but significant findings were obtained only with multiscale SDiff$_q$.

## 2. Materials and Methods

### 2.1. Experimental Protocol

Male Wistar rats (210 g on average) were obtained from the Animal Care Facility at the Campus of Ribeirão Preto of the University of São Paulo. The animals' usage was according to the Ethical Principles in Animal Research adopted by the National Council for the Control of Animal Experimentation, approved by the Local Animal Ethical Committee from the School of Medicine of Ribeirão Preto of the University of São Paulo.

The study divided animals into trained ($n = 13$) and sedentary groups ($n = 12$). Since animals could have distinct initial physical conditioning, they were individually tested for maximum velocity ($V_{max}$). For the $V_{max}$ test, the animals were placed on a treadmill, with no inclination, and the speed was increased in steps of 3 m/s every 3 min. The stage where the animal fatigued, as well as the time spent on the incomplete stage were noted to calculate the $V_{max}$ of each rat [34].

The trained group underwent a physical training protocol on the treadmill with no inclination for 9 consecutive weeks, 5 days per week. The training protocol consisted of a medium intensity training that initiated at 50% of $V_{max}$ for 20 min and ended, at the ninth week, at 70% of $V_{max}$ for 60 min (Adapted from [35]). At the fifth week, the trained group underwent a new $V_{max}$ test to adjust the training protocol as some animals acquire physical conditioning quicker than others. The sedentary group followed the same protocol, but the treadmill was kept off.

### 2.2. Data Acquisition and Processing

Two to three days after the end of the physical training protocol, rats were anesthetized with a mixture of ketamine and xylazine (50 and 10 mg/kg, ip) and implanted with subcutaneous electrodes for ECG recordings. Two days after surgery, with the animals conscious and under free movement conditions, the electrodes were connected to a bioelectric amplifier (Animal BioAmp FE136, ADInstruments, Bella Vista, Australia), and ECG recordings were acquired (2 kHz) by an IBM/PC coupled to an analog-to-digital interface (ML866 PowerLab 4/30, ADInstruments, Bella Vista, Australia).

ECG was recorded during one hour, so that multiscale measurements could be confidently estimated from HRV series. ECG recordings were processed using computer software (LabChart Pro, ADInstruments, Bella Vista, Australia) that creates HRV series as the sequence of R-R intervals, i.e., the time interval between adjacent R waves. All ECG recordings were carefully inspected, and missing beat detections and artifacts were manually corrected. HRV series are 20,000 beats in length, on average.

### 2.3. Multiscale Sample Entropy

Multiscale sample entropy (MSE) is a widely-known procedure to quantify the irregularity of time series within a time-scale range [36,37]. The MSE algorithm consists of creating multiple scaled versions of the original time series and calculating sample entropy (SampEn) from each scaled time series.

Consider a time series given by $u(1), u(2), \ldots, u(N)$. Let $x_m(i)$ be the set of consecutive samples in $u$ from $i$ to $i + m - 1$, i.e., $x_m(i) = [u(i), u(i + 1), u(i + 2), \ldots, u(i + m - 1)]$. Thus, SampEn is defined as [38]:

$$SampEn(m, r, N) = -\ln \frac{U^{m+1}(r)}{U^m(r)} \tag{1}$$

where:

$$U^m(r) = \frac{1}{N - m} \sum_{i=1}^{N-m} U_i^m \tag{2}$$

$$U_i^m = \frac{[\# \text{ of } x_m \mid d[x_m(i), x_m(j)] \leq r]}{N - m - 1} \tag{3}$$

and:

$$U^{m+1}(r) = \frac{1}{N - m} \sum_{i=1}^{N-m} U_i^{m+1} \tag{4}$$

$$U_i^{m+1} = \frac{[\# \text{ of } x_{m+1} \mid d[x_{m+1}(i), x_{m+1}(j)] \leq r]}{N - m - 1}. \tag{5}$$

The distance function $d$ is given by:

$$d[x_m(i), x_m(j)] = \max_{k=1,\ldots,m} (|u(i + k - 1) - u(j + k - 1)|). \tag{6}$$

In Equations (3) and (5), $1 \leq j \leq N - m$, $j \neq i$. In SampEn equations, $m$ is the pattern length or embedding dimension and $r$ is the tolerance factor assumed for similarity between samples.

To estimate MSE, multiple scaled versions of $u$ are created by a coarse-graining procedure, where each element $j$ in a $\tau$-scaled series is defined by:

$$u^\tau(j) = \frac{1}{\tau} \sum_{i=(j-1)\tau+1}^{j\tau} u(i), \qquad 1 \leq j \leq N/\tau. \tag{7}$$

Next, SampEn is calculated from each scaled time series $u^\tau$, resulting in a curve of entropy versus scale. It is worth noting that the higher the time scale ($\tau$), the slower the dynamics that the scaled time series is representing. Importantly, the tolerance factor ($r$) of SampEn is kept fixed for all time scales ($\tau$) in MSE.

In the present study, we calculated MSE with the most widely-used parameter setting, i.e., $m = 2$ and $r = 15\%$ of the original time series standard deviation. The maximum scale calculated was $\tau = 20$.

## 2.4. Multiscale Dispersion Entropy

Multiscale dispersion entropy (MDE) is similar to MSE and also quantifies the complexity of time series [39]. However, instead of calculating SampEn for each scaled time series, dispersion entropy (DispEn) is used to estimate irregularity.

Consider the same time series given before ($u$). First, $u$ is filtered by a normal cumulative distribution function (NCDF) with mean $\mu$ and standard deviation $\sigma$, resulting in a filtered time series $u_f$, which ranges from 0 to 1. This procedure is intended to better treat outliers. Next, $u_f$ is mapped into $c$ classes (1 to $c$), according to $z^c(j) = \text{round}(c * u_f(j) + 0.5)$, a function that linearly maps the range $[0, 1]$ to $[1, c]$.

Now, let $y_m(i)$ be the set of consecutive samples in $z^c$ from $i$ to $i + m - 1$, i.e., $y_m(i) = [z^c(i), z^c(i + 1), z^c(i + 2), \ldots, z^c(i + m - 1)]$, $i = 1, 2, \ldots, N - m + 1$. Each vector $y_m(i)$ represents a dispersion pattern. Considering that each value in $y_m$ can assume one of the $c$ possible classes, there will be $c^m$ potential dispersion patterns.

The probability of occurrence of each dispersion pattern $y_m(i)$ in $z^c$ can be calculated as the number of times the pattern $y_m(i)$ appears on $z^c$, divided by the total number of patterns in $z^c$ (i.e., $N - m + 1$). This procedure will result in a probability distribution for all possible dispersion patterns, $p[y_m(i)]$. Finally, the DispEn is defined as the Shannon entropy of $p[y_m(i)]$ [22]:

$$DispEn(m, c) = - \sum_{i=1}^{c^m} p[y_m(i)] \, \log(p[y_m(i)])$$ (8)

MDE uses the same coarse-graining procedure of MSE. Thus, MDE estimation consists of the creation of scaled versions of the original time series using Equation (7) and the calculation of DispEn from each scaled time series. However, the NCDF function applied to each scaled version is the same as that applied to the first scale, i.e., the original time series. This procedure has a similar effect of keeping $r$ fixed at all time scales in MSE and can be achieved choosing the same $\mu$ and $\sigma$ of the NCDF function at all scales.

Parameters of MDE were set as $m = 2$, $c = 6$ and maximum time scale $\tau = 20$. NCDF was generated with $\mu$ and $\sigma$ as the mean and standard deviation of the original time series, respectively.

## 2.5. Multiscale SDiff$_q$

An alternative proposal for multiscale complexity measurement is the multiscale SDiff$_q$ analysis [26]. Although still inspired by MSE in the sense of multiscale analysis, multiscale SDiff$_q$ do not use the entropy values over scales directly to characterize complexity. Instead, differences of entropy between the time series and its uncorrelated version, i.e., surrogate data, are used to represent the complexity. The difference of entropy is evaluated for a range of $q$-values, which is a parameter derived from nonadditive mechanical statistics [40,41]. The so-called nonadditive $q$-entropy has three regimes, namely classic additive when $q = 1$, sub-additive when $q > 1$ and super-additive when $q < 1$.

SDiff$_q$ accounts for the difference between the SampEn$_q$ of a given time series and the mean SampEn$_q$ of a set of surrogate series. SampEn$_q$ is a generalization of SampEn inspired by nonadditive statistics, which introduces the nonadditive parameter $q$ to SampEn. Its equation is given by [42]:

$$SampEn_q(m, r, N) = \log_q U^m(r) - \log_q U^{m+1}(r)$$ (9)

where $\log_q$ is defined as [43]:

$$\log_q(x) = \frac{x^{1-q} - 1}{1 - q}, \ [x \in \mathbb{R}_+^*; q \in \mathbb{R}; \log_1(x) = \log(x)]$$ (10)

and $[Z]_+ = \max\{Z, 0\}$. The definitions of $U^m(r)$ and $U^{m+1}(r)$ are the same as presented in Equations (2) and (4) for SampEn.

To calculate SDiff$_q$, one has to follow the steps:

- From a given time series $u$, $S$ surrogate series are generated from $u$. The surrogate series is obtained by simply shuffling $u$ [44];
- Next, values $A = U^m(r)$ and $B = U^{m+1}(r)$ are calculated from $u$;
- Values of $U^m(r)$ and $U^{m+1}(r)$ are also calculated from each surrogate instance, obtaining their mean values $C = \overline{U^m(r)}$ and $D = \overline{U^{m+1}(r)}$;
- Finally, SDiff$_q$ is defined by Equation (11) below:

$$
\begin{aligned}
SDiff_q &= \log_q(A) - \log_q(B) - [\log_q(C) - \log_q(D)] \\
&= \log_q(A) + \log_q(D) - \log_q(B) - \log_q(C).
\end{aligned}
\tag{11}
$$

Both SampEn$_q$ and SDiff$_q$ are parametrized in $q$ so that they represent a curve of entropy, or entropy difference, as a function of $q$. From SDiff$_q$ curves, three attributes ($q$-attributes) are obtained to characterize the time series dynamics, namely SDiff$_{q_{max}}$, $q_{max}$ and $q_{zero}$. The SDiff$_{q_{max}}$ represents the maximum value for SDiff$_q$ in the range of $q$. The $q_{max}$ and $q_{zero}$ represent the $q$-value where SDiff$_q$ finds its maximum and zero values, respectively. $q_{max}$ is the $q$ parameter that gives the largest entropic separation between the actual time series and its surrogate versions, whereas $q_{zero}$ is the $q$ parameter where original and shuffled dynamics have the same entropy. For more details on the calculation of $q$-attributes, please refer to [26,42].

The extension of SDiff$_q$ to a multiscale measurement is straightforward. Scaled versions of the original time series are created using the same coarse-graining procedure of MSE, given by Equation (7). Then, for each scaled time series, the SDiff$_q$ curve is calculated and $q$-attributes are obtained, so that it ends up with multiscale $q$-attributes.

Multiscale SDiff$_q$ parameters were set with the same values chosen for MSE, i.e., $m = 2, r = 0.15$ and maximum time scale $\tau = 20$. The number of surrogate instances generated for each time scale was $S = 20$, and the nonadditive $q$ parameter ranged from $-2$ to $2$ to estimate the $q$-attributes.

It is worth emphasizing the fact that $q$-attributes represent the SDiff$_q$ behavior. Furthermore, the $q$ parameter comes with the power law equation proposed for nonadditive entropy ($q$-entropy) [40,43]. Therefore, one can say that $q_{max}$ and $q_{zero}$ indicate where this power law results in maximum entropy differences regarding surrogates and where this difference is null (zero-crossing), respectively.

*2.6. Statistical Analysis*

We assessed mean MSE, MDE and multiscale $q$-attributes values in two range segments: short (1 to 5) and long (6 to 20) time scales. Those variables were checked for normality by the Shapiro–Wilk test. Differences between trained and sedentary groups were verified by Student's *t*-test or the Mann–Whitney rank sum test when required. Significance was assumed when $p < 0.05$.

**3. Results**

The curve profiles of MSE and MDE were very similar for both trained and sedentary rats (Figure 1A,B). Likewise, no difference was found between the groups in the mean values of MSE and MDE grouped by short (1 to 5) and long (6 to 20) time scales (Figure 1C,D), although for higher scales, there was a tendency of increasing differences among groups (Figure 1A,B).

The curve profiles of $q_{max}$ and $q_{zero}$ were very similar to each other (Figure 2B,C), which in turn were also very similar to MSE and MDE (Figure 1A,B), regardless of the experimental group. On the other hand, those curves are entirely different from the profile of SDiff$_{q_{max}}$ (Figure 2A). For $q_{max}$ and $q_{zero}$, the curve values decrease for the first two or three scales; after that, they start to increase (Figure 2B,C). However, in the case of SDiff$_{q_{max}}$, values increase for, approximately, the first six scales, and then, the values are virtually stable (Figure 2A). A significant difference was found between trained and sedentary rats in the mean $q_{max}$ at long time scales (6–20) (Figure 2E). No difference was observed among groups in the mean SDiff$_{q_{max}}$ (Figure 2D) or mean $q_{zero}$ (Figure 2F).

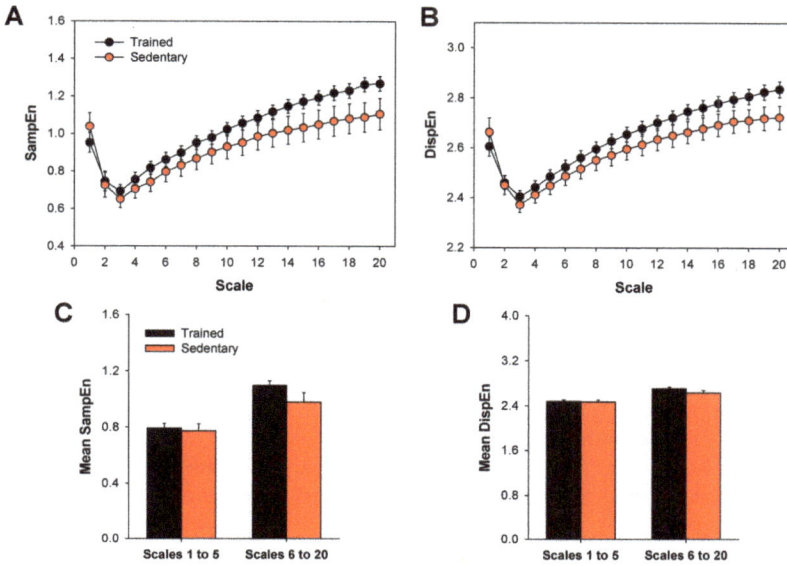

**Figure 1.** MSE or MDE did not detect differences between HRV complexity from trained and sedentary rats. Curve profiles are presented for MSE (**A**) and MDE (**B**), obtained from trained and sedentary groups. Bar graphs show mean entropy values obtained from MSE (**C**) and MDE (**D**) curves, grouped by short (1 to 5) and long (6 to 20) time scales. MSE: multiscale sample entropy; MDE: multiscale dispersion entropy; SampEn: sample entropy; DispEn: dispersion entropy; HRV: heart rate variability. Bars represent the mean ± standard error.

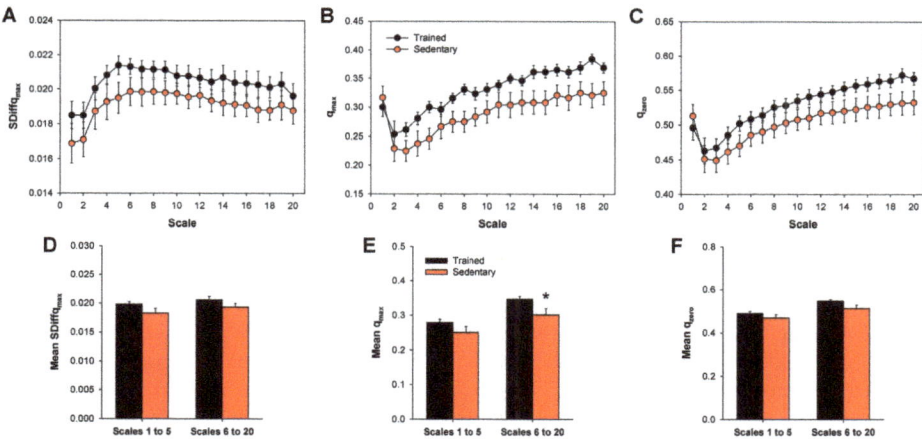

**Figure 2.** Multiscale $q$-attributes calculated from HRV series of trained and sedentary rats. Curve profiles are presented for $\text{SDiff}_{q_{max}}$ (**A**), $q_{max}$ (**B**) and $q_{zero}$ (**C**), obtained from trained and sedentary rats. Bar graphs show mean $q$-attributes values, obtained from $\text{SDiff}_{q_{max}}$ (**D**), $q_{max}$ (**E**) and $q_{zero}$ (**F**), grouped by short (1 to 5) and long (6 to 20) time scales. $\text{SDiff}_{q_{max}}$: maximal $\text{SDiff}_q$; $q_{max}$: $q$ value where $\text{SDiff}_q$ is maximal; $q_{zero}$: $q$ value where $\text{SDiff}_q$ is zero; HRV: heart rate variability. Bars represent the mean ± standard error. * $p < 0.05$ when compared to the trained group.

## 4. Discussion

The characterization of system physiological complexity from a univariate variable, such as HRV, is a hard task. Previous studies have reported on MSE as a powerful tool to assess the complexity of HRV [37,45–48]. Many studies have proposed and evaluated modifications in MSE, given its success in characterizing complex dynamics. Some of them are based on the replacement of sample entropy by another entropy measurement, such as MDE, attempting to improve the accuracy of MSE in specific situations. In the present study, we applied MSE and MDE to account for the complexity changes due to physical training in rats. However, neither MSE nor MDE were able to detect any difference between HRV complexity from trained and sedentary rats.

On the other hand, multiscale SDiff$_q$ is a recent proposal of complexity measurements ($q$-attributes), inspired by MSE, but with a different theoretical background. It relies on nonadditive statistics and uses the difference of $q$-entropy between the actual and surrogate HRV time series to characterize the complexity. Interestingly, from all the multiscale measurements studied, only $q_{max}$ was able to distinguish the complexity of HRV between trained and sedentary animals. Moreover, the difference was found only at long time scales (6 to 20). Recent studies have pointed out that short time scales of MSE are more associated with the vagal control of HRV, whereas long time scales seem to be more related (although not exclusively) to the sympathetic control of HRV [46,49], reinforcing the existence of long-term memory in the components of the autonomic nervous system. Extending this interpretation to SDiff$_q$, one could say that the difference between sedentary and trained HRV is more related to differences in the sympathetic control. This seems a reasonable assumption, given that (1) $q$-attributes use the same coarse-graining procedure of MSE to create the scaled time series and (2) physical training promotes, among other benefits, a lower sympathetic activity and modulation [50,51].

Even though there is a significant difference in $q_{max}$ between trained and sedentary groups, the difference is not huge. An interesting question to ask is: how much is changed in the physiological complexity with physical training? Another question would be: how do the interactions between physiological systems change in a physically trained animal? One has to bear in mind that all those multiscale measurements represent a general view of the system function. In other words, those complexity measurements extract the overall complexity of the system, which is the result of several mechanisms contributing to the homeostasis. Considering that the sedentary animals are healthy, a tremendous increase would not be expected in the complexity after physical training, given that most of the regulatory mechanisms are supposed to be already working at a high complexity level. Therefore, results suggest that systemic changes induced by physical training increase the system complexity to a slightly higher level.

The ability of those multiscale measurements to quantify the overall system complexity of HRV is a distinguishing feature. Many classical HRV indices seek to extract information related to the sympathetic or vagal autonomic modulation, not to mention that they are all linear models. Those indices are usually very sensitive to the environment and behavioral conditions and cannot represent the physiological complexity [4]. For example, during one hour of ECG recording, the rat may explore, sleep, groom, dig and other typical rat behaviors. All those situations will change the autonomic balance, and it is difficult to say what is the real sympathetic and vagal modulation of the rat during the whole one-hour period. On the other hand, applying the multiscale complexity measurements during the whole period, it was possible to identify that the dynamics of HRV has higher complexity in the trained rat compared to the sedentary one, even though the rat can change its physiological state several times during the recording. It is worth noting that all multiscale approaches were also applied to differential HRV series, but no difference was found between trained and sedentary animals, for any measurement [52].

The classical concept of entropy, e.g., SampEn and DispEn, relies on the quantification of the irregularity of a given series. The more irregular (unpredictable) the series, the higher the entropy. Thus, the entropy of any series is supposed to be lower than the entropy of its shuffled version (surrogate), even though the correlation properties of the dynamics were broken when samples are shuffled. However, with $q$-entropy, it is possible to achieve the same entropy values for both situations ($q_{zero}$). Therefore, if we consider the classical entropy ($q = 1$), surrogate data will always be assigned to a higher entropy value, but if we consider $q$ near 0.5 (super-additive), the two dynamics will be assigned the same $q$-sample entropy. More interestingly, there are some values of $q$ where the actual dynamics is assigned higher entropy regarding its surrogate (also for super-additive $q$). Hence, $q_{max}$ can be interpreted as the nonadditive parameter that maximizes the complex properties present in the actual dynamics.

In summary, results with multiscale SDiff$_q$ confirmed previous findings that $q_{max}$ and $q_{zero}$ provide similar, although not equivalent information, which is quite different from SDiff$_{q_{max}}$. Furthermore, MSE, MDE, $q_{max}$ and $q_{zero}$ presented very similar curve profiles, despite their different theoretical definitions, and $q_{max}$ was the only measurement that detected differences in the physiological complexity after physical training. There is no doubt that MSE represents a relevant tool for complexity analysis. This study reinforces that multiscale SDiff$_q$ is an alternative tool for characterizing the complexity of HRV time series, which can add information in some situations where MSE is not accurate enough. Multiscale SDiff$_q$ could also be used to help to characterize the complexity of HRV time series in different pathophysiological conditions, as well as in situations where the signal source is other than HRV.

**Acknowledgments:** This work was supported by CNPq (Grant 113) and CAPES (Grant PNPD20131672) and FAPESP (Grant 2017/05163-6).

**Author Contributions:** F.S.F., R.F.J. and L.E.V.S. conceived of the study. F.S.F. designed and performed the experiments. F.B., R.F.J. and L.E.V.S. collected the data. F.B., L.E.V.S. and L.O.M.J. analyzed the data. L.E.V.S. drafted the paper. All authors revised and approved the paper.

**Conflicts of Interest:** The authors declare no conflict of interest.

## References

1. Boccara, N. *Modeling Complex Systems*; Springer: New York, NY, USA, 2004.
2. Baranger, M. *Chaos, Complexity, and Entropy: A Physics Talk for Non-Physicists*; New England Complex Systems Institute: Cambridge, MA, USA, 2001.
3. Goldberger, A. Giles f. Filley lecture. Complex systems. *Proc. Am. Thorac. Soc.* **2006**, *3*, 467–471.
4. Goldberger, A.L.; Peng, C.K.; Lipsitz, L.A. What is physiologic complexity and how does it change with aging and disease? *Neurobiol. Aging* **2002**, *23*, 23–26.
5. Seely, A.J.E.; Macklem, P.T. Complex systems and the technology of variability analysis. *Crit. Care* **2004**, *8*, R367–R384.
6. Ahmad, S.; Ramsay, T.; Huebsch, L.; Flanagan, S.; McDiarmid, S.; Batkin, I.; McIntyre, L.; Sundaresan, S.R.; Maziak, D.E.; Shamji, F.M.; et al. Continuous multi-parameter heart rate variability analysis heralds onset of sepsis in adults. *PLoS ONE* **2009**, *4*, e6642.
7. Arab, C.; Dias, D.P.M.; de Almeida Barbosa, R.T.; de Carvalho, T.D.; Valenti, V.E.; Crocetta, T.B.; Ferreira, M.; de Abreu, L.C.; Ferreira, C. Heart rate variability measure in breast cancer patients and survivors: A systematic review. *Psychoneuroendocrinology* **2016**, *68*, 57–68.
8. Peng, C.K.; Havlin, S.; Stanley, H.E.; Goldberger, A.L. Quantification of Scaling Exponents and Crossover Phenomena In Nonstationary Heartbeat Time-series. *Chaos* **1995**, *5*, 82–87.
9. Ivanov, P.C.; Amaral, L.A.N.; Goldberger, A.L.; Havlin, S.; Rosenblum, M.G.; Struzik, Z.R.; Stanley, H.E. Multifractality in human heartbeat dynamics. *Nature* **1999**, *399*, 461–465.
10. Porta, A.; Casali, K.R.; Casali, A.G.; Gnecchi-Ruscone, T.; Tobaldini, E.; Montano, N.; Lange, S.; Geue, D.; Cysarz, D.; Van Leeuwen, P. Temporal asymmetries of short-term heart period variability are linked to autonomic regulation. *Am. J. Physiol. Regul. Integr. Comp. Physiol.* **2008**, *295*, R550–R557.
11. Costa, M.D.; Peng, C.K.; Goldberger, A.L. Multiscale analysis of heart rate dynamics: Entropy and time irreversibility measures. *Cardiovasc. Eng.* **2008**, *8*, 88–93.

12. Porta, A.; Tobaldini, E.; Guzzetti, S.; Furlan, R.; Montano, N.; Gnecchi-Ruscone, T. Assessment of cardiac autonomic modulation during graded head-up tilt by symbolic analysis of heart rate variability. *Am. J. Physiol. Heart Circ. Physiol.* **2007**, *293*, H702–H708.

13. Costa, M.D.; Davis, R.; Goldberger, A. Heart Rate Fragmentation: A Symbolic Dynamical Approach. *Front. Physiol.* **2017**, *8*, 827.

14. Bashan, A.; Bartsch, R.P.; Kantelhardt, J.W.; Havlin, S.; Ivanov, P.C. Network physiology reveals relations between network topology and physiological function. *Nat. Commun.* **2012**, *3*, 702.

15. Hou, F.Z.; Wang, J.; Wu, X.C.; Yan, F.R. A dynamic marker of very short-term heartbeat under pathological states via network analysis. *EPL (Europhys. Lett.)* **2014**, *107*, 58001.

16. Chen, C.; Jin, Y.; Lo, I.L.; Zhao, H.; Sun, B.; Zhao, Q.; Zheng, J.; Zhang, X.D. Complexity Change in Cardiovascular Disease. *Int. J. Biol. Sci.* **2017**, *13*, 1320–1328.

17. Xiong, W.; Faes, L.; Ivanov, P.C. Entropy measures, entropy estimators, and their performance in quantifying complex dynamics: Effects of artifacts, nonstationarity, and long-range correlations. *Phys. Rev. E* **2017**, *95*, 062114.

18. Humeau-Heurtier, A. The Multiscale Entropy Algorithm and Its Variants: A Review. *Entropy* **2015**, *17*, 3110–3123.

19. Bandt, C.; Pompe, B. Permutation entropy: A natural complexity measure for time series. *Phys. Rev. Lett.* **2002**, *88*, 174102.

20. Chen, W.; Wang, Z.; Xie, H.; Yu, W. Characterization of surface EMG signal based on fuzzy entropy. *IEEE Trans. Neural Syst. Rehabil. Eng.* **2007**, *15*, 266–272.

21. Li, P.; Liu, C.; Li, K.; Zheng, D.; Liu, C.; Hou, Y. Assessing the complexity of short-term heartbeat interval series by distribution entropy. *Med. Biol. Eng. Comput.* **2015**, *53*, 77–87.

22. Rostaghi, M.; Azami, H. Dispersion Entropy: A Measure for Time-Series Analysis. *IEEE Signal Process. Lett.* **2016**, *23*, 610–614.

23. Rényi, A. On measures of entropy and information. In *Proceedings of the Fourth Berkeley Symposium on Mathematical Statistics and Probability, Volume 1: Contributions to the Theory of Statistics*; The Regents of the University of California: Berkeley, CA, USA, 1961.

24. Manis, G.; Aktaruzzaman, M.; Sassi, R. Bubble Entropy: An Entropy Almost Free of Parameters. *IEEE Trans. Biomed. Eng.* **2017**, *64*, 2711–2718.

25. Hsu, C.F.; Wei, S.Y.; Huang, H.P.; Hsu, L.; Chi, S.; Peng, C.K. Entropy of Entropy: Measurement of Dynamical Complexity for Biological Systems. *Entropy* **2017**, *19*, 550.

26. Silva, L.E.V.; Cabella, B.C.T.; Neves, U.P.d.C.; Murta Junior, L.O. Multiscale entropy-based methods for heart rate variability complexity analysis. *Phys. A Stat. Mech. Its Appl.* **2015**, *422*, 143–152.

27. Amaral, L.S.d.B.; Silva, F.A.; Correia, V.B.; Andrade, C.E.F.; Dutra, B.A.; Oliveira, M.V.; de Magalhães, A.C.M.; Volpini, R.A.; Seguro, A.C.; Coimbra, T.M.; et al. Beneficial effects of previous exercise training on renal changes in streptozotocin-induced diabetic female rats. *Exp. Biol. Med.* **2016**, *241*, 437–445.

28. Faleiros, C.M.; Francescato, H.D.C.; Papoti, M.; Chaves, L.; Silva, C.G.A.; Costa, R.S.; Coimbra, T.M. Effects of previous physical training on adriamycin nephropathy and its relationship with endothelial lesions and angiogenesis in the renal cortex. *Life Sci.* **2017**, *169*, 43–51.

29. Weippert, M.; Behrens, K.; Rieger, A.; Kumar, M.; Behrens, M. Effects of breathing patterns and light exercise on linear and nonlinear heart rate variability. *Appl. Physiol. Nutr. Metab.* **2015**, *40*, 762–768.

30. Soares-Miranda, L.; Sandercock, G.; Vale, S.; Silva, P.; Moreira, C.; Santos, R.; Mota, J. Benefits of achieving vigorous as well as moderate physical activity recommendations: Evidence from heart rate complexity and cardiac vagal modulation. *J. Sports Sci.* **2011**, *29*, 1011–1018.

31. Karavirta, L.; Costa, M.D.; Goldberger, A.L.; Tulppo, M.P.; Laaksonen, D.E.; Nyman, K.; Keskitalo, M.; Häkkinen, A.; Häkkinen, K. Heart rate dynamics after combined strength and endurance training in middle-aged women: Heterogeneity of responses. *PLoS ONE* **2013**, *8*, e72664.

32. Goulopoulou, S.; Fernhall, B.; Kanaley, J.A. Hemodynamic responses and linear and non-linear dynamics of cardiovascular autonomic regulation following supramaximal exercise. *Eur. J. Appl. Physiol.* **2009**, *105*, 525–531.

33. Platisa, M.M.; Mazic, S.; Nestorovic, Z.; Gal, V. Complexity of heartbeat interval series in young healthy trained and untrained men. *Physiol. Meas.* **2008**, *29*, 439–450.

34.  Kuipers, H.; Verstappen, F.; Keizer, H.; Geurten, P.; Van Kranenburg, G. Variability of aerobic performance in the laboratory and its physiologic correlates. *Int. J. Sports Med.* **1985**, *6*, 197–201.

35.  Silva, K.A.d.S.; Luiz, R.S.; Rampaso, R.R.; de Abreu, N.P.; Moreira, E.D.; Mostarda, C.T.; De Angelis, K.; Teixeira, V.d.P.C.; Irigoyen, M.C.; Schor, N. Previous exercise training has a beneficial effect on renal and cardiovascular function in a model of diabetes. *PLoS ONE* **2012**, *7*, e48826.

36.  Costa, M.; Goldberger, A.L.; Peng, C.K. Multiscale Entropy Analysis of Complex Physiologic Time Series. *Phys. Rev. Lett.* **2002**, *89*, 068102.

37.  Costa, M.; Goldberger, A.L.; Peng, C.K. Multiscale entropy analysis of biological signals. *Phys. Rev. E* **2005**, *71*, 021906.

38.  Richman, J.S.; Moorman, J.R. Physiological time-series analysis using approximate entropy and sample entropy. *Am. J. Physiol. Heart Circ. Physiol.* **2000**, *278*, H2039–H2049.

39.  Azami, H.; Rostaghi, M.; Abasolo, D.; Escudero, J. Refined Composite Multiscale Dispersion Entropy and Its Application to Biomedical Signals. *IEEE Trans. Biomed. Eng.* **2017**, *64*, 2872–2879.

40.  Tsallis, C. Possible generalization of Boltzmann-Gibbs statistics. *J. Stat. Phys.* **1988**, *52*, 479–487.

41.  Tsallis, C. *Introduction to Nonextensive Statistical Mechanics*; Springer: New York, NY, USA, 2009.

42.  Silva, L.E.V.; Murta, L.O. Evaluation of physiologic complexity in time series using generalized sample entropy and surrogate data analysis. *Chaos* **2012**, *22*, 043105.

43.  Borges, E.P. A possible deformed algebra and calculus inspired in nonextensive thermostatistics. *Phys. A Stat. Mech. Its Appl.* **2004**, *340*, 95–101.

44.  Theiler, J.; Eubank, S.; Longtin, A.; Galdrikian, B.; Doyne Farmer, J. Testing for nonlinearity in time series: The method of surrogate data. *Phys. D Nonlinear Phenom.* **1992**. *58*, 77–94.

45.  Silva, L.E.V.; Lataro, R.M.; Castania, J.A.; da Silva, C.A.A.; Valencia, J.F.; Murta, L.O.; Salgado, H.C.; Fazan, R.; Porta, A. Multiscale entropy analysis of heart rate variability in heart failure, hypertensive, and sinoaortic-denervated rats: Classical and refined approaches. *Am. J. Physiol. Regul. Integr. Comp. Physiol.* **2016**, *311*, R150–R156.

46.  Silva, L.E.V.; Silva, C.A.A.; Salgado, H.C.; Fazan, R. The role of sympathetic and vagal cardiac control on complexity of heart rate dynamics. *Am. J. Physiol. Heart Circ. Physiol.* **2017**, *312*, H469–H477.

47.  Silva, L.E.V.; Rodrigues, F.L.; de Oliveira, M.; Salgado, H.C.; Fazan, R. Heart rate complexity in sinoaortic-denervated mice. *Exp. Physiol.* **2015**, *100*, 156–163.

48.  Ho, Y.L.; Lin, C.; Lin, Y.H.; Lo, M.T. The Prognostic Value of Non-Linear Analysis of Heart Rate Variability in Patients with Congestive Heart Failure—A Pilot Study of Multiscale Entropy. *PLoS ONE* **2011**, *6*, e18699.

49.  Bari, V.; Valencia, J.F.; Vallverdú, M.; Girardengo, G.; Marchi, A.; Bassani, T.; Caminal, P.; Cerutti, S.; George, A.L.; Brink, P.A.; et al. Multiscale complexity analysis of the cardiac control identifies asymptomatic and symptomatic patients in long QT syndrome type 1. *PLoS ONE* **2014**, *9*, e93808.

50.  Pardo, Y.; Merz, N.; Bairey, C.; Velasquez, I.; Paul-Labrador, M.; Agarwala, A.; Peter, C.T. Exercise conditioning and heart rate variability: Evidence of a threshold effect. *Clin. Cardiol.* **2000**, *23*, 615–620.

51.  Mueller, P.J. Exercise training and sympathetic nervous system activity: evidence for physical activity dependent neural plasticity. *Clin. Exp. Pharmacol. Physiol.* **2007**, *34*, 377–384.

52.  Liu, C.; Gao, R. Multiscale Entropy Analysis of the Differential RR Interval Time Series Signal and Its Application in Detecting Congestive Heart Failure. *Entropy* **2017**, *19*, 251.

*entropy*

MDPI

*Article*

# Entropy Analysis of Short-Term Heartbeat Interval Time Series during Regular Walking

Bo Shi [1], Yudong Zhang [2], Chaochao Yuan [3], Shuihua Wang [4] and Peng Li [3,*]

[1]  Department of Medical Imaging, Bengbu Medical College, Bengbu 233030, China; shibo@bbmc.edu.cn
[2]  Department of Informatics, University of Leicester, Leicester LE1 7RH, UK; yudongzhang@ieee.org
[3]  School of Control Science and Engineering, Shandong University, Jinan 250061, China;
    chaochao.yuan@philips.com
[4]  School of Computer Science and Technology, Nanjing Normal University, Nanjing 210023, China;
    shuihuawang@ieee.org
*   Correspondence: pli@sdu.edu.cn; Tel.: +86-159-6968-8120

Received: 18 September 2017; Accepted: 21 October 2017; Published: 24 October 2017

**Abstract:** Entropy measures have been extensively used to assess heart rate variability (HRV), a noninvasive marker of cardiovascular autonomic regulation. It is yet to be elucidated whether those entropy measures can sensitively respond to changes of autonomic balance and whether the responses, if there are any, are consistent across different entropy measures. Sixteen healthy subjects were enrolled in this study. Each subject undertook two 5-min ECG measurements, one in a resting seated position and another while walking on a treadmill at a regular speed of 5 km/h. For each subject, the two measurements were conducted in a randomized order and a 30-min rest was required between them. HRV time series were derived and were analyzed by eight entropy measures, i.e., approximate entropy (ApEn), corrected ApEn (cApEn), sample entropy (SampEn), fuzzy entropy without removing local trend (FuzzyEn-g), fuzzy entropy with local trend removal (FuzzyEn-l), permutation entropy (PermEn), conditional entropy (CE), and distribution entropy (DistEn). Compared to resting seated position, regular walking led to significantly reduced CE and DistEn (both $p \leq 0.006$; Cohen's $d = 0.9$ for CE, $d = 1.7$ for DistEn), and increased PermEn ($p < 0.0001$; $d = 1.9$), while all these changes disappeared after performing a linear detrend or a wavelet detrend ($<\sim0.03$ Hz) on HRV. In addition, cApEn, SampEn, FuzzyEn-g, and FuzzyEn-l showed significant decreases during regular walking after linear detrending (all $p < 0.006$; $0.8 < d < 1$), while a significantly increased ApEn ($p < 0.0001$; $d = 1.9$) and a significantly reduced cApEn ($p = 0.0006$; $d = 0.8$) were observed after wavelet detrending. To conclude, multiple entropy analyses should be performed to assess HRV in order for objective results and caution should be paid when drawing conclusions based on observations from a single measure. Besides, results from different studies will not be comparable unless it is clearly stated whether data have been detrended and the methods used for detrending have been specified.

**Keywords:** exercise; short-term heart rate variability (HRV); complexity; entropy; approximate entropy (ApEn); conditional entropy (CE); distribution entropy (DistEn); fuzzy entropy (FuzzyEn); permutation entropy (PermEn); sample entropy (SampEn)

## 1. Introduction

Reduced heart rate variability (HRV), a sign of impaired cardiovascular autonomic control [1], has been associated with elevated risk for cardiovascular disease in the general population [2–4], and increased mortality in patients with various circulatory system diseases [5–8]. HRV, by definition, indicates the tiny fluctuations of the time intervals between consecutive normal sinus heartbeats. It can easily be extracted from the electrocardiographic (ECG) recordings. Since the measurement is simple, non-invasive, and cost-efficient, HRV has emerged as a promising tool for assessing risk for cardiovascular diseases and monitoring disease progression.

In common clinical settings, HRV is usually measured under well-controlled conditions (e.g., resting supine or seated position) over a short period (e.g., 5–30 min). Ambulatory HRV monitoring, however, appears to attracted increasing attention nowadays [9]. The long-term ambulatory measurement facilitates the track of HRV changes with activities of free living (e.g., exercise) [10]. The exercise-evoked HRV changes could potentially provide disease-related information [11] but may easily be overlooked by single laboratory assessments that usually do not last long. A couple of previous studies have examined acute HRV changes induced by different activity patterns, e.g., intense exercise or low-intensity exercise, isometric or dynamic exercise [12–21].

Regulated by a feed-back control network that involves balanced spontaneity (as a result of the spontaneous depolarization and repolarization of the sinoatrial node) and adaptability (as a consequence of the regulation of the autonomic nervous system [ANS]), HRV is accepted to be nonlinear in nature [22]. Nonlinear methods can potentially better capture the tiny but physiologically important changes in HRV which, on the contrary, cannot be caught by traditional linear methods. Amongst a vast number of nonlinear approaches, several entropy measures derived from the theory of chaos have witnessed their broad suitability in especially short-term HRV analysis. Those established entropy measures include approximate entropy (ApEn) [23], sample entropy (SampEn) [24], fuzzy entropy (FuzzyEn) [25], permutation entropy (PermEn) [26], conditional entropy (CE) [27], and distribution entropy (DistEn) [28], etc. Different entropy measures likely capture different dynamical properties [29]. However, to our knowledge, there are no published studies that have examined whether those entropy measures respond to exercise in the same way and which entropy measure responds to the stimuli more sensitively.

Therefore, in this study we aimed to test how different entropies of HRV change during exercise. In particular, we focused on the effect of common daily exercise. To imitate daily exercise in the laboratory, walking at a regular speed of 5 km/h on a treadmill was used as a proxy. To examine the within-subject changes, each participant undertook a walking protocol and a rest protocol. The next section explains in detail the subjects, experimental protocols, and analysis methods. Experimental results are summarized in the Results section, followed by discussions in the Discussion section.

## 2. Materials and Methods

### 2.1. Subjects

Sixteen healthy college students (four females/12 males; age: $20.1 \pm 0.6$ years old (mean $\pm$ standard deviation unless otherwise indicated)) were enrolled in this study. Health status was confirmed by questionnaires on the subjects' cardiovascular disease history, neurological disorders, and diabetes. Subjects should not be taking medications with known effects on the ANS within two weeks before participation. Subjects were asked to have adequate sleep during the night before coming to the laboratory, and not to have performed vigorous exercise during the test day and the day before.

### 2.2. Protocols

All tests were performed in a quiet, temperature-controlled ($23 \pm 1$ °C) measurement room. After a 30-min rest to stabilize the cardiovascular system, each participant underwent two 5-min ECG measurement protocols: (i) a "*Rest*" protocol during which the participant was in a resting seated position; and (ii) a "*Walk*" protocol during which the participant kept walking at a speed of 5 km/h on a treadmill (ZR11, Reebok, Canton, MA, USA). A 30-min rest was scheduled between the two tests. To minimize possible training effect, eight randomly selected participants (two females) undertook the *Rest* protocol first and the remaining eight participants undertook the *Walk* protocol first. ECG was recorded using a Holter (DiCare-mlCP, Dimetek Digital Medical Tech., Ltd., Shenzhen, China) with a sampling frequency of 200 Hz. Standard unipolar chest lead V5 was applied.

### 2.3. Extraction of HRV Time-Series

After data collection, ECGs were imported into a self-designed MATLAB program for all subsequent analyses. First, all data recordings were visually inspected for quality issues. After a thorough visual check, we confirmed that all the recordings were with high signal quality. Then, R peaks were detected based on a template-matching procedure [30] followed by a second-round visual inspection to remove incorrectly identified peaks (either false positive or false negative) and ectopic beats. We confirmed that no ectopic beats occurred in those subjects. HRV time-series were finally constructed by the consecutive R-R intervals.

### 2.4. Entropy Measures

#### 2.4.1. Algorithm of ApEn

ApEn evaluates the irregularity of time-series by measuring the unpredictability of fluctuation patterns, i.e., the more repetitive patterns the more predictable (less irregular) the time-series. For a time-series of N points $\{u(i), 1 \le i \le N\}$, ApEn can be calculated using following steps [23]:

1.   State space reconstruction

Form $(N - m\tau)$ vectors $\mathbf{X}_m(i)$ by $\mathbf{X}_m(i) = \{u(i), u(i+\tau), \cdots, u(i+(m-1)\tau)\}, 1 \le i \le N - m\tau$. Here $m$ indicates the embedding dimension and $\tau$ the time delay.

2.   Ranking similar vectors

Define the distance between $\mathbf{X}_m(i)$ and $\mathbf{X}_m(j)$ ($1 \le i, j \le N - m\tau$) by $d_{i,j} = \max(|u(i+k) - u(j+k)|, 0 \le k \le m - 1)$. For a given $i$, calculate the percentage of the vectors $\mathbf{X}_m(j)$ that are within $r$ of $\mathbf{X}_m(i)$ (i.e., $d_{i,j} \le r$):

$$C_i^m(r) = \frac{N_i^m(r)}{N - m\tau}. \tag{1}$$

where $N_i^m(r)$ indicating the number of vectors $\mathbf{X}_m(j)$ that are within $r$ of $\mathbf{X}_m(i)$.

Define $\Phi^{(m)}(r)$ the average of the percentages over $1 \le i \le N - m\tau$ after logarithmic transform, i.e., $\Phi^{(m)}(r) = \sum_{i=1}^{N-m\tau} \ln[C_i^m(r)]/(N - m\tau)$. Repeat steps (1) and (2) to calculate $\Phi^{(m+1)}(r)$ for dimension $(m + 1)$. Here, $r$ indicates the threshold parameter.

3.   Calculation

The ApEn value of the time-series $\{u(i)\}$ can be calculated by:

$$\mathrm{ApEn}(m, \tau, r) = \Phi^{(m)}(r) - \Phi^{(m+1)}(r). \tag{2}$$

ApEn is accepted to be a biased estimator since it allows self-matches (i.e., the distance between $\mathbf{X}_m(i)$ and itself) [24]. In order to reduce the bias, a corrected ApEn (cApEn) algorithm has been proposed [31]. Briefly, if limiting the number of vectors to $N - (m + 1)\tau$ for dimension $m$, Equation (2) can be rewritten as $-\frac{1}{N-(m+1)\tau} \sum_{i=1}^{N-(m+1)\tau} \ln \frac{N_i^{m+1}(r)}{N_i^m(r)}$, which is exactly the formula for cApEn. In addition, when $N_i^{m+1}(r) = 1$ or $N_i^m(r) = 1$ that implies the occurrence of self-match, the ratio $\frac{N_i^{m+1}(r)}{N_i^m(r)}$ in cApEn formula should be substituted with $\frac{1}{N-(m+1)\tau}$.

#### 2.4.2. Algorithm of SampEn

SampEn is mathematically the negative natural logarithm of the conditional probability that two vectors (in the state space representation) that are similar for $m$ points (i.e., the distance between them is within $r$) remain similar at the next point [24]. The following algorithm can be used to determine the SampEn value of a time-series of N points $\{u(i), 1 \le i \le N\}$:

1. State space reconstruction

Form $(N - m\tau)$ vectors $\mathbf{X}_m(i)$ by $\mathbf{X}_m(i) = \{u(i), u(i+\tau), \cdots, u(i+(m-1)\tau)\}, 1 \leq i \leq N - m\tau$. Here $m$ indicates the embedding dimension and $\tau$ the time delay.

2. Ranking similar vectors

Define the distance between $\mathbf{X}_m(i)$ and $\mathbf{X}_m(j)$ $(1 \leq i, j \leq N - m\tau, i \neq j)$ by $d_{i,j} = \max(|u(i+k) - u(j+k)|, 0 \leq k \leq m - 1)$. Denote $A_i^{(m)}(d,r)$ the average number of vectors $\mathbf{X}_m(j)$ within $r$ of $\mathbf{X}_m(i)$ (i.e., $d_{i,j} \leq r$) for all $j = 1, 2, \cdots, N - m\tau$ and $j \neq i$ to exclude self-matches. Similarly, we define $A_i^{(m+1)}(d,r)$ to rank the similarity between vectors with the next point added in the comparison. Here, $r$ indicates the threshold parameter.

3. Calculation

The SampEn value of the time-series $\{u(i)\}$ can be calculated by:

$$\text{SampEn}(m, \tau, r) = -\ln \frac{\sum_{i=1}^{N-m\tau} A_i^{(m+1)}(d,r)}{\sum_{i=1}^{N-m\tau} A_i^{(m)}(d,r)}. \tag{3}$$

### 2.4.3. Algorithm of FuzzyEn

FuzzyEn is methodologically the same to SampEn except that the average number of vectors $\mathbf{X}_m(j)$ that are within $r$ of $\mathbf{X}_m(i)$ (in step 2 of the algorithm of SampEn) is replaced with the average degree of membership. Specifically, for a given fuzzy membership function $e^{-\ln(2)(x/r)^2}$, $A_i^{(m)}(d,r) = \sum_{j=1, j \neq i}^{N-m\tau} e^{-\ln(2)(d_{i,j}/r)^2}$ is applied in FuzzyEn [25]. In addition, in the original FuzzyEn algorithm [25], the local mean of the corresponding vector is removed before calculating the distance, i.e., $d_{i,j} = \max(|[u(i+k) - \overline{u}_i] - [u(j+k) - \overline{u}_j]|, 0 \leq k \leq m - 1)$, wherein $\overline{u}_i$ and $\overline{u}_j$ are the local means for vectors $\mathbf{X}_m(i)$ and $\mathbf{X}_m(j)$, respectively (i.e., $\overline{u}_i = m^{-1} \sum_{k=0}^{m-1} u(i+k)$ and $\overline{u}_j = m^{-1} \sum_{k=0}^{m-1} u(j+k)$). In this way, FuzzyEn evaluates the similarity between vectors based on mainly their shape. However, the memory effect of the autonomic regulation may be manifested in the low frequency component which cannot be captured after removing the local trend. Thus, here we calculated two FuzzyEn values that are with and without local trend removal, respectively, and denoted these two versions by FuzzyEn-l and FuzzyEn-g. In a previous study, a fuzzy measure entropy was developed by combining (i.e., linearly adding them up) those two versions [32]. Since the underlying meanings of the two different approaches could clearly be uncovered by each individual version, here we did not apply this combined measure.

### 2.4.4. Algorithm of PermEn

PermEn evaluates the diversity of ordinal patterns within a time-series [26]. First, a permutation vector $\pi$ can be obtained by resorting the state-space vectors $\mathbf{X}_m(i) = \{u(i), u(i+\tau), \cdots, u(i+(m-1)\tau)\}, 1 \leq i \leq N - m\tau$, in an increasing order ($\pi$ is defined by the index of elements in $\mathbf{X}_m(i)$ when resorting it). Note that the orders of two equal values are defined according to the orders of appearance. Here, we denote the frequency of each $\pi_j, 1 \leq j \leq m!$ as $p_j(m, \tau)$. Then, the PermEn of the time-series $\{u(i), 1 \leq i \leq N\}$ can be calculated by:

$$\text{PermEn}(m, \tau) = -\frac{1}{\log_2 m!} \sum_{j=1}^{m!} p_j(m, \tau) \log_2 [p_j(m, \tau)]. \tag{4}$$

### 2.4.5. Algorithm of CE

CE evaluates the information carried by a new sampling point given the previous samples by estimating the Shannon entropy of the vectors with length $m$ and vectors with the new sampling point added (i.e., with length $m + 1$) [27,33]. Specifically, for a time-series of $N$ points $\{u(i), 1 \leq i \leq N\}$, CE can be calculated by the following steps:

1.  Coarse-graining

    The full range of dynamics is divided into a fixed number of $\xi$ values labelled from zero to $\xi - 1$. The coarse-graining resolution thus equals $[\max(u) - \min(u)]/\xi$. It renders $u(i)$ sequences of symbols $\hat{u}(i), i = 1, 2, \ldots, N$. Here $\xi$ indicates the quantization level.

2.  State space reconstruction

    Form $\mathbf{X}_m(i)$ and $\mathbf{X}_{m+1}(j)$ by:

    $$\mathbf{X}_m(i) = [\hat{u}(i), \hat{u}(i - \tau), \ldots, \hat{u}(i - (m-1)\tau)], \tag{5}$$

    $$\mathbf{X}_{m+1}(j) = [\hat{u}(j), \mathbf{X}_m(j - \tau)], \tag{6}$$

    respectively, where $(m-1)\tau + 1 \leq i, j \leq N$.

3.  Encoding

    The vectors $\mathbf{X}_m(i)$ and $\mathbf{X}_{m+1}(j)$ can be codified in decimal format as:

    $$\{\mathbf{X}_m(i)\}_{10} = \hat{u}(i)\xi^{m-1} + \hat{u}(i - \tau)\xi^{m-2} + \cdots + \hat{u}(i - (m-1)\tau)\xi^0 = w_i, \tag{7}$$

    $$\{\mathbf{X}_{m+1}(j)\}_{10} = \hat{u}(j)\xi^m + \{\mathbf{X}_m(j - \tau)\}_{10} = z_j, \tag{8}$$

    thus rendering each sequence of vectors $\mathbf{X}_m(i)$ and $\mathbf{X}_{m+1}(j)$ series of integer numbers $w_i$ and $z_j$ with $w_i$ ranging from zero to $(\xi - 1) \sum_{i=1}^{m-1} \xi^i$, and $z_j$ ranging from zero to $(\xi - 1) \sum_{j=1}^{m} \xi^j$.

4.  Probability estimation

    Estimate the probability of each possible value for $w_i$ and $z_j$ by the corresponding frequency.

5.  Calculation

    Define CE by:
    $$CE(m, \tau) = SE(z_j) - SE(w_i) + perc(m)SE(1), \tag{9}$$

    where $SE(\cdot)$ calculates the Shannon entropy of a specific distribution, $perc(m)$ is the percentage of patterns $w_i$ found only once in the data set, $SE(1)$ the Shannon entropy of the quantized series $\hat{u}(i)$.

### 2.4.6. Algorithm of DistEn

Instead of quantifying only the probability of "similar vectors" in the state-space that has been applied in SampEn, DistEn takes full advantage of the state-space representation of the time-series by quantifying the distribution characteristics of the inter-vector distances. For the time-series $\{u(i), 1 \leq i \leq N\}$, DistEn can be estimated as follows [28]:

1.  State space reconstruction

    Form $(N - (m-1)\tau)$ vectors $\mathbf{X}(i)$ by $\mathbf{X}(i) = \{u(i), u(i + \tau), \cdots, u(i + (m-1)\tau)\}, 1 \leq i \leq N - (m-1)\tau$. Here, $m$ indicates the embedding dimension and $\tau$ the time delay.

2.  Distance matrix construction

    Calculate the inter-vector distances (distances between all possible combinations of $\mathbf{X}(i)$ and $\mathbf{X}(j)$) by $d_{i,j} = \max(|u(i + k) - u(j + k)|, 0 \leq k \leq m - 1)$ for all $1 \leq i, j \leq N - m$. The distance matrix is denoted as $\mathbf{D} = \{d_{i,j}\}$.

3. Probability density estimation

Estimate the empirical probability density function of the distance matrix **D** by the histogram approach with a fixed bin number of $B$. The probability of each bin can be denoted as $\{p_t, t = 1, 2, \cdots, B\}$. Note here elements with $i = j$ in **D** are excluded in the estimation. Besides, since **D** is always symmetric about the main diagonal, the estimation can be performed on only the diagonal matrix (with the main diagonal excluded).

4. Calculation

The DistEn value of the time-series $\{u(i)\}$ can be calculated by:

$$\text{DistEn}(m, \tau, B) = -\frac{1}{\log_2(B)} \sum_{t=1}^{B} p_t \log_2(p_t). \tag{10}$$

*2.5. Entropy Analysis of HRV Time-Series*

Before all calculations, the HRV time-series were first normalized by subtracting the corresponding mean and then dividing the results by the corresponding standard deviation (SD), i.e., z-scored. Table 1 summarizes the assignments of input parameters for different entropies.

**Table 1.** Assignments of input parameters.

| Interpretation | $m$ Embedding Dimension | $\tau$ Time Delay | $r$ [1] Threshold Value | $B$ Bin Number | $\xi$ Quantization Level | References |
|---|---|---|---|---|---|---|
| ApEn | 2 | 1 | $0.2 \times$ SD | - | - | [24] |
| cApEn | 2 | 1 | $0.2 \times$ SD | - | - | [31] |
| SampEn | 2 | 1 | $0.2 \times$ SD | - | - | [24] |
| FuzzyEn-g | 2 | 1 | $0.2 \times$ SD | - | - | [25,32] |
| FuzzyEn-l | 2 | 1 | $0.2 \times$ SD | - | - | [25] |
| PermEn | 3 | 1 | - | - | - | [34] |
| CE | 2 | 1 | - | - | 6 | [27,33] |
| DistEn | 2 | 1 | - | 512 | - | [28,35] |

[1] SD = 1 after normalizing the signals.

To explore the possible influence of nonstationary trend, we performed a linear detrending [31] and a wavelet detrending, separately, on the HRV time-series and repeated all those calculations. To perform the wavelet detrending, HRV were first evenly resampled to 4 Hz by spline interpolation. A 6-level wavelet decomposition using the coif5 wavelet was then conducted. The approximation coefficients on the 6th level were reconstructed to the original scale and were non-evenly "recovered" by spline interpolation which resulted in the final trend that would be subtracted. The 6-level decomposition was used so that the frequency band of the trend would be less than ~0.03 Hz.

*2.6. Statistical Analysis*

All results were first subjected to the Shapiro-Wilk $W$ test to examine the normality. The null hypothesis of this test is that the data under examined follow a normal distribution. A $p$ value of less than 0.05 rejects the null hypothesis and thus indicates a non-normal distribution. For a specific entropy measure, paired $t$-test would be used to examine the difference between *Rest* and *Walk* protocols, if the Shapiro-Wilk $W$ test suggested a normal distribution for that measure; Wilcoxon signed-rank test of each pair would be applied if otherwise. In addition, Cohen's $d$ static was calculated for statistically significant observations to examine the effect size of the corresponding measure responding to the stimuli of regular walking, no matter the measure was normally distributed or not. An effect size $d > 0.8$ was considered large and was considered very large if $d > 1.2$ [36]. To further check the performance of those entropy measures, bivariate Pearson correlation analyses between each two entropy measures were explored under *Rest* and *Walk* conditions, separately. Bonferroni criterion was used to correct for multiple comparisons. Bonferroni corrected $p < 0.05$ was considered statistically significant. All the statistical analyses were performed using the JMP software (Pro 13, SAS Institute, Cary, NC, USA).

## 3. Results

An average of 375 (SD: 46; min: 307; max: 485) RR intervals were obtained from the 16 participants during the *Rest* protocol. During the *Walk* protocol, the average length of HRV was 576 (SD: 49; min: 500; max: 699; $p < 0.0001$ vs. *Rest*). The Shapiro-Wilk $W$ tests suggested normality for ApEn, cApEn, and SampEn (all $p > 0.2$) while it refuted the normal distribution hypothesis for the rest measures (all $p < 0.05$) except DistEn, for which $p = 0.07$. We here still considered DistEn non-normally distributed partly because the distribution, as visually checked, was less likely to follow a normal distribution and partly because of the relatively small sample size. Therefore, paired *t*-test was applied to examine the differences in ApEn, cApEn, and SampEn between *Rest* and *Walk* protocols, whereas for the rest five measures, Wilcoxon signed-rank test of each pair was applied.

Figure 1 shows the pair-wise changes of the eight entropy measures of HRV time-series before detrending between *Rest* and *Walk* conditions, with the corresponding mean (or median if non-normally distributed) and SD (or the 1st and 3rd quartiles) specified by short bars. Since eight tests were performed, here a $p$ value of $\leq 0.006$ ($0.05/8 \approx 0.006$) was considered statistically significant using the Bonferroni criterion. The results show no significant changes between the *Walk* and *Rest* conditions in ApEn ($p = 0.1$), cApEn ($p > 0.1$), and SampEn ($p > 0.1$) as suggested by the paired *t*-test, and no significant changes in FuzzyEn-g and FuzzyEn-l (both $p = 0.008$), either, as indicated by the Wilcoxon signed-rank test. By contrast, CE and DistEn reduce significantly under *Walk* condition as indicated by the Wilcoxon signed-rank test (all $p \leq 0.006$). Large or even very large effect sizes are observed (i.e., $d = 0.90$ for CE; $d = 1.7$ for DistEn). PermEn however increases significantly under *Walk* condition ($p < 0.0001$) with very large effect size ($d = 1.9$).

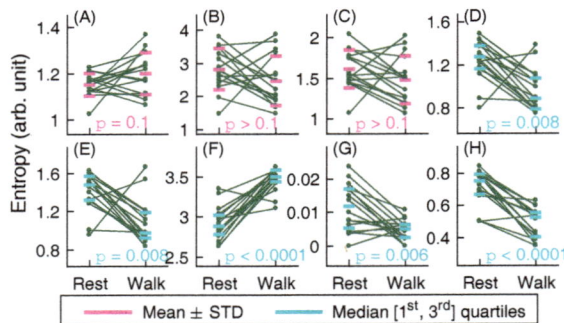

**Figure 1.** The entropies of raw short-term heartbeat interval series. In order to show the changes, results from the same participants were connected by lines. (**A**) ApEn; (**B**) cApEn; (**C**) SampEn; (**D**) FuzzyEn-g; (**E**) FuzzyEn-l; (**F**) PermEn; (**G**) CE; (**H**) DistEn.

The results after linear detrending are shown in Figure 2. There is still no significant change in ApEn ($p > 0.1$). Moreover, the observed changes in PermEn, CE, and DistEn become not significant (all $p > 0.05$). However, the results show significantly reduced cApEn, SampEn, FuzzyEn-g, and FuzzyEn-l during *Walk* condition (all $p < 0.006$) with large effect sizes (all $0.8 < d < 1$). Figure 3 shows the results after wavelet detrending. Surprisingly, ApEn shows a significant increase during *Walk* condition ($p < 0.0001$; $d = 1.9$). Similar to the result after linear detrending, cApEn decreases significantly ($p = 0.006$; $d = 0.8$). However, all the rest entropy measures do not indicate significant changes (all $p \geq 0.02$).

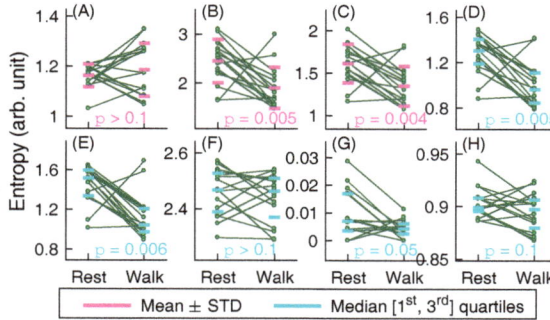

**Figure 2.** The entropies of short-term heartbeat interval series after linear detrending. In order to show the changes, results from the same participants were connected by lines. (**A**) ApEn; (**B**) cApEn; (**C**) SampEn; (**D**) FuzzyEn-g; (**E**) FuzzyEn-l; (**F**) PermEn; (**G**) CE; (**H**) DistEn.

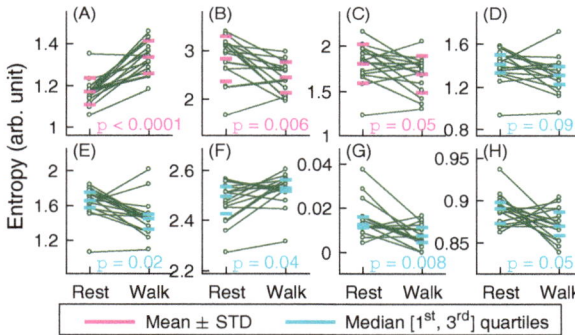

**Figure 3.** The entropies of short-term heartbeat interval series after wavelet detrending. In order to show the changes, results from the same participants were connected by lines. (**A**) ApEn; (**B**) cApEn; (**C**) SampEn; (**D**) FuzzyEn-g; (**E**) FuzzyEn-l; (**F**) PermEn; (**G**) CE; (**H**) DistEn.

The bivariate Pearson correlation analysis results are showed in Figure 4 (which shows the test significance $p$) and Figure 5 (which shows the Pearson $r$). Results are summarized below.

1. For raw HRV time-series without detrending (upper panels in Figures 4 and 5)

Under *Rest* condition, cApEn is positively correlated with SampEn, FuzzyEn-g, and FuzzyEn-l; SampEn is positively correlated with FuzzyEn-g and FuzzyEn-l; FuzzyEn-g is positively correlated with FuzzyEn-l; PermEn is negatively correlated with DistEn; no significant correlations are observed between all other pairs.

Under *Walk* condition, ApEn is positively correlated with FuzzyEn-g and FuzzyEn-l; c-ApEn is positively correlated with SampEn; FuzzyEn-g is positively correlated with FuzzyEn-l; no significant correlations are observed between the rest pairs.

Since many correlation results change during walking, those entropy measures are less likely to reproduce each other. Their responses to change of conditions may also be different. In order to illustrate this assumption, the bivariate correlation analysis between the pair-wise changes of entropy measures from *Rest* to *Walk* was also performed and the results are superimposed on those for *Rest* and *Walk* conditions in Figures 4 and 5. The results are similar to those under *Walk* condition, except that the difference of PermEn is negatively correlated with the difference of DistEn, which is the same to that under *Rest* condition.

2.  For HRV time-series after linear detrending (middle panels in Figures 4 and 5)

The results are similar to those for raw HRV data, except that the negative correlation between PermEn and DistEn disappears. The results for *Rest-Walk* differences are the same to those for *Rest* condition. However, under *Walk* condition, in addition to those significant pairs for *Rest* condition, ApEn also shows positive correlations with cApEn, SampEn, FuzzyEn-g, and FuzzyEn-l.

3.  For HRV time-series after wavelet detrending (lower panels in Figures 4 and 5)

The results are exactly the same to those for HRV data after linear detrending except some changes in $p$ and $r$ values.

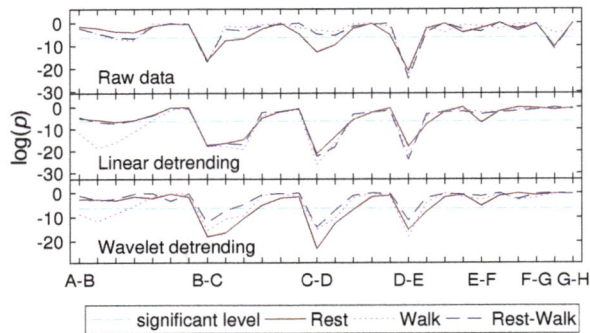

**Figure 4.** Bivariate correlation analysis results. The abscissa shows between which two entropy measures the correlation analysis was performed. The codes A to H mean ApEn, cApEn, SampEn, FuzzyEn-g, FuzzyEn-l, PermEn, CE, and DistEn, respectively. The label 'A-B' thus indicates the correlation between ApEn and cApEn, and so do the rest labels. The $p$ values are shown in logarithmic scale in the ordinate, such that a significant test result is obtained if the corresponding $p$ value is less than the significant level, which is $\log\left(0.05/C_8^2\right)$ after Bonferroni correction.

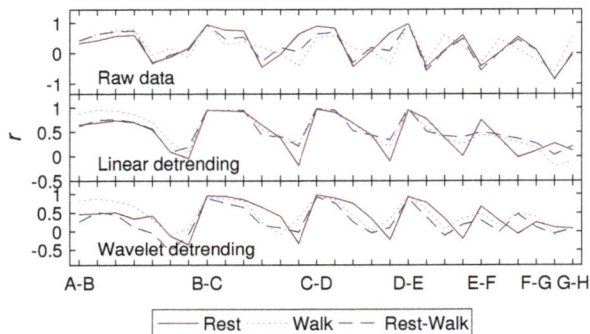

**Figure 5.** Bivariate correlation analysis results. Results are shown in the same way as has applied in Figure 4, except that the ordinate is showing the Pearson $r$.

## 4. Discussion

Based on eight well-established entropy measures, we studied the within-subject changes of HRV during walking with a regular speed of 5 km/h on a treadmill (the *Walk* protocol) compared to those in a resting seated position (the *Rest* protocol). We also explored the potential effects of nonstationary linear, or very low-frequency trend. Our main findings are summarized in Table 2.

**Table 2.** Summary of findings.

| HRV vs. Measures [1] | A | B | C | D | E | F | G | H |
|---|---|---|---|---|---|---|---|---|
| Raw | - [2] | - | - | - | - | ↑ [3] | ↓ [4] | ↓ |
| After linear detrending | - | ↓ | ↓ | ↓ | ↓ | - | - | - |
| After wavelet detrending | ↑ | ↓ | - | - | - | - | - | - |

[1] A: ApEn; B: cApEn; C: SampEn; D: FuzzyEn-g; E: FuzzyEn-l; F: PermEn; G: CE; H: DistEn; [2] No statistical significance; [3] Statistically significant increase; [4] Statistically significant decrease.

Even though the significant observations in PermEn, CE, and DistEn disappeared after linear or wavelet detrending, the changing directions were unchanged (i.e., walking leads to increased PermEn while decreased CE and DistEn, see Figures 1–3). The insignificant results might be due to lack of power because only 16 subjects were enrolled. This is one of our study limitations. However, the within-subject design we applied may help improve the power. We are planning to enroll more participants and use a field protocol to further investigate the effects of daily activity.

Those findings suggest that the information that PermEn, CE, and DistEn capture may easily be "masked" by nonstationary trend which makes them less able to probe the changes in dynamics that are beyond the trend. Besides, PermEn is considered to be highly sensitive to noise [37] while few knowledge is known regarding the robustness of CE and DistEn against noise. The relative contribution of noise is supposed to be augmented after trend removal (which reduces signal power and thus reduces the signal-to-noise ratio). Noise in RR intervals may come from the tiny deviations between the detected and real R peaks (random noise) because of fixed rate sampling or spikes because of false positive or false negative detection, or ectopic beats. As we have mentioned in Method section, we conducted a thorough visual inspection regarding detection error and confirmed that no ectopic beats occurred. So only random noise may be considered as one of the factors that may affect our results. Besides, we note that the sampling frequency of the device we used is relatively low (200 Hz) which is considered to be another study limitation. The low sampling rate may lead to increased noise power that exacerbates the adverse effects on entropy measures.

SampEn, FuzzyEn-g, and FuzzyEn-l showed significant decrease during walking after linear detrending while the results became not significant again after wavelet detrending. Though seemingly erratic, the results are actually consistent to some extent as all the changing directions remain the same (Figures 1–3). These findings imply that a linear detrend may be considered prior to performing the SampEn, FuzzyEn-g, and FuzzyEn-l analyses. The cApEn decreased significantly during walking after both linear and wavelet detrending (Figures 2 and 3), which also suggests performing a detrend before using this method, irrelative to detrend methods. The ApEn only showed a significant difference between resting and walking conditions after wavelet detrending, suggesting ApEn highly sensitive to nonstationary trend. As a result, removal of nonstationary, very low-frequency trend should be performed before ApEn analysis.

However, there is little knowledge on whether or not the very low frequency trend contains useful physiological information. Those suggestions, as described above, are thus purely observation-based. Comprehensive studies on how nonstationary trend affects entropy analysis and further physiological investigations on the meaning of HRV trend are warranted. Anyway, the interpretation does differ from each other if different strategies are applied. For example, the insignificant ApEn observation on raw HRV and HRV after linear detrending might be due to the biasness of the ApEn algorithm as it includes self-matches [24], but ApEn became capable after wavelet detrending which seems to refute the possibility of biasness. However, considering the methodology, the number of similar vectors indeed increases after removing the trend component which thus reduces the weight of self-matches.

One interesting finding is that the changing direction of ApEn was opposite of cApEn and the other SampEn-based measures (i.e., SampEn, FuzzyEn-g, and FuzzyEn-l). Methodologically, both ApEn and SampEn assess the creation of information, or the unpredictability, in a time series [24]. The cApEn is basically a modified version of ApEn such that it is supposed to capture similar properties [31]. So does FuzzyEn which is actually a modified version of SampEn based on fuzzy logic [25,32]. However, the results come out to be that cApEn performs more similarly as SampEn does, instead of ApEn itself. This is also supported by the correlation analysis which showed that cApEn was always highly correlated to SampEn (i.e., *r* value is very close to 1). The correction algorithm in cApEn may actually be equivalent to what SampEn applies. ApEn thus captures certain property hidden in the fluctuations that is different from SampEn-based measures.

The other interesting finding is that PermEn also shows different changing direction as compared to SampEn-based measures, CE, and DistEn. PermEn estimates the diversity of fluctuation patterns which may reflect the randomness [26]. CE functions like SampEn except that it estimates the average amount of information based on encoded time-series [27]. DistEn assesses actually the diversity of vectors in the state-space representation of a time-series [28]. Based on these methodological differences, the properties captured by different entropy measures may actually be intrinsically different or be different aspects of the complexity of HRV.

Overall, several other reasons may also contribute to the observed discrepancies across these entropy measures:

- All calculations were based on fixed input parameters which may not work well all the time. In other words, some results might not be completely true because improper parameters were applied. In the current study, we did not repeat our analyses using different combinations of parameters partly because that it would make things rather cumbersome for real application. Furthermore, there is no solid way to find out a proper choice for each individual case even though different combinations are able to be traversed. Previous studies have explored how to define parameters but the proposed approaches are mostly achieved retrospectively by maximizing the pre-hypothesized group differences [38–40]. However, it is not necessarily be always true that those hypothesized group differences exist.
- From the perspective of the underlying physiological mechanisms, it is still yet to be determined which branch in the ANS (i.e., sympathetic or vagal nerves) actually becomes dominant during walking at such a relatively lower but regular (for typical populations) speed. Some studies indicated that vagal withdrawal is the dominant mechanisms during lower intensity, dynamic exercise while others also points to a sympathetic HR modulation even at the onset of exercise [13,15]. Studies also suggested that the relative role of the two drives may depend on the exercise intensity [41]. It has been hypothesized that the withdrawal of parasympathetic (vagal) modulation might already be obvious during low intensity exercise, whereas the sympathetic increase may present at higher intensity exercises [17]. In addition, it is also controversial whether sympathetic drive to the heart or vagal withdrawal is the main contributor of HR complexity [21,31,42,43], let alone each specific complexity measure.

In term of sensitivity to the stimuli of regular walking, different measures actually indicate different sensitivity regarding different strategies applied for detrending. When raw HRV without detrending is used, PermEn and DistEn suggest the best sensitivity. The cApEn, SampEn, FuzzyEn-g, and FuzzyEn-l show comparable sensitivity for HRV after linear detrending. When wavelet detrending is applied, ApEn, however, suggests the best performance.

CE was not correlated with any other entropies under both *Rest* and *Walk* conditions, suggesting that it may capture a unique HRV nonlinear property. However, as shown in Figure 1G, CE under *Rest* condition distributed more dispersedly even though an overall significant reduction was displayed.

*Entropy* **2017**, *19*, 568

Besides, many CE results actually researched the theoretical lower limit and this floor effect might introduce some bias in the estimation of effect size. This lack of robustness may thus deserve further elucidations. By contrast, DistEn showed correlation with SampEn and PermEn under *Rest* condition while under *Walk* condition no significant correlations were shown. Besides, no significant correlations between DistEn and others were observed for HRV after linear or wavelet detrending during both conditions. These results also render DistEn a unique role in analyzing HRV.

As a brief conclusion, when applying entropy analyses to short-term HRV data, we suggest: (1) using PermEn or DistEn on raw short-term HRV data; (2) performing linear detrend before applying SampEn-based measures; and (3) removing the very-low frequency trend before ApEn analysis. In order to make results comparable and to help better interpret different observations, whether or not detrend is performed as well as what detrending method is applied should be clearly specified.

With the rapid advances of technology and reduced cost, the use of wearable devices that are able to monitor heart rate continuously will likely become rather commonplace. Our current study shows that regular walking may acutely affect the commonly applied HRV entropies. The effect of daily activities should therefore be taken into consideration when interpreting results from long-time ambulatory recordings. Device developers may consider including event reporters in those devices for users to track their activities throughout the day, which can be a useful reference for data analyses. In our future studies, field protocols will be designed to examine the effects of real daily activities.

**Acknowledgments:** This work was supported by the Key Program on Natural Scientific Research from the Department of Education of Anhui Province, China (No. KJ2016A470), the National Natural Science Foundation of China (No. 61601263), and the Natural Science Foundation of Shandong Province (No. ZR2015FQ016).

**Author Contributions:** P.L. and B.S. conceptualized the study. B.S. collected the data. All authors analyzed the data, interpreted the results, and drafted the manuscript. P.L. significantly revised and finalized the manuscript. All authors read and approved the final manuscript.

**Conflicts of Interest:** The authors declare no conflict of interest.

## References

1.  Hayano, J.; Sakakibara, Y.; Yamada, A.; Yamada, M.; Mukai, S.; Fujinami, T.; Yokoyama, K.; Watanabe, Y.; Takata, K. Accuracy of assessment of cardiac vagal tone by heart rate variability in normal subjects. *Am. J. Cardiol.* **1991**, *67*, 199–204. [CrossRef]
2.  Ziegler, D.; Voss, A.; Rathmann, W.; Strom, A.; Perz, S.; Roden, M.; Peters, A.; Meisinger, C.; Group, K.S. Increased prevalence of cardiac autonomic dysfunction at different degrees of glucose intolerance in the general population: The kora s4 survey. *Diabetologia* **2015**, *58*, 1118–1128. [CrossRef] [PubMed]
3.  Felber Dietrich, D.; Schindler, C.; Schwartz, J.; Barthelemy, J.C.; Tschopp, J.M.; Roche, F.; von Eckardstein, A.; Brandli, O.; Leuenberger, P.; Gold, D.R.; et al. Heart rate variability in an ageing population and its association with lifestyle and cardiovascular risk factors: Results of the sapaldia study. *Europace* **2006**, *8*, 521–529. [CrossRef] [PubMed]
4.  Tsuji, H.; Venditti, F.J., Jr.; Manders, E.S.; Evans, J.C.; Larson, M.G.; Feldman, C.L.; Levy, D. Reduced heart rate variability and mortality risk in an elderly cohort. The framingham heart study. *Circulation* **1994**, *90*, 878–883. [CrossRef] [PubMed]
5.  Drawz, P.; Babineau, D.; Brecklin, C.; He, J.; Kallem, R.; Soliman, E.; Xie, D.; Appleby, D.; Anderson, A.; Rahrnan, M.; et al. Heart rate variability is a predictor of mortality in chronic kidney disease: A report from the cric study. *Am. J. Nephrol.* **2013**, *38*, 517–528. [CrossRef] [PubMed]
6.  Ziegler, D.; Zentai, C.P.; Perz, S.; Rathmann, W.; Haastert, B.; Döring, A.; Meisinger, C.; Group, K.S. Prediction of mortality using measures of cardiac autonomic dysfunction in the diabetic and nondiabetic population: The monica/kora augsburg cohort study. *Diabetes Care* **2008**, *31*, 556–561. [CrossRef] [PubMed]
7.  Dekker, J.M.; Crow, R.S.; Folsom, A.R.; Hannan, P.J.; Liao, D.; Swenne, C.A.; Schouten, E.G. Low heart rate variability in a 2-minute rhythm strip predicts risk of coronary heart disease and mortality from several causes: The aric study. Atherosclerosis risk in communities. *Circulation* **2000**, *102*, 1239–1244. [CrossRef] [PubMed]

8.  Liao, D.; Cai, J.; Rosamond, W.D.; Barnes, R.W.; Hutchinson, R.G.; Whitsel, E.A.; Rautaharju, P.; Heiss, G. Cardiac autonomic function and incident coronary heart disease: A population-based case-cohort study. The aric study. Atherosclerosis risk in communities study. *Am. J. Epidemiol.* **1997**, *145*, 696–706. [CrossRef] [PubMed]

9.  Akintola, A.A.; van de Pol, V.; Bimmel, D.; Maan, A.C.; van Heemst, D. Comparative analysis of the equivital eq02 lifemonitor with holter ambulatory ecg device for continuous measurement of ecg, heart rate, and heart rate variability: A validation study for precision and accuracy. *Front. Physiol.* **2016**, *7*, 391. [CrossRef] [PubMed]

10. Kristiansen, J.; Korshoj, M.; Skotte, J.H.; Jespersen, T.; Sogaard, K.; Mortensen, O.S.; Holtermann, A. Comparison of two systems for long-term heart rate variability monitoring in free-living conditions—A pilot study. *Biomed. Eng. Online* **2011**, *10*, 27. [CrossRef] [PubMed]

11. Morris, C.J.; Purvis, T.E.; Hu, K.; Scheer, F.A. Circadian misalignment increases cardiovascular disease risk factors in humans. *Proc. Natl. Acad. Sci. USA* **2016**, *113*, E1402–E1411. [CrossRef] [PubMed]

12. Taylor, K.A.; Wiles, J.D.; Coleman, D.D.; Sharma, R.; O'Driscoll, J.M. Continuous cardiac autonomic and haemodynamic responses to isometric exercise. *Med. Sci. Sports Exerc.* **2017**. [CrossRef] [PubMed]

13. White, D.W.; Raven, P.B. Autonomic neural control of heart rate during dynamic exercise: Revisited. *J. Physiol.* **2014**, *592*, 2491–2500. [CrossRef] [PubMed]

14. Weippert, M.; Behrens, M.; Rieger, A.; Behrens, K. Sample entropy and traditional measures of heart rate dynamics reveal different modes of cardiovascular control during low intensity exercise. *Entropy* **2014**, *16*, 5698–5711. [CrossRef]

15. Fisher, J.P. Autonomic control of the heart during exercise in humans: Role of skeletal muscle afferents. *Exp. Physiol.* **2014**, *99*, 300–305. [CrossRef] [PubMed]

16. Goya-Esteban, R.; Barquero-Perez, O.; Sarabia-Cachadina, E.; de la Cruz-Torres, B.; Naranjo-Orellana, J.; Rojo-Alvarez, J.; Murray, A. Heart rate variability non linear dynamics in intense exercise. *Comput. Cardiol.* **2012**, *39*, 177–180.

17. Boettger, S.; Puta, C.; Yeragani, V.K.; Donath, L.; Muller, H.J.; Gabriel, H.H.; Bar, K.J. Heart rate variability, qt variability, and electrodermal activity during exercise. *Med. Sci. Sports Exerc.* **2010**, *42*, 443–448. [CrossRef] [PubMed]

18. Leicht, A.S.; Sinclair, W.H.; Spinks, W.L. Effect of exercise mode on heart rate variability during steady state exercise. *Eur. J. Appl. Physiol.* **2008**, *102*, 195–204. [CrossRef] [PubMed]

19. Princi, T.; Accardo, A.; Peterec, D. Linear and non-linear parameters of heart rate variability during static and dynamic exercise in a high-performance dinghy sailor. *Biomed. Sci. Instrum.* **2004**, *40*, 311–316. [PubMed]

20. Cottin, F.; Durbin, F.; Papelier, Y. Heart rate variability during cycloergometric exercise or judo wrestling eliciting the same heart rate level. *Eur. J. Appl. Physiol.* **2004**, *91*, 177–184. [CrossRef] [PubMed]

21. Tulppo, M.; Makikallio, T.; Takala, T.; Seppanen, T.; Huikuri, H. Quantitative beat-to-beat analysis of heart rate dynamics during exercise. *Am. J. Physiol. Heart Circ. Physiol.* **1996**, *271*, H244–H252.

22. Sugihara, G.; Allan, W.; Sobel, D.; Allan, K.D. Nonlinear control of heart rate variability in human infants. *Proc. Natl. Acad. Sci. USA* **1996**, *93*, 2608–2613. [CrossRef] [PubMed]

23. Pincus, S.M. Approximate entropy as a measure of system complexity. *Proc. Natl. Acad. Sci. USA* **1991**, *88*, 2297–2301. [CrossRef] [PubMed]

24. Richman, J.S.; Moorman, J.R. Physiological time-series analysis using approximate entropy and sample entropy. *Am. J. Physiol. Heart Circ. Physiol.* **2000**, *278*, H2039–H2049. [PubMed]

25. Chen, W.; Zhuang, J.; Yu, W.; Wang, Z. Measuring complexity using fuzzyen, apen, and sampen. *Med. Eng. Phys.* **2009**, *31*, 61–68. [CrossRef] [PubMed]

26. Bandt, C.; Pompe, B. Permutation entropy: A natural complexity measure for time series. *Phys. Rev. Lett.* **2002**, *88*, 174102. [CrossRef] [PubMed]

27. Porta, A.; Baselli, G.; Liberati, D.; Montano, N.; Cogliati, C.; Gnecchi-Ruscone, T.; Malliani, A.; Cerutti, S. Measuring regularity by means of a corrected conditional entropy in sympathetic outflow. *Biol. Cybern.* **1998**, *78*, 71–78. [CrossRef] [PubMed]

28. Li, P.; Liu, C.; Li, K.; Zheng, D.; Liu, C.; Hou, Y. Assessing the complexity of short-term heartbeat interval series by distribution entropy. *Med. Biol. Eng. Comput.* **2015**, *53*, 77–87. [CrossRef] [PubMed]

29. Li, P.; Karmakar, C.; Yan, C.; Palaniswami, M.; Liu, C. Classification of five-second epileptic eeg recordings using distribution entropy and sample entropy. *Front. Physiol.* **2016**, *7*, 136. [CrossRef] [PubMed]

30. Li, P.; Liu, C.; Zhang, M.; Che, W.; Li, J. A real-time qrs complex detection method. *Acta Biophys. Sin.* **2011**, *27*, 222–230. [CrossRef]

*Entropy* **2017**, *19*, 568

31. Porta, A.; Gnecchi-Ruscone, T.; Tobaldini, E.; Guzzetti, S.; Furlan, R.; Montano, N. Progressive decrease of heart period variability entropy-based complexity during graded head-up tilt. *J. Appl. Physiol. (1985)* **2007**, *103*, 1143–1149. [CrossRef] [PubMed]

32. Liu, C.; Li, K.; Zhao, L.; Liu, F.; Zheng, D.; Liu, C.; Liu, S. Analysis of heart rate variability using fuzzy measure entropy. *Comput. Biol. Med.* **2013**, *43*, 100–108. [CrossRef] [PubMed]

33. Li, P.; Li, K.; Liu, C.; Zheng, D.; Li, Z.-M.; Liu, C. Detection of coupling in short physiological series by a joint distribution entropy method. *IEEE Trans. Biomed. Eng.* **2016**, *63*, 2231–2242. [CrossRef] [PubMed]

34. Makowiec, D.; Kaczkowska, A.; Wejer, D.; Żarczyńska-Buchowiecka, M.; Struzik, Z. Entropic measures of complexity of short-term dynamics of nocturnal heartbeats in an aging population. *Entropy* **2015**, *17*, 1253–1272. [CrossRef]

35. Karmakar, C.; Udhayakumar, R.K.; Li, P.; Venkatesh, S.; Palaniswami, M. Stability, consistency and performance of distribution entropy in analysing short length heart rate variability (hrv) signal. *Front. Physiol.* **2017**, *8*, 720. [CrossRef] [PubMed]

36. Sawilowsky, S.S. New effect size rules of thumb. *J. Mod. Appl. Stat. Methods* **2009**, *8*, 597–599. [CrossRef]

37. Porta, A.; Bari, V.; Marchi, A.; De Maria, B.; Castiglioni, P.; di Rienzo, M.; Guzzetti, S.; Cividjian, A.; Quintin, L. Limits of permutation-based entropies in assessing complexity of short heart period variability. *Physiol. Meas.* **2015**, *36*, 755–765. [CrossRef] [PubMed]

38. Liu, C.; Liu, C.; Shao, P.; Li, L.; Sun, X.; Wang, X.; Liu, F. Comparison of different threshold values r for approximate entropy: Application to investigate the heart rate variability between heart failure and healthy control groups. *Physiol. Meas.* **2011**, *32*, 167–180. [CrossRef] [PubMed]

39. Lu, S.; Chen, X.; Kanters, J.K.; Solomon, I.C.; Chon, K.H. Automatic selection of the threshold value r for approximate entropy. *IEEE Trans. Biomed. Eng.* **2008**, *55*, 1966–1972. [PubMed]

40. Li, P.; Liu, C.; Wang, X.; Li, L.; Yang, L.; Chen, Y.; Liu, C. Testing pattern synchronization in coupled systems through different entropy-based measures. *Med. Biol. Eng. Comput.* **2013**, *51*, 581–591. [CrossRef] [PubMed]

41. Aubert, A.E.; Seps, B.; Beckers, F. Heart rate variability in athletes. *Sports Med.* **2003**, *33*, 889–919. [CrossRef] [PubMed]

42. Porta, A.; Castiglioni, P.; Bari, V.; Bassani, T.; Marchi, A.; Cividjian, A.; Quintin, L.; Di Rienzo, M. K-nearest-neighbor conditional entropy approach for the assessment of the short-term complexity of cardiovascular control. *Physiol. Meas.* **2013**, *34*, 17–33. [CrossRef] [PubMed]

43. Porta, A.; Guzzetti, S.; Furlan, R.; Gnecchi-Ruscone, T.; Montano, N.; Malliani, A. Complexity and nonlinearity in short-term heart period variability: Comparison of methods based on local nonlinear prediction. *IEEE Trans. Biomed. Eng.* **2007**, *54*, 94–106. [CrossRef] [PubMed]

*Article*

# Sample Entropy of the Heart Rate Reflects Properties of the System Organization of Behaviour

Anastasiia V. Bakhchina [1,2,*], Karina R. Arutyunova [1], Alexey A. Sozinov [1], Alexander V. Demidovsky [3] and Yurii I. Alexandrov [1,4]

[1] Institute of Psychology of Russian Academy of Sciences, Laboratory of Neural Bases of Mind Named after V.B. Shvyrkov, 129366 Moscow, Russia; arutyunova@inbox.ru (K.R.A.); alesozinov@yandex.ru (A.A.S.); yuraalexandrov@yandex.ru (Y.I.A.)

[2] Department of Psychophysiology, National Research University Nizhny Novgorod State University Named after N.I. Lobachevsky, 603950 Nizhny Novgorod, Russia

[3] Computer Science Department, National Research University Higher School of Economics, 603014 Nizhny Novgorod, Russia; ademidovskij@hse.ru or monadv@yandex.ru

[4] Department of Psychology, National Research University Higher School of Economics, 101000 Moscow, Russia

* Correspondence: nastya18-90@mail.ru or nastya189065@gmail.com; Tel.: +7-964-638-8360

Received: 27 April 2018; Accepted: 5 June 2018; Published: 8 June 2018

**Abstract:** Cardiac activity is involved in the processes of organization of goal-directed behaviour. Each behavioural act is aimed at achieving an adaptive outcome and it is subserved by the actualization of functional systems consisting of elements distributed across the brain and the rest of the body. This paper proposes a system-evolutionary view on the activity of the heart and its variability. We have compared the irregularity of the heart rate, as measured by sample entropy (SampEn), in behaviours that are subserved by functional systems formed at different stages of individual development, which implement organism-environment interactions with different degrees of differentiation. The results have shown that SampEn of the heart rate was higher during performing tasks that included later acquired knowledge (foreign language vs. native language; mathematical vocabulary vs. general vocabulary) and decreased in the stress and alcohol conditions, as well as at the beginning of learning. These results are in line with the hypothesis that irregularity of the heart rate reflects the properties of a set of functional systems subserving current behaviour, with higher irregularity corresponding to later acquired and more complex behaviour.

**Keywords:** heart rate variability; irregularity; sample entropy; functional systems; individual development; stress; alcohol administration; learning

## 1. Introduction

Studies of the physiological bases of behaviour do not often take into account the processes that occur outside the anatomical borders of the brain. The mental processes, such as perceptions, thoughts, and feelings, for a long time have been viewed without considering the physiological state of the body. At the current stage of the development of cognitive sciences more attention is paid to the analyses of "embodied cognition" and whole-organism integration that underpin behaviour and psychological processes [1–4]; however, the mechanisms of such integration are still a debatable question.

From the physiological perspective, the interest in the "mind-body" relationship is becoming obvious as the number of studies on autonomic regulation of processes involving internal organs is increasing [5,6]. Some previous studies were focused on how different subcortical structures modulate the activity of internal organs but now more attention is paid to the corticovisceral coordination [7–9]. In general, it appears important to develop an integrated approach to study the psychophysiological bases of behaviour organization that would take into account the significance of the impact of the body's physiological states. In this direction, the phenomenon of heart rate variability is of interest.

The heart beats with complicated oscillations in coherence with the functioning of the entire body. Heart rate variability (HRV) denotes the change in the time intervals between adjacent heartbeats and it is necessary to regulate the transport of resources through the body in order to adapt to external challenges and achieve optimal performance [10]. Therefore, the cardiovascular activity, and particularly, heart rate, are viewed in psychophysiological studies as related to relevant behavioural processes.

From the physiological perspective, the primary origin of HRV is related to the activity of the sympathetic and parasympathetic subdivisions of the autonomic nervous system (ANS) [11]. Higher HRV at rest supports a better performance at a number of tasks [12]. Several theories [13] explain why large HRV is associated with physical fitness and youth, while physical and psychological stress is associated with decreased HRV. However, such purely physiological models are limited in explaining the HRV related to behaviour. Moreover, although it is clear how the activity of ANS is reflected in HRV, the regulation of heart rate by cortical structures is still an unanswered question [14–16]. Here, we address this problem from the positions of the system-evolutionary theory and propose a model explaining HRV in relation to neuronal processes in the brain.

The system-evolutionary theory [3,4,17] was built upon the fundamentals of P.K. Anokhin's theory of functional systems [18], which suggests that morphologically different components of the body and brain comprise functional systems, in which their co-operative activity leads to achieving adaptive results within the organism-environment relations. P.K. Anokhin emphasized that breathing and blood flow observed during performing different behavioural acts may seem the same, but they are actually different, because their characteristics depend on the behavioural result that is being achieved by means of performing this particular act [19]. The system-evolutionary theory considerably extends this approach and proposes to view neurons and other body cells, as "organisms within the organism", i.e., the activity of each cell is aimed at satisfying its metabolic needs by means of interacting with the environment and other cells [3]. Neurons and body cells joint cooperative activity leads to the adaptive result that is a new relation between an organism and the environment. Each novel way of adaptive organism-environment interaction underlies the formation of a new functional system represented by the cells that were active when achieving this adaptive outcome. The subsequent actualization of this functional system underlies the realization of this behavioural act, which can be considered as an element of subjective experience. Thus, a functional system is understood as a dynamic organization of activity of components across different anatomical localizations, both in the brain and the rest of the body, which provides the achievement of an adaptive result for the whole organism.

When considered within the framework of the system-evolutionary theory, HRV originates in cooperation of the heart with the other components of actualized functional systems, including neuronal groups. This approach suggests that HRV depends on the behavioural results that were achieved during the organism-environment interactions and, therefore, may differ between sequential behavioural acts. The current study is aimed at analyzing and examining this dependency and its implications.

It has been shown [2,3,20] that a newly formed complex instrumental behaviour is based on the simultaneous activation of the corresponding "new" functional system with the older systems that had been formed at previous stages of individual development. The older systems become involved in many behavioural patterns as they belong to the elements of individual experience that are common for various acts. Therefore, the system organization of behaviour represents the history of behavioural development. Multiple systems are involved in behaviour, and each of them was formed at a certain stage of development of the current behaviour [21].

The structure of experience becomes more complex and differentiated during individual development. The formation of new functional systems results in growing complexity and the degree of differentiation of organism-environment relations. A new functional system does not replace previously formed systems, but instead, it is "superimposed" on them [20]. Individual development can be considered as the process of increasing differentiation along with the number of learned behaviors [22,23]. Therefore, behaviour formed at later stages of individual development provides a more differentiated (more detailed and accurate) organism-environment relation and supported by a larger set of systems and links between them. From this view, heart activity depends on the system characteristics of current behaviour and HRV is hypothesized to be higher when an individual performs recently formed behaviour than behaviour formed at earlier stages of development. This is the first hypothesis that is explored in our study.

Psychological and physiological stress along with other factors, such as alcohol intoxication, lead to a temporary increase of the role of earlier formed systems in the organization of behaviour [24,25]. This phenomenon of "system dedifferentiation" is usually accompanied by increased emotional arousal, decreased cognitive control, less detailed perception and performance, preference of intuitive strategies in decision making to rational ones, etc. On the neuronal level, it is shown that alcohol decreases the number of activated cortical neurons specialized in relation to later formed systems [26]. This implies that HRV would decrease in the context of system dedifferentiation, which is the second hypothesis that is explored in this study.

Thus, the goal of the study was to examine the interrelations between the systems supporting behaviour and the HRV during the performance of this behaviour. We hypothesized that a lower HRV is observed during performing early-formed behaviour in comparison with behaviour formed more recently, and HRV decreases in the contexts of system dedifferentiation.

## 2. Materials and Methods

### 2.1. Ethical Statement

All of the participants gave written informed consent to take part in the study after receiving an explanation of the procedures.

The study was conducted in accordance with the Declaration of Helsinki. The Ethics Committee of the Federal State-Financed Institution, Institute of Psychology, and Russian Academy of Sciences (Moscow) approved the experimental protocols and the specific consent procedure used in this study, and assessed it as safe for the participants' psychic and physical health. All of the participants were native Russian-speakers and were paid for their participation.

### 2.2. Heart Rate Measurement & Heart Rate Variability Analyses

In all of the experiments, RR-intervals (the time periods between consecutive heartbeats) were measured using a miniature ECG sensor (HxM; sampling rate 250 Hz, Zephyr Technology, Annapolis, MD, USA, www.zephyranywhere.com). Participants wore a special chest belt with two plastic electrodes that were located in the first and second chest leads. Batch data transmission from the sensor to a mobile device was carried out through the wireless protocol Bluetooth. Connecting, data transmission and storage on the mobile device were performed using custom software "HR-Reader" [developed by Kozhevnikov V.V.] for Android OS.

Acquired sequences of RR-intervals were pre-processed before proceeding to the analysis. The sequences with abnormal beats and any artifacts (ectopic beats, coughs, and motion artifacts) were excluded from the analyses. The artifacts were identified as RR-intervals that did not satisfy the condition $|RR_i - RR_{i-1}| < 0.7 \times (RR_i - RR_{i-1})/2$. Thus, we only analyzed sequences that were free from artifacts.

For estimation of heart rate irregularity we used the sample entropy (SampEn) as a set of measures of system irregularity reporting on similarity in time series. SampEn can be applied to relatively short and noisy data; it is largely independent of record length and displays relative consistency under circumstances. SampEn (m, r, N) is precisely the negative natural logarithm of the conditional probability that two vectors that are similar for m points remain similar at the next point, where self-matches are not included in calculating the probability (1) [27].

$$SampEn(m, \; r, \; N) = -\ln\frac{A}{B}, \tag{1}$$

The parameter N is the length of the time series, m is the length of vectors to be compared, and r is the tolerance for accepting matches. A is the number of pairs of vectors (x) for m points that satisfy the condition d[xm(i), xm(j)] ≤ r, and B is the number of pairs of vectors (x) for (m+1) points that satisfy the condition d[xm(i), xm(j)] ≤ r. Thus, a low value of SampEn reflects a high degree of regularity. The parameters m and r were fixed: m = 2, r = 0.5 × SDNN (SDNN—standard deviation of RR-intervals). It had been shown previously that higher values of r were accompanied by lower dispersion of SampEn [27]. While a lot of studies use r = 0.2 × SDNN, we chose to use a larger value of r to avoid extra dispersion in our small samples. It is important to note that SampEn still reliably distinguishes the processes with different degrees of order when calculated with r = 0.5 × SDNN [27]. Calculations were made using the cross-platform software that contained a set of algorithms implemented in Python and PyQt5 as the UI rendering framework. The source code is available in the open source repo (https://github.com/demid5111/approximate-enthropy) under the MIT license.

Additionally the time domain indexes of HRV (mean (av-RR, ms) and standard deviation (SDNN, ms) of RR-intervals) were calculated.

It is accepted that irregularity of heart rate (measured by SampEn) and its variability (measured by SDNN) can correlate, but these correlations are not linear [28]. Therefore, we can face such modes of heart activity which differ in irregularity and have the same variability, and vice versa. For example, SDNN, as an index of HRV, is modulated by breathing, while SampEn is independent of the characteristics of respiratory arrhythmia. Thus, SampEn and SDNN are both HRV indexes partly complementing each other. In this work, we used SampEn as an index of irregularity, SDNN as an index of variability, and av-RR as an index of frequency of the heart rate.

### 2.3. Experimental Protocols

The study included five experiments. In the beginning of each experiment, the participants filled questionnaires about their demographics, health, and well-being during the past several days prior to the experiment. Then, the participants were informed about the details of experimental procedures.

#### 2.3.1. Experiment 1

The aim of Experiment 1 was to compare HRV during the realization of behaviour formed at earlier stages of individual development with behaviour formed later in life (see Figure 1). Participants performed a set of linguistic tasks either in native language (earlier formed), or foreign language (later formed). It had been shown previously that foreign-language processing reduces the impact of intuition and/or increases the impact of deliberation on individuals' choices [29]. Using a foreign language affects moral judgment by blunting emotional reactions that are associated with violation of moral rules [30]. It has been shown in EEG studies that the latencies of event-related potentials are longer when performing tasks in foreign language as compared to performing them in native language, which has been viewed as an indicator of different degrees of automaticity and quantity of the subprocesses that are involved in sentence comprehension [31]. Thus, solving tasks using a foreign language can be considered as a later formed behaviour based on the activation of a wider distributed neuronal subserving in the brain, as opposed to solving tasks using a native language [32].

Participants (N = 29, 25 females, 18 to 26 years old, mean = 20, median = 20) were native Russian speakers recruited at the Nizhny Novgorod Linguistic State University, named after N.A. Dobrolubov, where they were studying German as a foreign language. For some participants (N = 11), German was a primary foreign language and they spent 6–18 (mean = 12.2; median = 13) years studying it, while for others (N = 18) it was a secondary foreign language which they studied for 1–5 years (mean = 2.6; median = 2). We compared the number of mistakes, response times (ms), and heart rate indexes during performing the tests in these two groups of participants. The number of mistakes in the German-language test was lower in the group who had been studying German as a primary foreign language for a longer period of time (6–18 years of studying: Med = 5, Q1 = 2, Q3 = 10; 1–5 years of studying: Med = 12, Q1 = 6, Q3 = 13; U = 50, Z = 2.17, $p < 0.03$, Mann-Whitney test). Therefore, we analyzed the data for these two groups separately.

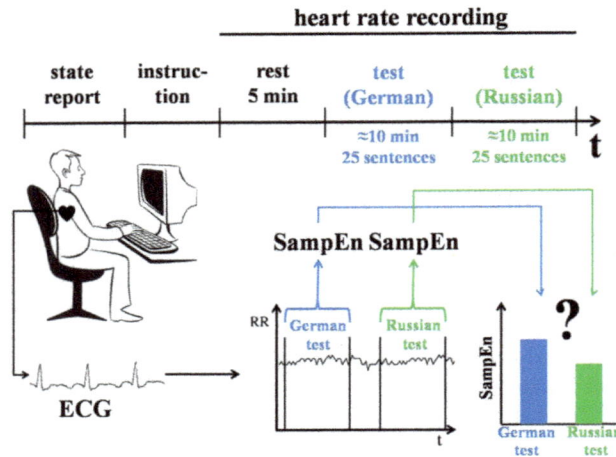

**Figure 1.** Design of Experiment 1. Participants performed two computer tests: one in a foreign language (German) and the other one in a native language (Russian). Participants' heart rate was recorded. Sample entropy (SampEn), mean (av-RR), and standard deviation (SDNN) values were calculated for RR-intervals and compared between the periods of performing the German and Russian tests.

The participants performed two tests: one in German and one in Russian. The order of the tests was counterbalanced. Both tests consisted of 25 sentences that were presented on a computer screen one at a time in a randomized order. Each sentence had a missing word, which was always a noun. The task was to complete the sentences. The time to perform the task was not limited.

Each participant was seated in a quiet room approximately 50 cm from a computer screen. Sentences were presented in white letters against a black background in the center of the screen. A standard computer keyboard was used. To type in a missing word, participants had to press the spacebar and then press the appropriate keys on the keyboard, followed by pressing the Enter key to proceed to the next sentence. The tests were presented using custom experiment software. The cross-platform software was implemented in Java to support conducting experiments with wide configuration capabilities. The source code is available in the open source repo (https://bitbucket.org/ademidovskij/test_me_words) under the MIT license.

Participants' test performance, response times (in ms), and heart rate were recorded. Using Wilcoxon test, we compared SampEn, av-RR, and SDNN values calculated for RR-intervals during the periods of performing the language tests in German and Russian.

### 2.3.2. Experiment 2

The aim of Experiment 2 was to compare HRV during behaviour that was formed at earlier or later stages of individual development (see Figure 2). As opposed to Experiment 1, in this case, we used a linguistic task presented in the participants' native language (Russian) to avoid inter-lingual differences. The task included words that were typically learnt at a different age [33]. It has been shown that early-acquired words are recalled, read, and recognized faster than later acquired words [34]. Early-acquired adjectives are also rated by participants as more emotional [22]. Therefore, we considered using early-acquired words as related to earlier formed and less detailed behaviour.

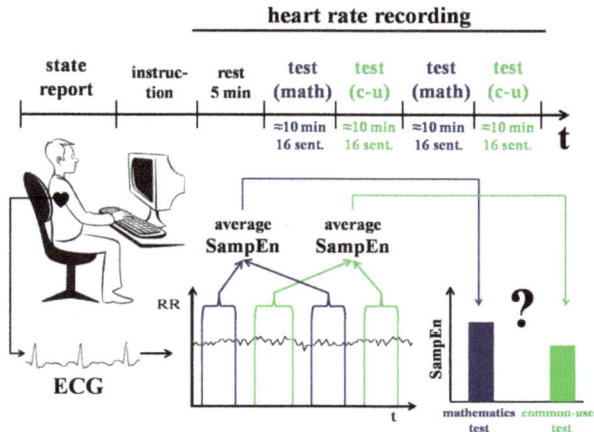

**Figure 2.** Design of Experiment 2. Participants were asked to complete sentences with missing words. Two groups of sentences included words with different age of acquisition: early acquired commonly used words ("c-u") and later acquired mathematical terms ("math"). Participants' heart rate was recorded while performing the task. SampEn, av-RR, and SDNN values were calculated for RR-intervals and compared between the periods of using earlier and later acquired words.

Participants (N = 35, 5 females, 23 to 37 years old, mean = 27.78, median = 28) were professional mathematicians with work experience of 1–10 years (median = 4.84 years). They were presented with two types of tests. The first test consisted of 32 sentences, which contained mathematical terms. These terms are usually learned at an undergraduate level (age of acquisition is 18–19 years old). For example, "A normal is a vector that is perpendicular to a given object". The second test consisted of 32 sentences, which contained commonly used words that are familiar from childhood (age of acquisition is 5–6 years old). For example, "Plasticine is a material for modelling figures". Sentences from the two tests had equal linguistic characteristics, such as the number of words, syllables, letters and Fog's index (a value of text's complexity). Each sentence had a missing word and the experimental task was to complete these sentences. The time to perform the task was not limited. The task was organized in four sets, 16 items each: two of them contained mathematical terms and the two others contained commonly used words. The presentation of the sets was counterbalanced. The sentences in each set were mixed and presented one at a time in a randomized order.

Each participant was seated in a quiet room approximately 50 cm from a computer screen. Sentences were presented in white letters against a black background in the center of the screen. An underscored gap indicated a place for a missing word. A standard computer keyboard was used. To type a missing word, the participants had to press the spacebar and then press the appropriate keys on the keyboard, followed by pressing the Enter key to proceed to the next sentence. The tests were presented using custom experiment software. The cross-platform software was implemented in Java to support conducting experiments with wide configuration capabilities. The source code is available in the open source repo (https://bitbucket.org/ademidovskij/test_me_words) under the MIT license.

Participants' test performance, response times (in ms), and heart rate were recorded. SampEn, av-RR, and SDNN values were calculated for RR-intervals during performing both sets of sentences.

We averaged SampEn, av-RR, and SDNN values for the sentences containing mathematical terms and the sentences containing commonly used words, and then these averaged SampEn, av-RR, and SDNN values were compared using Wilcoxon test.

### 2.3.3. Experiment 3

The aim of Experiment 3 was to study the dynamics of HRV in the conditions of system dedifferentiation that was induced by alcohol administration (see Figure 3). We compared the dynamics of HRV after drinking an alcoholic beverage with the dynamics of HRV after drinking a non-alcoholic beverage. As mentioned above (see Introduction), alcohol administration decreases the activity of neurons that are specialized in relation to later formed behaviour and it does not significantly affect the activity of neurons specialized in relation to earlier formed behaviour [32,35,36]. Therefore, behaviour becomes less differentiated and detailed after alcohol administration.

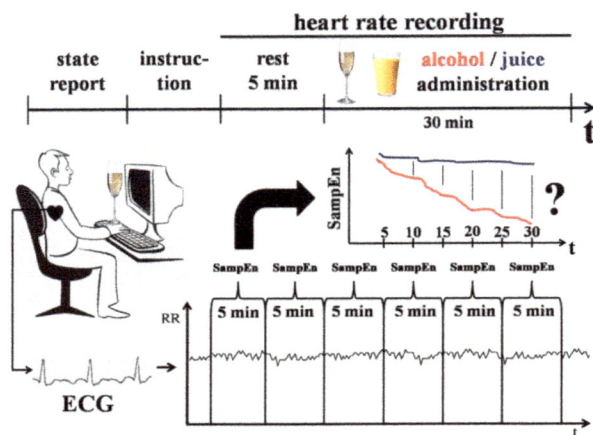

**Figure 3.** Design of Experiment 3. Heart rate was recorded during a 30 min period while participants were drinking a beverage and watching a video. SampEn was calculated for sequential 5 min sections of RR-intervals. The dynamics of SampEn, av-RR, and SDNN values in the control (juice + water) and experimental (juice + alcohol) conditions were compared.

Participants (N = 25, 5 females, 23 to 35 years old, mean = 26.82, median = 28) took part in the experiment twice, once in an alcohol condition and once in a control condition, with the time interval of 1–2 months. The order of the alcohol and control conditions was counterbalanced. In the beginning of each experiment, the participants were weighed and given a glass of beverage, which they were asked to drink within 30 min while watching an emotionally neutral video (BBC, "Planet Earth"). An alcoholic drink was given in the experimental condition and a non-alcoholic drink was given in the control condition.

The dose of alcohol was 1 g of ethanol (medical ethanol, 96%) to each 1 kg of a participant's weight. A measured amount of alcohol was mixed with apple juice to achieve an overall drink volume of 750 mL. In the control condition, the participants drank apple juice mixed with water in the same proportion as in the alcohol condition. We measured breath alcohol content (BrAC, mg/L) using AlcoDigital AL7000 Pro Breathalyzer in the beginning and in the end of the experiment. BrAC was always equal to zero in the control condition. In the alcohol condition, the average level of BrAC before consuming an alcoholic drink was equal to zero and at the end of the experiment BrAC was equal to $0.72 \pm 0.11$ mg/L.

Heart rate was recorded during all experimental sessions. The SampEn, av-RR, and SDNN were calculated for sequential 5 min sections of RR-intervals during the drinking phase. The dynamics of sequential SampEn, av-RR, and SDNN values was compared in the alcohol and the control conditions using the Wilcoxon test.

### 2.3.4. Experiment 4

The aim of Experiment 4 was to study the dynamics of HRV in the condition of system dedifferentiation induced by stress (see Figure 4). The state of stress is usually characterized by high emotional arousal, reduced attention to details [37], preference to the most familiar behavioural strategies and habits [38], as well as making decisions that are typical for earlier stages of development [39]. Under stress, individuals tend to choose intuitive explanations, rather than rational [40]. From the neurobiological perspective, the biochemical diversity in the brain comes to three basic cascades during stress [41]. Overall, stress can be described as a state of temporal system dedifferentiation, which arises as a bias to earlier ways of adaptation [42].

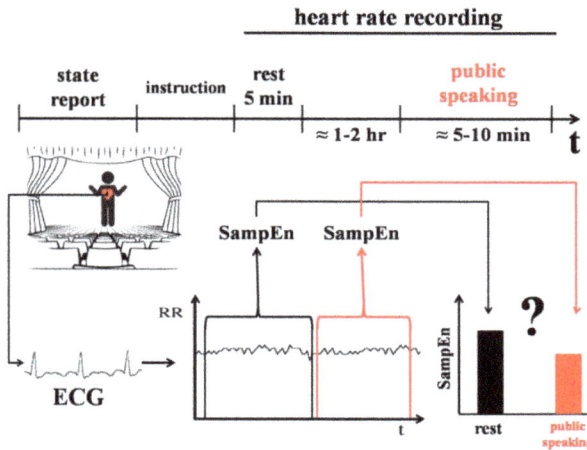

**Figure 4.** Design of Experiment 4. Public speaking was used as a model of social stress. Heart rate was recorded for 5 min at rest 1–2 h prior to public speaking and for the duration of public speaking which lasted 5–10 min. SampEn, av-RR, and SDNN were calculated for RR-intervals at rest and the period of stress and were compared between the conditions.

Public speaking is part of the Trier Social Stress Test [43], which is widely used to model social stress. Participants (N = 13, 7 females, 21 to 30 years old, mean = 24.14, median = 24) were students of the Nizhny Novgorod State University, named after N.I. Lobachevski, who were taking part in academic conferences with oral reports.

The participants' heart rate was recorded when they spoke in public (5–10 min) and while they rested (5 min) 1 or 2 h before their speech. SampEn, av-RR, and SDNN were calculated for the rest and stress sections of RR-intervals and were compared between the conditions using Wilcoxon test.

### 2.3.5. Experiment 5

The aim of Experiment 5 was to analyze the dynamics of HRV during the process of learning. The beginning of learning a new task usually involves novelty and mismatch between the existing individual experience and current organism-environment relations. The same is observed when an individual experiences stress or intense emotions [44]. Moreover, a similar pattern of hormonal changes, i.e., an increase in cortisol levels and concentration of endogenous opioids, is observed during periods of learning, stress, and experiencing emotions [45]. Thus, we viewed the beginning stages of learning as a period of system dedifferentiation, which could be manifested in specific changes of HRV.

Participants (N = 35, 14 females, 18 to 38 years old, mean = 25.4, median = 25) had to learn to play a computer game. The task was to figure out the rules of the game. Before participants started to play, they were briefly familiarized with the apparatus. Then, they read an instruction, which translates as follows: "You are invited to play a computer game. The game has several levels and your general aim is to pass as many levels as you can. You don't know the rules so you have to figure them out by yourself. All events of the game will take place on the screen and the only thing you will need is a computer mouse. You have unlimited time. You can stop playing if you feel bored or get tired. If you are ready to start, click "Play"". Participants played the game until they felt that they wanted to stop or that they achieved the end of the game by passing the last level.

The computer game had been developed using "A-Ware" software (Sozinov A.A., Bokhan A.I.) and was used previously to study behavioural characteristics and strategies in situations of unpredictability, novelty, and mismatching during forming new experience [46]. The design of the game involved a target object that was hidden within a playing field on the screen. The task was to find the target object and to draw a rectangle around it with the cursor as closely to its borders as possible. Good game performance was rewarded with points and players could see their points on the screen after each trial. The maximum possible number of points was 10,000. Once a player achieved 10,000 points in three sequential trials, she/he progressed to the next level. The game consisted of eight levels. The number of target objects (or their size) increased by one with each level passed.

The participant's heart rate was recorded during the game (Figure 5). SampEn, av-RR, and SDNN values were calculated for sequential RR-intervals with a running window (window length—was 100 points, the shift of running—10 points) from the start to the end of the game. To control for the factor of motivation, participants were given an opportunity to stop the game at any time. Therefore, the number of levels that participants passed was different (median = 3). Thus, we analyzed the dynamics of SampEn, av-RR, and SDNN at the beginning (about 150 s from the start) of the first level, which was passed by 26 participants. The heart rate indexes for the first section of RR-intervals from the start of the game were compared with following sections, using the Wilcoxon test.

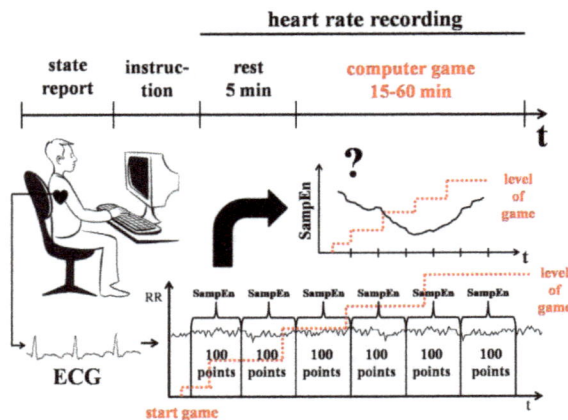

**Figure 5.** Design of Experiment 5. Heart rate was recorded while participants were learning to play a computer game. SampEn, av-RR, and SDNN were calculated for sequential 100 points sections of RR-intervals from the start of the game. The average dynamics of SampEn, av-RR, and SDNN values during the game were analyzed.

## 2.4. Data Analyses

Statistical analyses were performed using STATISTICA10 software (http://documentation.statsoft.com/; STA999K347150-W; Expires 21.12.2021). For data used in statistical analyses, see Supplementary Material.

Distributions of all the variables were tested for normality by Shapiro-Wilk's test. We calculated the medians and interquartile ranges and used nonparametric tests (Wilcoxon test and Mann-Whitney test) to compare distributions different from the normal distribution. In other cases, we calculated mean values and standard errors, and used parametric tests (*t*-test) to compare distributions. To evaluate the dynamics of variables we used Friedman test. To check the correlations between variables, we used Spearman's rank correlation coefficient. Significance was assumed when $p < 0.05$.

## 3. Results

### 3.1. Dynamics of HRV During Performing a Linguistic Task Using Native and Foreign Languages (Experiment 1)

We compared SampEn, SDNN and av-RR values during the periods of performing the linguistic task in native (Russian) and foreign (German) languages in two groups of participants: those who had been learning German for 1–5 years, and those who had been learning it for more than five years. For descriptive statistics and results, see Table 1.

**Table 1.** Heart rate indexes during performing a language task in native (Russian) and foreign (German) languages for groups of participants who had been learning the foreign language (German) for 1–5 years and 6–18 years [1].

| HR Index | Group | Language | Med | Q1 | Q3 | Comparison |
|---|---|---|---|---|---|---|
| SampEn | 1–5 years | German | 1.30 | 1.14 | 1.36 | $T = 7, Z = 3.42, p < 0.01$ |
| | | Russian | 1.12 | 1.08 | 1.17 | |
| | 6–18 years | German | 1.17 | 1.07 | 1.38 | $T = 16, Z = 1.51, p = 0.13$ |
| | | Russian | 1.07 | 0.95 | 1.19 | |
| av-RR (ms) | 1–5 years | German | 705 | 647 | 759 | $T = 78, Z = 0.33, p = 0.74$ |
| | | Russian | 738 | 641 | 765 | |
| | 6–18 years | German | 713 | 584 | 749 | $T = 20, Z = 1.16, p = 0.25$ |
| | | Russian | 695 | 579 | 756 | |
| SDNN (ms) | 1–5 years | German | 49.39 | 40.12 | 57.50 | $T = 79, Z = 0.28, p = 0.78$ |
| | | Russian | 48.81 | 42.25 | 62.55 | |
| | 6–18 years | German | 48.05 | 37.20 | 52.97 | $T = 25, Z = 0.71, p = 0.48$ |
| | | Russian | 47.97 | 33.91 | 57.92 | |

[1] Median values (Med) and quartiles (Q1 and Q3) are presented along with the results of Wilcoxon test comparisons between the periods of using native and foreign languages.

In the group of participants who had been studying German for 1–5 years, SampEn was significantly lower when using Russian language as compared to using German language (Figure 6). Av-RR and SDNN did not differ between the two conditions in this group. In the group of participants who had been studying German for 6–18 years, SampEn, av-RR, and SDNN did not differ between the two conditions.

The analysis of behavioral performance and response time (ms) showed that when the participants of both groups were using foreign language, they made more mistakes and responded slower than when they were using native language (Table 2).

These results show that irregularity of the heart rate is lower when participants perform a task in their native language than when they perform the same task in a foreign language, if they have been learning it for a comparatively short period of time (1–5 years). Irregularity of the heart rate did not differ between the conditions when native and foreign languages were used in participants who had been learning the foreign language for a considerable period of time (6–18 years). It is important to note that irregularity of the heart rate during behaviour, but not its frequency or variability, was associated with the age when this behaviour had been formed.

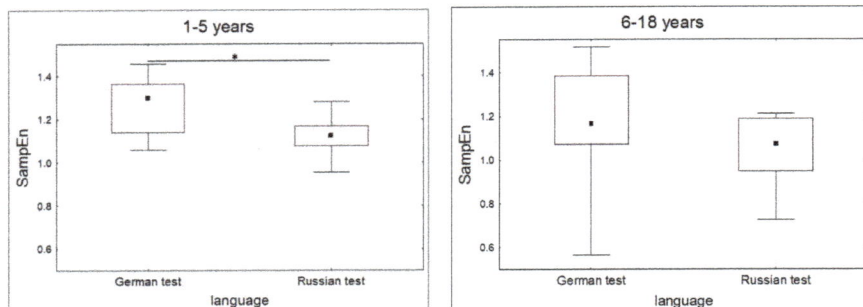

**Figure 6.** Median values (quartiles and range) of SampEn compared between the periods of performing the foreign (German) and native (Russian) language tasks by participants who had been learning the foreign language (German) for 1–5 years (**left** graph) and 6–18 years (**right** graph). * $p < 0.01$, Wilcoxon test.

**Table 2.** Task performance in native (Russian) and foreign (German) languages for groups of participants who had been learning the foreign language (German) for 1–5 years and 6–18 years [1].

| Variables | Group | Language | Med | Q1 | Q3 | Comparison |
|---|---|---|---|---|---|---|
| number of mistakes | 1–5 years | German | 12 | 6 | 13 | $T = 0, Z = 3.62, p < 0.01$ |
| | | Russian | 0 | 0 | 0 | |
| | 6–18 years | German | 5 | 2 | 10 | $T = 0, Z = 2.93, p < 0.01$ |
| | | Russian | 0 | 0 | 0 | |
| response time (ms) | 1–5 years | German | 8166 | 6615 | 10,953 | $T = 0, Z = 3.62, p < 0.01$ |
| | | Russian | 4271 | 3787 | 5191 | |
| | 6–18 years | German | 6803 | 3565 | 8412 | $T = 4, Z = 2.58, p < 0.01$ |
| | | Russian | 3300 | 2579 | 3835 | |

[1] Median values (Med) and quartiles (Q1 and Q3) are presented along with the results of Wilcoxon test comparisons between the periods of using native and foreign languages.

### 3.2. Dynamics of HRV When Using Mathematical and General Vocabulary (Experiment 2)

We compared SampEn, SDNN, and av-RR values during the periods when the participants were using mathematical and general vocabulary. For descriptive statistics and the results of comparisons, see Table 3. SampEn was significantly lower during performing the task involving general vocabulary than the task involving mathematical vocabulary (Figure 7). Av-RR and SDNN did not differ significantly between the two tasks.

**Table 3.** Heart rate indexes during performing a task using mathematical and everyday vocabulary [1].

| HR Index | Test | Med | Q1 | Q3 | Comparison |
|---|---|---|---|---|---|
| SampEn | mathematical terms | 1.09 | 0.98 | 1.15 | $t = 7.70, p < 0.01, t\text{-test}$ |
| | general vocabulary | 0.97 | 0.86 | 1.04 | |
| av-RR (ms) | mathematical terms | 803 | 739 | 916 | $t = -1.79, p = 0.08, t\text{-test}$ |
| | general vocabulary | 814 | 754 | 922 | |
| SDNN (ms) | mathematical terms | 57.30 | 45.41 | 76.68 | $T = 309, Z = 0.09, p = 0.92,$ Wilcoxon test |
| | general vocabulary | 56.66 | 44.52 | 78.32 | |

[1] Median values (Med) and quartiles (Q1 and Q3) are presented along with the results of Wilcoxon and *t*-test pair-wise comparisons between the periods of using mathematical and everyday vocabulary.

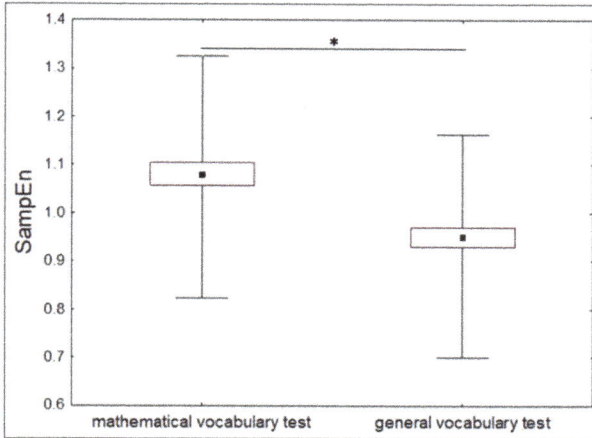

**Figure 7.** Mean values (±standard errors; range) of SampEn as compared between the intervals of performing a task using mathematical vocabulary and everyday vocabulary. * $p < 0.01$, $t$-test.

The participants made more mistakes when using mathematical vocabulary (median = 6; Q1 = 4; Q3 = 8) as compared to general vocabulary (median = 2; Q1 = 2; Q3 = 3) (Wilcoxon test T = 0, Z = 4.7, $p < 0.001$). Response times were longer (Wilcoxon test T = 21, Z = 4.73, $p < 0.001$) when using mathematical vocabulary (median = 4488 ms; Q1 = 3254 ms; Q3 = 6502 ms) than general vocabulary (median = 3902 ms; Q1 = 2779 ms; Q3 = 5336 ms).

No significant correlation was observed between the HRV indexes and the behavioural characteristics (see Table 4). Thus, the differences in SampEn between the tasks could not be accounted for by the differences in performance.

These results indicate that the irregularity of the heart rate is lower when individuals are using knowledge acquired earlier in their development than knowledge acquired later, i.e., heart rate irregularity is associated with the age the current behaviour was formed. We did not observe such association for the frequency or variability of the heart rate.

**Table 4.** Spearman correlation coefficients between the heart rate indexes and behavioural characteristics when using mathematical and general vocabulary.

| Mathematical Vocabulary | SampEn | av-RR (ms) | SDNN (ms) |
|---|---|---|---|
| number of mistakes | 0.07 | 0.06 | −0.09 |
| response time (ms) | −0.27 | 0.13 | −0.32 |
| **General vocabulary** | **SampEn** | **av-RR (ms)** | **SDNN (ms)** |
| number of mistakes | 0.09 | 0.16 | 0.01 |
| response time (ms) | 0.11 | −0.28 | 0.12 |

*3.3. Dynamics of HRV During Alcohol Administration (Experiment 3)*

We tested whether the dynamics of heart rate indexes during alcohol administration consisted of non-random changes. Values of SampEn, SDNN, and av-RR for each 5-min section of RR-intervals were compared pair-wisely in the alcohol and control conditions. The results of these comparisons are shown in Tables 5–7.

**Table 5.** SampEn values for consecutive 5 min sections of the heart rate during the alcohol and control conditions [1].

| Statistics | Conditions | Time Period after the Start of Drinking | | | | | | Friedman Test |
| | | 5 min | 10 min | 15 min | 20 min | 25 min | 30 min | |
|---|---|---|---|---|---|---|---|---|
| Med | alcohol | 0.97 | 0.96 | 0.94 | 0.90 | 0.88 | 0.85 | $X^2 = 13.62, p < 0.03$ |
| | control | 1.07 | 1.02 | 1.09 | 1.04 | 1.11 | 1.04 | $X^2 = 1.60, p = 0.90$ |
| Q1 | alcohol | 0.88 | 0.89 | 0.82 | 0.83 | 0.82 | 0.75 | |
| | control | 0.89 | 0.89 | 0.92 | 0.98 | 0.97 | 0.95 | |
| Q3 | alcohol | 1.06 | 1.07 | 1.02 | 1.05 | 1.07 | 0.99 | |
| | control | 1.14 | 1.13 | 1.18 | 1.12 | 1.18 | 1.15 | |
| | Wilcoxon test | $T = 121$ $Z = 1.12$ $p = 0.26$ | $T = 127$ $Z = 0.96$ $p = 0.34$ | $T = 71$ $Z = 2.46$ $p < 0.05$ | $T = 46$ $Z = 3.14$ $p < 0.01$ | $T = 79$ $Z = 2.25$ $p < 0.05$ | $T = 73$ $Z = 2.41$ $p < 0.05$ | |

[1] Median values (Med) and quartiles (Q1 and Q3) are shown along with the results of Friedman test and Wilcoxon pair-wise comparisons between the alcohol and control conditions.

**Table 6.** Av-RR values (ms) for consecutive 5 min sections of the heart rate during the alcohol and control conditions [1].

| Statistics | Conditions | Time Period after the Start of Drinking | | | | | | Friedman Test |
| | | 5 min | 10 min | 15 min | 20 min | 25 min | 30 min | |
|---|---|---|---|---|---|---|---|---|
| Med | alcohol | 716 | 734 | 729 | 718 | 717 | 715 | $X^2 = 23.52, p < 0.01$ |
| | control | 770 | 796 | 804 | 829 | 832 | 811 | $X^2 = 40.63, p < 0.01$ |
| Q1 | alcohol | 680 | 676 | 649 | 659 | 657 | 632 | |
| | control | 697 | 736 | 745 | 752 | 739 | 748 | |
| Q3 | alcohol | 782 | 797 | 785 | 792 | 794 | 764 | |
| | control | 870 | 921 | 898 | 903 | 940 | 916 | |
| | Wilcoxon test | $T = 87$ $Z = 2.25$ $p < 0.05$ | $T = 57$ $Z = 3.01$ $p < 0.01$ | $T = 37$ $Z = 3.52$ $p < 0.01$ | $T = 16$ $Z = 4.05$ $p < 0.01$ | $T = 25$ $Z = 3.82$ $p < 0.01$ | $T = 24$ $Z = 3.85$ $p < 0.01$ | |

[1] Median values (Med) and quartiles (Q1 and Q3) are shown along with the results of Friedman test and Wilcoxon pair-wise comparisons between the alcohol and control conditions.

**Table 7.** SDNN values (ms) for consecutive 5 min sections of the heart rate during the alcohol and control conditions [1].

| Statistics | Conditions | Time Period after the Start of Drinking | | | | | | Friedman Test |
| | | 5 min | 10 min | 15 min | 20 min | 25 min | 30 min | |
|---|---|---|---|---|---|---|---|---|
| Med | alcohol | 70.99 | 67.78 | 70.26 | 69.99 | 67.34 | 65.42 | $X^2 = 8.90, p = 0.11$ |
| | control | 68.99 | 80.66 | 73.62 | 78.15 | 78.09 | 74.90 | $X^2 = 4.69, p = 0.45$ |
| Q1 | alcohol | 54.32 | 52.34 | 46.61 | 46.03 | 48.47 | 49.35 | |
| | control | 54.55 | 65.01 | 59.90 | 58.90 | 69.42 | 58.87 | |
| Q3 | alcohol | 80.97 | 76.46 | 76.33 | 80.93 | 76.20 | 70.47 | |
| | control | 100.77 | 92.83 | 97.11 | 108.10 | 96.74 | 120.74 | |
| | Wilcoxon test | $T = 136$ $Z = 1.01$ $p = 0.32$ | $T = 68$ $Z = 2.73$ $p < 0.01$ | $T = 111$ $Z = 1.64$ $p = 0.101$ | $T = 62$ $Z = 2.88$ $p < 0.01$ | $T = 64$ $Z = 2.83$ $p < 0.01$ | $T = 59$ $Z = 2.96$ $p < 0.01$ | |

[1] Median values (Med) and quartiles (Q1 and Q3) are shown along with the results of Friedman test and Wilcoxon pair-wise comparisons between the alcohol and control conditions.

SampEn values decreased significantly during the 30 min of drinking in the alcohol condition, while there was no registered change in the dynamics of SampEn in the control condition (Figure 8). In general, the SampEn values were lower in the alcohol condition as compared to the control condition at 15, 20, 25, and 30 min periods after the beginning of the experiment.

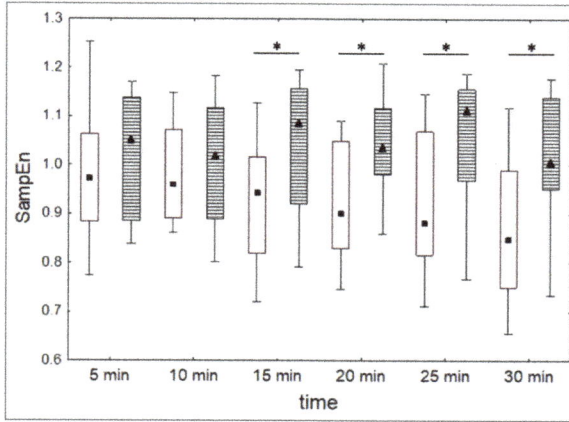

**Figure 8.** Median values (quartiles and range) of SampEn for consecutive 5 min sections of the heart rate during the alcohol (square points) and control (triangular points) conditions. *—Wilcoxon pair-wise comparisons, $p < 0.05$.

In the alcohol condition, av-RR increased during the period between 5 and 10 min after the beginning of drinking, followed by a significant decrease at the end of the experiment. In the control condition, av-RR significantly increased during drinking throughout the experiment. In general, av-RR was significantly lower in the alcohol condition as compared to the control condition.

No non-random changes were observed in the dynamics of SDNN values in either alcohol or control conditions. SDNN was significantly lower in the alcohol condition as compared to the control condition at 10, 20, 25, and 30 min periods after the beginning of the experiment.

All in all, the results have shown that alcohol administration, as a factor inducing the system dedifferentiation, reduced HRV and irregularity of the heart rate while increasing frequency of the heart rate. The dynamics of SampEn during alcohol administration was more consequent and linear than the dynamics of SDNN.

### 3.4. Dynamics of HRV During Public Speaking (Experiment 4)

We compared SampEn, SDNN, and av-RR values between the periods of rest (1–2 h before public speaking) and stress (during public speaking). The results are summarized in Table 8. SampEn was significantly lower during public speaking than at rest (Figure 9). The social stress condition also involved a decrease of SDNN and av-RR. Thus, social stress, as a factor inducing the system dedifferentiation, was accompanied by reduced irregularity of the heart rate, which was also observed in the condition of alcohol administration.

**Table 8.** Heart rate indexes at rest and during social stress [1].

| HR Index | Period | Med | Q1 | Q3 | Wilcoxon Test |
|----------|--------|------|-------|-------|---------------|
| SampEn | rest | 0.77 | 0.64 | 1.07 | $T = 16, Z = 2.06, p < 0.05$ |
| | stress | 0.58 | 0.42 | 0.67 | |
| av-RR (ms) | rest | 585 | 505 | 686 | $T = 8, Z = 2.62, p < 0.05$ |
| | stress | 470 | 411 | 531 | |
| SDNN (ms) | rest | 56.14 | 47.99 | 69.28 | $T = 0, Z = 3.18, p < 0.01$ |
| | stress | 43.26 | 27.19 | 54.11 | |

[1] Median values (Med) and quartiles (Q1 and Q3) are presented along with the results of Wilcoxon pair-wise comparisons between the periods of rest and social stress during speaking in public.

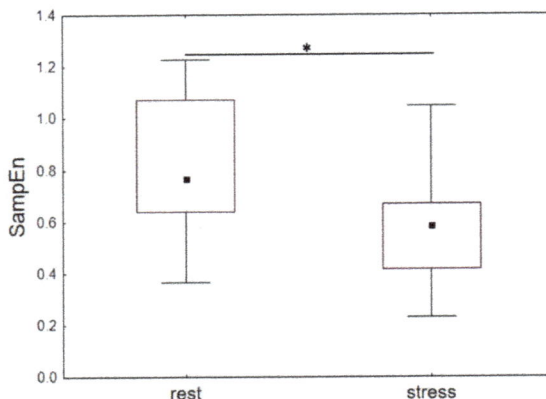

**Figure 9.** Median values (quartiles and range) of SampEn at rest and during social stress. * $p < 0.05$, Wilcoxon test.

### 3.5. Dynamics of HRV During Learning to Play a Computer Game (Experiment 5)

We analyzed the dynamics of SampEn, av-RR, and SDNN at the beginning of the first level of the game. The heart rate indexes were calculated for the initial 100 RR-intervals followed by five subsequent sections of RR-intervals with a step in 10 points from the beginning of the game: 1–100, 10–110, 20–120, 30–130, 40–140, and 50–150 points. The indexes were compared between the first section (1–100 points) and five subsequent sections (10–110, 20–120, 30–130, 40–140, and 50–150 points). For descriptive statistics, see Table 9.

**Table 9.** Heart rate indexes (means and standard errors) calculated for the initial 150 points of playing a computer game.

| HR Index | Point of Time from the Start of Playing the Game | | | | | |
|---|---|---|---|---|---|---|
| | 100 | 110 | 120 | 130 | 140 | 150 |
| SampEn | $0.78 \pm 0.02$ | $0.73 \pm 0.02$ | $0.75 \pm 0.02$ | $0.74 \pm 0.02$ | $0.75 \pm 0.02$ | $0.76 \pm 0.02$ |
| av-RR (ms) | $789 \pm 19.02$ | $793 \pm 19.73$ | $792 \pm 20.12$ | $792 \pm 20.24$ | $791 \pm 20.21$ | $791 \pm 20.06$ |
| SDNN (ms) | $58.91 \pm 4.66$ | $52.19 \pm 4.01$ | $50.79 \pm 3.84$ | $50.04 \pm 3.71$ | $49.88 \pm 3.81$ | $49.93 \pm 3.52$ |

SampEn calculated for the section of 1–100 points was significantly higher than the SampEn calculated for the section of 10–110 points (t = 2.42, $p < 0.05$, t-test dependent samples) (Figure 10). No other significant difference in SampEn was observed between the sections of RR-intervals during the initial 150 points of learning to play the game. Av-RR that was calculated for the section of 1–100 points was significantly lower than 10–110 points (t = −2.11, $p < 0.05$, t-test dependent samples). No other significant difference in av-RR was observed between the sections of RR-intervals during the initial 150 points of learning to play the game. SDNN for the section of 1–100 points was significantly higher than sections of 10–110 (t = 3.46, $p < 0.01$), 20–120 (t = 3.53, $p < 0.01$), 30–130 (t = 3.11, $p < 0.01$), 40–140 (t = 3.01, $p < 0.01$), and 50–150 (t = 2.74, $p < 0.05$, t-test dependent samples) points from the beginning of learning to play the game.

These results showed that irregularity of the heart rate as well as its frequency decreased at the beginning of learning. Variability of the heart rate also decreased at the beginning of learning and was lower than at any further point of leaning analyzed in our study.

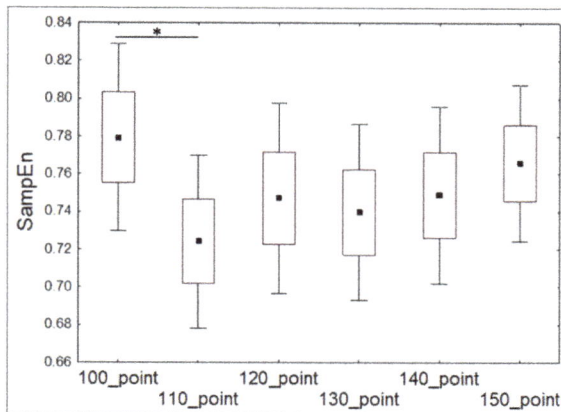

**Figure 10.** Mean values (±standard errors; range) of SampEn calculated for the initial 150 points of playing a computer game. *—$p < 0.05$, $t$-test for dependent samples.

## 4. Discussion

The aim of the current study was to explore how the system organization of behaviour can be reflected in the irregularity of the heart rate, as measured by SampEn. In experiments reported in this paper, we show that behaviour based on early-formed experience, related to native language as opposed to foreign language, and general vocabulary as opposed to abstract mathematical vocabulary, is associated with a lower irregularity of the heart rate. From the positions of system-evolutionary theory, this can be viewed as a result of increasing of system differentiation in the organization of later formed behaviour. Behaviour that is formed at later stages of individual development is based on a larger set of functional systems that are represented in the brain by neuronal groups specialized in relation to different acts comprising this behaviour, each formed at a certain point of an organism's life [2,17,20,47]. Therefore, the later formed behaviour is considered as more complex and detailed. Our results show that the regularity of the physiological processes underlying such behaviour decreases and their dynamics become more complex.

Heart rate irregularity, as a measure of non-linear dynamics of functional systems that are involved in the organization of current behaviour, is higher when this behaviour was formed later in individual development, but such differences were not observed for frequency of the heart rate, which implies that the higher heart rate irregularity cannot be explained by differences in the intensity of cognitive load engaging more internal recourses when performing later-formed behaviors. Therefore, we argue that this effect is related to characteristics of the functional systems that are actualized in the current behaviour.

It was shown that later-formed behaviour is subserved by the activity of a higher quantity of neural networks [20,21,47]. Therefore, we assume that when an individual is performing a later formed behaviour, his heart is involved in a more complex non-linear activity in order to achieve an adaptive coordination with the activity of neuronal elements of functional systems. This results in a decreased regularity of the heart rate.

We also show that irregularity of the heart rate is decreased during stress, alcohol administration, and at the beginning of learning. From the positions of system-evolutionary theory, this can be viewed as a result of system dedifferentiation in the organization of behaviour. In the context of system dedifferentiation, the number of actualized functional systems declines as the organization of behaviour becomes less complex [42], which is reflected in the reduced irregularity of the dynamics of physiological processes, as shown in our study.

The decrease in heart rate irregularity during public speaking shown in our study is in line with similar results obtained for examination stress [48] and hemorrhagic shock [49]. A state of shock can be considered as a higher degree of stress; heart rate irregularity that is observed in the state of shock is four times lower than heart rate irregularity during stress. This suggests that a decrease in heart rate irregularity may be quantitatively related to the degree of stress and system dedifferentiation induced by the stress.

We have not found any other studies that would show the dynamics of heart rate irregularity during alcohol administration, yet a decrease in heart rate variability, as measured by SDNN and HF (power of the spectrum of heart rate in the high-frequency range), during alcohol administration had been demonstrated previously [28].

A decrease of heart rate irregularity that was observed during the 30 min of drinking in the alcohol condition comports with the dynamics of ethanol absorption for the initial 20–40 min of alcohol intake reported in other studies [50]. At this initial stage of absorption, concentration of ethanol in the bloodstream grows before it gradually transitions to the next stage of ethanol elimination, when the concentration of ethanol in the bloodstream starts reducing. It is reasonable to assume that in a case of a longer experiment with alcohol administration (about 3 h), one can expect a gradual increase in heart rate irregularity, which would be slower than the initial decrease, as the stage of elimination lasts longer than the stage of absorption. It is worth noting that a similar decrease of heart rate irregularity, as measured by approximate entropy (ApEn), has been shown in experiments with cocaine administration [51], which can be viewed as a possible factor of the system dedifferentiation. Moreover, the cocaine-induced decrease in heart rate irregularity was dose-dependent. Therefore, we assume that decreasing heart rate irregularity followed after alcohol administration reflects and may be quantitatively related to the processes of system dedifferentiation.

The dynamics of heart rate irregularity at the beginning of learning can be viewed along with data on the pattern of hormonal changes, i.e., an increase in cortisol levels and concentration of endogenous opioids [45], typical for novel situations involving a mismatch between the existing individual experience and current organism-environment relations. The initial stages of learning can be described as a state of high emotional arousal, accompanied by a reduction in the actualization of later-formed systems, when an individual goes back to the states of earlier developmental stages [23], which could be viewed as an adaptive mechanism. Experiments with mathematical modelling have shown that without such a reduction in differentiation of the system organization of behaviour at the beginning of learning, it took animates a significantly longer time to find ways of solving a new task [42]. Thus, a decrease of heart rate irregularity during the initial stages of learning can also be viewed as a phenomenon that is associated with the processes of system dedifferentiation.

In addition, this view is supported by the results that were obtained in the studies of heart rate irregularity and emotional arousal. G. Valenza and coauthors [52] compared heart rate indexes during viewing the IAPS pictures, which elicited different levels of emotional arousal. They showed that the heart rate irregularity, as measured by ApEn, was lower when participants were viewing the pictures with a non-zero level of arousal than emotionally neutral pictures. I. Grossman and coauthors found a positive correlation between heart rate variability and the number of decisions based on wise reasoning about political and social problems [53].

It is important to note that we have not found any significant changes in the dynamics of av-RR or SDNN when comparing them between behaviors that are formed at earlier and later stages of individual development (Experiment 1 and 2). These results imply that different modes of heart activity manifested in the dynamics of SampEn cannot be explained by the difference in intensity of cognitive load, which demands unequal amounts of internal resources, cognitive control, and efforts when performing early- and later-formed behaviors. Typical decreases of av-RR and SDNN (Experiment 3 and 4) were observed during stress and alcohol administration, which had been reported in many previous studies. However, we believe that decreasing SampEn in these situations can also be explained by the dynamics of functional systems. For example, in Experiment 5 we observed decreasing SampEn and SDNN, accompanied by increasing av-RR. Thus, decreasing SampEn is not always observed when av-RR and SDNN are also decreasing. This suggests that SampEn reflects not only the basic states of general well-being of the organism, but can also be used to study the system processes orchestrating behaviour.

In summary, we argue that heart rate variability originates from complex system processes subserving behaviour, and it would be inaccurate to view it simply as a result of excitatory and inhibitory commands from the structures of the central nervous system. In the framework of system-evolutionary theory, behaviour is considered as an individual's activity at the whole-organism level. Therefore, changes in activity of the heart are viewed as a result of its coordination with other elements of functional systems actualized in the current behaviour, particularly with changes in the sets of activated neurons distributed across the cerebral cortex and sub-cortical structures and specialized in relation to this behaviour. This coordination is subserved via a number of afferent and efferent nerve fibers of the parasympathetic and sympathetic autonomic nervous system, as well as neurons of the intracardiac nervous system tightly-distributed in the myocardium [54].

The results of this study allow for us to assume a positive correlation between the number of functional systems and inter-system connections subserving current behaviour, which become more complex in individual development, and heart rate irregularity observed during this behaviour (Figure 11). The more functional systems actualized in the behaviour, the more elements the heart coordinates its activity with in order to achieve a common whole-organism outcome, which is reflected in higher heart rate irregularity.

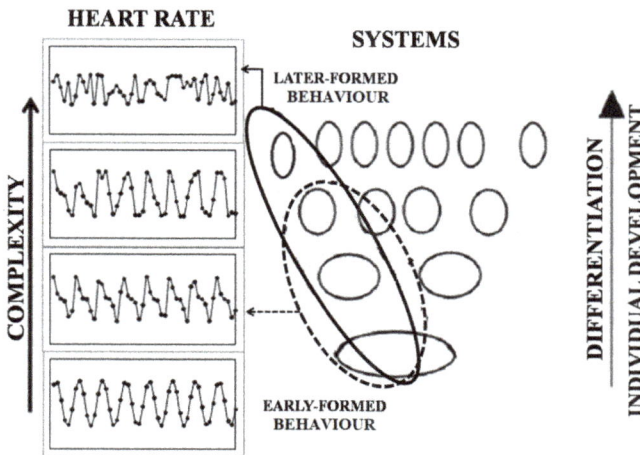

**Figure 11.** Heart rate irregularity and the number of functional systems actualized in current behaviour. The ovals depict functional systems formed at different stages of individual development. The selected groups of ovals illustrate combinations of functional systems that provide realization of earlier (dotted line) and later (solid line) formed behaviour. Heart rate irregularity, as measured by SampEn, increases along with behavioural differentiation in the process of individual development.

Integrating results of the five experiments, the study proposes that SampEn of the heart rate can be used as a new tool to study psychological processes and organization of behaviour. Many previous studies had demonstrated the sensitivity of SampEn to physiological and behavioural dynamics, which can be applied for diagnostics of optimal and extremal functional states [28,55], as well as various clinical pathologies [28,56]. Our study has demonstrated the potential of using SampEn in psychological research in order to study the system organization of behaviour.

An important output of this study is that the dynamics of functional systems supporting current behavior is reflected not only in activity of the brain, but also in the activity of the rest of the body. Functional systems subserving behaviour are not entirely neuronal systems but they also consist of elements located in different parts of the body. All elements of functional systems, neuronal and non-neuronal, continuously change their activity in order to achieve an effective cooperation resulting in an adaptive outcome relevant for the whole organism.

## 5. Conclusions

Heart rate irregularity is higher when an individual performs behaviour formed at later stages of his development (e.g., using foreign language vs. native language and later acquired words vs. earlier acquired words). The processes of system dedifferentiation (a reduction of the number of functional systems subserving current behaviour) that were observed during stress, after alcohol administration, and at the beginning stages of learning are accompanied by a decrease in heart rate irregularity. Thus, irregularity of the heart rate reflects characteristics of the system organization of behaviour.

**Supplementary Materials:** The supplementary materials are available online at http://www.mdpi.com/1099-4300/20/6/449/s1.

**Author Contributions:** A.V.B. and Y.I.A. conceived and designed the experiments; A.V.B., A.A.S. and K.R.A. performed the experiments; A.V.B., A.A.S. and A.V.D. analyzed the data; A.V.D. contributed analysis tools; A.V.B., Y.I.A. and K.R.A. wrote the paper.

**Acknowledgments:** The research has been supported by the Russian Science Foundation under Grant [number 14-28-00229] for Institute of Psychology Russian Academy of Sciences partly—the fifth experiment; and partly by Russian Foundation for Basic Research under Grant [number 16-36-60044 mol_a_dk]—the first, second, and fourth experiments.

**Conflicts of Interest:** The authors declare no conflict of interest.

## References

1. Matheson, H.E.; Barsalou, L.W. Embodiment and Grounding in Cognitive Neuroscience. In *Stevens' Handbook of Experimental Psychology and Cognitive Neuroscience*, 4th ed.; Wixted, J.T., Thompson-Schill, S.L., Eds.; John Wiley & Sons: Hoboken, NJ, USA, 2018; Volume 3, pp. 1–27, ISBN 978-1-119-1701-67.

2. Alexandrov, Y.I.; Grechenko, T.N.; Gavrilov, V.V.; Gorkin, A.G.; Shevchenko, D.G.; Grinchenko, Y.V.; Aleksandrov, I.O.; Maksimova, N.E.; Bezdenezhnych, B.N.; Bodunov, M.V. Formation and realization of individual experience: A psychophysiological approach. In *Conceptual Advances in Brain Research. Conceptual Advances in Russian Neuroscience: Complex Brain Functions*; Miller, R., Ivanitsky, A.M., Balaban, P.V., Eds.; Harwood Academic Publishers: Amsterdam, The Netherlands, 2000; Volume 2, pp. 181–200, ISBN 978-9-0582-3021-8.

3. Shvyrkov, V.B. Neurophysiological study of animals' subjective experience. In *Machinery of the Mind*; John, R., Harmony, T., Eds.; Birkhäuser: Boston, MA, USA, 1990; pp. 337–350, ISBN 978-1-4757-1085-4.

4. Alexandrov, Y.I. Comparative description of consciousness and emotions in the framework of systemic understanding of behavioural continuum and individual development. In *Neuronal Bases and Psychological Aspects of Consciousness, Proceedings of the International School of Biocybernetics, Casamicciola, Napoli, Italy, 13–18 October 1997*; Teddei-Ferretti, C., Musio, C., Eds.; World Scientific: New York, NY, USA, 1999; Volume 8, pp. 220–235, ISBN 978-981-02-3597-0.

5. Lasrado, R.; Boesmans, W.; Kleinjung, J.; Pin, C.; Bell, D.; Bhaw, L.; McCallum, S.; Zong, H.; Luo, L.; Clevers, H.; et al. Lineage-dependent spatial and functional organization of the mammalian enteric nervous system. *Science* **2017**, *356*, 722–726. [CrossRef] [PubMed]

6. Armour, J.A. Cardiac neuronal hierarchy in health and disease. *Am. J. Physiol. Regul. Integr. Comp. Physiol.* **2004**, *287*, 262–271. [CrossRef] [PubMed]

7. Bagaev, V.; Aleksandrov, V. Visceral-related area in the rat insular cortex. *Auton. Neurosci.* **2006**, *125*, 16–21. [CrossRef] [PubMed]

8. Brüggemann, J.; Shi, T.; Apkarian, A.V. Viscero-somatic neurons in the primary somatosensory cortex (SI) of the squirrel monkey. *Brain Res.* **1997**, *756*, 297–300. [CrossRef]

9.   Taggart, P.; Critchley, H.; van Duijvendoden, S.; Lambiase, P.D. Significance of neuro-cardiac controll mechanisms governed by higher regions of the brain. *Auton. Neurosci.* **2016**, *199*, 54–65. [CrossRef] [PubMed]

10.  Obrist, P.A.; Webb, R.A.; Sutterer, J.R.; Howard, J.L. The cardiac-somatic relationship: Some reformulations. *Psychophysiology* **1970**, *6*, 569–587. [CrossRef] [PubMed]

11.  Shaffer, F.; McCraty, R.; Zerr, C.L. A healthy heart is not a metronome: An integrative review of the heart's anatomy and heart rate variability. *Front. Psychol.* **2014**, *5*, 1040. [CrossRef] [PubMed]

12.  Thayer, J.F.; Ahs, F.; Fredrikson, M.; Sollers, J.J.; Wager, T.D. A meta-analysis of heart rate variability and neuroimaging studies: Implications for heart rate variability as a marker of stress and health. *Neurosci. Biobehav. Rev.* **2012**, *36*, 747–756. [CrossRef] [PubMed]

13.  Thayer, J.F. What the Heart Says to the Brain (and vice versa) and Why We Should Listen. *Psychol. Top.* **2007**, *16*, 241–250.

14.  Matthews, S.C.; Paulus, M.P.; Simmons, A.N.; Nelesen, R.A.; Dimsdale, J.E. Functional subdivisions within anterior cingulate cortex and their relationships to autonomic nervous system function. *NeuroImage* **2004**, *22*, 1151–1156. [CrossRef] [PubMed]

15.  Mather, M.; Thayer, J.F. How heart rate variability affects emotion regulation brain networks. *Curr. Opin. Behav. Sci.* **2018**, *19*, 98–104. [CrossRef] [PubMed]

16.  Winkelmann, T.; Thayer, J.F.; Pohlack, S.; Nees, F.; Grimm, O.; Flor, H. Structural brain correlates of heart rate variability in a healthy young adult population. *Brain Struct. Funct.* **2017**, *222*, 1061–1068. [CrossRef] [PubMed]

17.  Alexandrov, Y.I. Psychophysiological regularities of the dynamics of individual experience and the "stream of consciousness". In *Neuronal Bases and Psychological Aspects of Consciousness, Proceedings of the International School of Biocybernetics, Casamicciola, Napoli, Italy, 13–18 October 1997*; Teddei-Ferretti, C., Musio, C., Eds.; World Scientific: New York, NY, USA, 1999; Volume 8, pp. 201–219, ISBN 978-981-02-3597-0.

18.  Anokhin, P.K. *Biology and Neurophysiology of Conditioned Reflex and Its Role in Adaptive Behavior*, 1st ed.; Pergamon Press: Oxford, UK, 1974; pp. 190–254, ISBN 978-0-08-021516-7.

19.  Anokhin, P.K. *Biology and Neurophysiology of Conditioned Reflex*, 1st ed.; Meditzina: Moscow, Russia, 1968; pp. 406–432.

20.  Aleksandrov, Y.I. Learning and Memory: Traditional and Systems Approaches. *Neurosci. Behav. Physiol.* **2006**, *36*, 969–985. [CrossRef] [PubMed]

21.  Gorkin, A.G.; Shevchenko, D.G. Distinctions in the neuronal activity of the rabbit limbic cortex under different training strategies. *Neurosci. Behav. Physiol.* **1996**, *26*, 103–112. [CrossRef] [PubMed]

22.  Kolbeneva, M.G.; Alexandrov, Y.I. Mental Reactivation and Pleasantness Judgment of Experience Related to Vision, Hearing, Skin Sensations, Taste and Olfaction. *PLoS ONE* **2016**, *11*, e0159036. [CrossRef] [PubMed]

23.  Alexandrov, Y.I.; Krylov, A.K.; Arutyunova, K.R. Activity during Learning and the Nonlinear Differentiation of Experience. *Nonlinear Dyn. Psychol. Life Sci.* **2017**, *21*, 391–405.

24.  Alexandrov, Y.I.; Grinchenko, Y.V.; Shevchenko, D.G.; Averkin, R.G.; Matz, V.N.; Laukka, S.; Korpusova, A.V. A subset of cingulate cortical neurons is specifically activated during alcohol-acquisition behaviour. *Acta Physiol. Scand.* **2001**, *171*, 87–97. [PubMed]

25.  Alexandrov, Y.I. How we fragment the world: The view from inside versus the view from outside. *Soc. Sci. Inf.* **2008**, *47*, 419–457. [CrossRef]

26.  Alexandrov, Y.I. Differentiation and development. In *Development Theory: Differential-Integration Paradigm*; Chuprikova, N.I., Ed.; Languages of Slavic Culture: Moscow, Russia, 2009; pp. 17–28. (In Russian)

27.  Richman, J.S.; Moorman, J.R. Physiological time-series analysis using approximate entropy and sample entropy. *Am. J. Physiol. Heart Circ. Physiol.* **2000**, *278*, 2039–2049. [CrossRef] [PubMed]

28.  Acharya, U.R.; Joseph, K.P.; Kannathal, N.; Lim, C.M.; Suri, J.S. Heart rate variability: A review. *Med. Biol. Eng. Comput.* **2006**, *44*, 1031–1051. [CrossRef] [PubMed]

29.  Costa, A.; Vives, M.-L.; Corey, J.D. On language processing shaping decision making. *Curr. Dir. Psychol. Sci.* **2017**, *26*, 146–151. [CrossRef]

30.  Hayakawa, S.; Tannenbaum, D.; Costa, A.; Corey, J.D.; Keysar, B. Thinking more or feeling less? Explaining the foreign-language effect on moral judgment. *Psychol. Sci.* **2017**, *28*, 1387–1397. [CrossRef] [PubMed]

31.  Hahne, A. What's Different in Second-Language Processing? Evidence from Event-Related Brain Potentials. *J. Psycholinguist. Res.* **2001**, *30*, 221–266. [CrossRef]

32.  Oscar-Berman, M. Alcohol-related ERP changes in cognition. *Alcohol* **1987**, *4*, 289–292. [CrossRef]

33. Walley, A.C.; Metsala, J.L. Young children's age-of-acquisition estimates for spoken words. *Mem. Cognit.* **1992**, *20*, 171–182. [CrossRef] [PubMed]

34. Izura, C.; Ellis, A.W. Age of acquisition effects in word recognition and production in first and second languages. *Psicológica* **2002**, *23*, 245–281.

35. Alexandrov, Y.I.; Grinchenko, Y.V.; Laukka, S.; Järvilehto, T.; Maz, V.N.; Korpusova, A.V. Effect of ethanol on hippocampal neurons depends on their behavioural specialization. *Acta. Physiol. Scand.* **1993**, *149*, 105–115. [CrossRef] [PubMed]

36. Alexandrov, L.I.; Alexandrov, Y.I. Changes of auditory-evoked potentials in response to behaviorally meaningful tones induced by acute ethanol intake in altricial nestlings at the stage of formation of natural behavior. *Alcohol* **1993**, *10*, 213–217. [CrossRef]

37. Schwabe, L.; Joels, M.; Roozendaal, B.; Wolf, O.T.; Oitzl, M.S. Stress effects on memory: An update and integration. *Neurosci. Biobehav. Rev.* **2011**, *36*, 1740–1749. [CrossRef] [PubMed]

38. Schwabe, L.; Wolrf, O.T. Stress prompts habit behavior in humans. *J. Neurosci.* **2009**, *22*, 7191–7198. [CrossRef] [PubMed]

39. Znamenskaja, I.I.; Markov, A.V.; Bahchina, A.V.; Aleksandrov, J.I. Attitude to outgroup members in stress: System dedifferentiation. *Psikhol. Zh.* **2016**, *37*, 44–58.

40. Yu, R. Stress potentiates decision biases: A stress induced deliberation-to-intuition (SIDI) model. *Neurobiol. Stress* **2016**, *3*, 83–95. [CrossRef] [PubMed]

41. Parin, S.B.; Bakhchina, A.V.; Polevaya, S.A. A neurochemical framework of the theory of stress. *Int. J. Psychophysiol.* **2014**, *94*, 230. [CrossRef]

42. Alexandrov, Y.I.; Svarnik, O.E.; Znamenskaya, I.I.; Kolbeneva, M.G.; Arutyunova, K.R.; Krylov, A.K.; Bulava, A.I. *Regression as the Stage of Development*, 1st ed.; IPRAS: Moscow, Russia, 2017; pp. 10–143, ISBN 978-5-9270-0354-9. (In Russian)

43. Allen, A.P.; Kennedy, P.J.; Cryan, J.F.; Dinana, T.G.; Clarke, G. Biological and psychological markers of stress in humans: Focus on the Trier Social Stress Test. *Neurosci. Biobehav. Rev.* **2014**, *38*, 94–124. [CrossRef] [PubMed]

44. Gray, J.A. Brain Systems that Mediate both Emotion and Cognition. *Cognit. Emot.* **1990**, *4*, 269–288. [CrossRef]

45. Wassum, K.M. Neurochemistry of Desire: Endogenous Opioid and Glutamate Involvement in Incentive Learning and Reward Seeking Actions. Ph.D. Thesis, University of California, Los Angeles, CA, USA, 2010; pp. 120–199.

46. Sozinov, A.A.; Bohan, A.I.; Alexandrov, Y.I. A software for assessment of new experience formation and problem solving under achievement or avoidance conditions. *Exp. Psychol.* **2018**, *11*, 75–91.

47. Aleksandrov, Y.I. Systemic psychophysiology. In *Russian Cognitive Neuroscience: Historical and Cultural Context*, 1st ed.; Forsythe, C., Zotov, M.V., Radvansky, G.A., Tsvetkova, L., Eds.; CreateSpace Independent Publishing: New York, NY, USA, 2015; pp. 65–100.

48. Melillo, P.; Bracale, M.; Pecchia, L. Nonlinear heart rate variability features for real-life stress detection. Case study: Students under stress due to university examination. *Biomed. Eng. Online* **2011**, *10*, 96. [CrossRef] [PubMed]

49. Batchinsky, A.I.; William, H.C.; Kuusela, T.; Cancio, L.C. Loss of complexity characterizes the heart response to experimental hemorrhagic shock in swine. *Crit. Care Med.* **2007**, *35*, 519–525. [CrossRef] [PubMed]

50. Swift, R. Direct measurement of alcohol and its metabolites. *Addiction* **2003**, *98*, 73–80. [CrossRef] [PubMed]

51. Newlin, D.B.; Wong, C.J.; Stapleton, J.M.; London, E.D. Intravenous cocaine decreases cardiac vagal tone, vagal index (derived in Lorenz Space), and heart period complexity (approximate entropy) in cocaine abusers. *Neuropsychopharmacology* **2000**, *23*, 560–568. [CrossRef]

52. Valenza, G.; Allegrini, P.; Lanata, A.; Scilingo, E.P. Dominant Lyapunov exponent and approximate entropy in heart rate variability during emotional visual elicitation. *Front. Neuroeng.* **2012**, *5*, 3. [CrossRef] [PubMed]

53. Grossman, I.; Balljinder, K.S.; Ciarrochi, J. A heart and a mind: Self-distancing facilitates the association between heart rate variability, and wise reasoning. *Front. Behav. Neurosci.* **2016**, *10*, 68. [CrossRef] [PubMed]

54. Wake, E.; Brack, K. Characterization of the intrinsic cardiac nervous system. *Auton. Neurosci. Basic Clin.* **2016**, *199*, 3–16. [CrossRef] [PubMed]

55. Kokonozi, A.K.; Michail, E.M.; Chouvarda, I.C.; Maglaveras, N.M. A Study of Heart Rate and Brain System Complexity and Their Interaction in Sleep-Deprived Subjects. *Comput. Cardiol.* **2008**, *35*, 969–971.

56. Al-Angari, H.M.; Sahakian, A.V. Use of Sample Entropy Approach to Study Heart Rate Variability in Obstructive Sleep Apnea Syndrome. *IEEE Trans. Biomed. Eng.* **2007**, *54*, 1900–1904. [CrossRef] [PubMed]

entropy

*Article*

# Research on Recognition Method of Driving Fatigue State Based on Sample Entropy and Kernel Principal Component Analysis

**Beige Ye \*, Taorong Qiu \*, Xiaoming Bai and Ping Liu**

Department of Computer, Nanchang University, Nanchang 330029, China;
baixm2006@163.com (X.B.); beibei541757635@163.com (P.L.)
\* Correspondence: 406130916148@email.ncu.edu.cn (B.Y.); Qiutaorong@ncu.edu.cn (T.Q.);
Tel.: +86-189-7910-0236 (T.Q.)

Received: 29 July 2018; Accepted: 10 September 2018; Published: 13 September 2018

**Abstract:** In view of the nonlinear characteristics of electroencephalography (EEG) signals collected in the driving fatigue state recognition research and the issue that the recognition accuracy of the driving fatigue state recognition method based on EEG is still unsatisfactory, this paper proposes a driving fatigue recognition method based on sample entropy (SE) and kernel principal component analysis (KPCA), which combines the advantage of the high recognition accuracy of sample entropy and the advantages of KPCA in dimensionality reduction for nonlinear principal components and the strong non-linear processing capability. By using support vector machine (SVM) classifier, the proposed method (called SE_KPCA) is tested on the EEG data, and compared with those based on fuzzy entropy (FE), combination entropy (CE), three kinds of entropies including SE, FE and CE that merged with KPCA. Experiment results show that the method is effective.

**Keywords:** driving fatigue; sample entropy; kernel principal component analysis; support vector machine

---

## 1. Introduction

Driving fatigue is a phenomenon in which, due to continuous driving, drivers' ability of perception, judgment and operation appear to decrease [1]. Drivers are prone to driving fatigue after long driving, and if they keep driving, their limbs will be stiff, their attention will decrease and their judgment will decline. Driving fatigue may cause people to become delirious, and they may be prone to traffic accidents [2]. Therefore, an effective driving fatigue state recognition method is the key to construct the dangerous driving state warning system.

At present, a series of studies has been conducted on the recognition of driving fatigue status at home and abroad. Guo et al. [3] explored the correlation between ECG indicators and driving fatigue state based on ECG signals and constructed the driving fatigue state recognition model combined with the SVM classifier. Yang et al. [4] conducted research on driving fatigue recognition on the basis of the fusion of eye movement and pulse information. Zhao et al. [5] applied functional brain networks to establish a fatigue recognition model based on EEG data and graph theory methods. Zhao et al. [6] constructed a driving fatigue recognition model based on the human eye feature by using a concatenated convolutional neural network. To judge whether the driver felt fatigue, Zhang et al. [7] conducted research on driving fatigue recognition on the feature extraction of the wavelet entropy of EEG signals. Moreover, Chai and Naik et al. [8] used entropy rate bound minimization as a source separation technique, the autoregressive (AR) modeling as the feature extraction algorithm and the Bayesian neural network as the classification algorithm for driving fatigue recognition; they combined independent component by entropy rate bound minimization analysis (ICA-ERBM) and EEG feature extraction components, which have not been explored previously for fatigue classification. Zeng et al. [9] proposed to use deep convolutional neural networks and deep residual learning to predict the drivers' mental states from EEG signals; they also developed two mental state classification models that are the architecture of our CNN-based EEG classier called EEG-Conv and combining EEG-Conv with residual learning called EEG-Conv-R. Chai and Ling et al. [10] combined the AR modeling feature extractor with a sparse-deep belief networks (sparse-DBN) classifier to constructed a driving fatigue recognition model, which have not been explored previously for EEG-based driving fatigue classification. Hu et al. [11] conducted a driving fatigue recognition model by the combination of the feature set that consisted of sample entropy, fuzzy entropy, approximate entropy and spectral entropy and the gradient boosting decision tree (GBDT) through the use of three different classifiers.

The literature mentioned above has enriched and extended the research of fatigue recognition from different perspectives. EEG signals, with the highest sensitivity in driving fatigue detection and recognition and the highly correlated relationship with the driver's mental state, have been studied more deeply [12]. There are many ways to analyze EEG signals, such as time domain analysis, frequency domain analysis, multidimensional statistical analysis and nonlinear analysis [13]. However, the recognition accuracy of the driving fatigue state obtained by using these methods is still not satisfactory after the feature extraction of EEG signals. PCA and SVM were used to obtain better recognition accuracy based on motion imagination EEG signals in [14]. However, because of the nonlinear characteristics of EEG data, the internal model of the PCA is linear, the same with the relationship among the principal components. PCA will lose its effectiveness when the principal components of the study object are nonlinear. Therefore, this paper proposes a driving fatigue recognition method based on sample entropy and kernel principal component analysis. There are several advantages to using the sample entropy kernel principal component analysis (SE_KPCA) method. On the one hand, the recognition of driving fatigue state based on sample entropy (SE) [15] is more accurate. On the other hand, the result of dimension reduction in KPCA [16] is more positive and has a strong non-linear processing capability. On the basis of the above, a driving fatigue state recognition model is constructed with the combination of the support vector machine (SVM) [17] algorithm to achieve effective recognition of the driver's fatigue state.

## 2. Sample Entropy

The sample entropy (SE) calculation process is described as follows, given the original signal of length $N$, denote it by $x(1), x(2), \ldots, x(N)$, and define the $m$-dimensional vector:

$$X_m(i) = \{x(i), x(i+1), \ldots, x(i+m-1)\}; 1 \leq i \leq N-m+1 \tag{1}$$

Calculate any two $m$-dimensional vectors:

$$D[X_m(i), X_m(j)] = max[x(i+k) - x(j+k)], 0 \leq k \leq m-1; i \neq j, \ i,j \leq N-m+1 \tag{2}$$

$D[X_m(i), X_m(j)]$ is the maximum difference between $X_m(i)$ and $X_m(j)$. Given a threshold $r$, calculate the total number of the maximum difference between any two elements that is less than the threshold:

$$C = \sum_{i=1}^{N-m} (D(i) < r) \tag{3}$$

Define a ratio:

$$B_i^m(r) = \frac{C}{N-m} \tag{4}$$

$B_i^m(r)$ is the ratio of $C$ to the total; calculate its mean:

$$\overline{B}^m(r) = \frac{1}{N-m+1} \sum_{i=1}^{N-m+1} B_i^m(r) \tag{5}$$

where $\overline{B}^m(r)$ is the proportion mean of the $m$-dimensional sequence. When the signal increases to $m + 1$-dimension, repeat Equation (1) to Equation (4), and calculate the proportion mean of the $m + 1$-dimensional sequence:

$$A^{m+1}(r) = \frac{C}{N-m} \sum_{i=1}^{N-m} B_i^{m+1}(r) \tag{6}$$

Get the sample entropy of the sequence:

$$SanmEn(m,r) = \lim_{N \Rightarrow \infty} \left\{ -ln(A^{m+1}(r)/\overline{B}^m(r)) \right\} \tag{7}$$

When $N$ is finite, Equation (7) can be expressed as follows:

$$SanmEn(m,r,N) = -ln(A^{m+1}(r)/\overline{B}^m(r)) \tag{8}$$

From Equation (8), it is known that the value of $SanmEn$ is related to $m$ and $r$. Pincus [18] pointed out that the value of $m$ is generally taken as two, when $r$ is set to be 0.1- to 0.25-times the standard deviation of the original EEG signal time series (0.1 to 0.25 SD; SD is the standard deviation). Thus, $m$ is set as two, and $r$ is set as 0.25 SD in this paper.

## 3. Principal Component Analysis and Kernel Principal Component Analysis

### 3.1. Basic Principles of PCA

Suppose that the $m$-times extracted data matrix of $n$ variables $X_i, X_2, \ldots, X_m$ is $X = (X_{pq})_{m*n}$. The main steps of PCA analysis [19,20] are as follows:

1. Calculate the sample mean and standard deviation for each indicator $X$:

$$\overline{X} = \frac{1}{n} \sum_{p=1}^{m} X_{pq}, \; S_q = \sqrt[2]{\frac{1}{N-1} \sum_{p=1}^{m} (X_{pq} - \overline{X_q})^2}, \; q = 1, 2, \ldots, n \tag{9}$$

2. Normalize $X_{pq}$ and calculate its normalization matrix:

$$Y_{pq} = \frac{X_{pq} - \overline{X_q}}{m}, \; p = 1, 2, \ldots, m, \; q = 1, 2, \ldots, n \tag{10}$$

3. Calculate the correlation coefficient matrix $R$ according to the obtained standardized matrix $Y = (Y_{pq})_{m*n}$:

$$r_{qk} = \frac{1}{m-1} \sum_{p=1}^{m} Y_{pq} * Y_{pk} \tag{11}$$

$$R = (r_{pq})_{mn}, \; r_{qq} = 1, \; r_{qk} = r_{kq} \tag{12}$$

4. Get the eigenvalue of $R$, denoted as $\lambda$. Suppose $\lambda_1 \geq \lambda_2 \geq \ldots \geq \lambda_n > 0$ and $l_1, l_2, \ldots, l_n$ are the corresponding feature vectors. Determine the range of $K$ according to the cumulative variance contribution as $CVC > 90\%$, and define the $CVC$ as:

$$CVC = \sum_{q=1}^{k} \lambda_q / \sum_{q=1}^{n} \lambda_q \tag{13}$$

the $K$ principal components are created, denoted as:

$$Z_q = l_q X \tag{14}$$

### 3.2. Basic Principles of KPCA

There are $M$ samples in the input space, denoted as $x_k (k = 1, 2, \ldots, M)$, $x_k \in R^N$, $\sum_{k=1}^{M} x_k = 0$. The nonlinear mapping function $\Phi$ is introduced to the algorithm, transforming the sample points in the input space $x_1, x_2, \ldots, x_M$ into sample points in the feature space as $\Phi(x_1), \Phi(x_2), \ldots, \Phi(x_M)$, and the hypothesis:

$$\sum_{k=1}^{M} \Phi(x_k) = 0 \tag{15}$$

Then, the covariance matrix in the feature space $F$ is defined as:

$$\overline{C} = \frac{1}{M} \sum_{j=1}^{M} \Phi(x_j) \Phi(x_j)^T \tag{16}$$

Therefore, the solving equation of PCA in the feature space is:

$$\lambda V = \overline{C} v \tag{17}$$

$\lambda$ is the eigenvalue, and $v \in F \backslash \{0\}$ is the eigenvector, so:

$$\lambda(\Phi(x_k) * v) = \Phi(x_k) * \overline{C} v, (k = 1, 2, \ldots, M) \tag{18}$$

Note that $v$ can be expressed linearly by $\Phi(x_i)(i = 1, 2, \ldots, M)$ in the above formula.

$$v = \sum_{i=1}^{M} a_i \Phi(x_i) \tag{19}$$

where $a_1, a_2, \ldots, a_N$ is constant. Define an $N * N$ matrix satisfying the Mercer condition, denoted as $K$:

$$K_{ij} = \Phi(x_i) * \Phi(x_j) \tag{20}$$

$K$ is called the nuclear matrix, which can be obtained from Equation (16) to Equation (19) as follows:

$$M\lambda a = Ka \tag{21}$$

The required eigenvalues and eigenvectors are obtained by solving the formula Equation (21). The projection of the test sample on the $F$-space vector $V^k$ is:

$$(V^k * \Phi(x)) = \sum_{i=1}^{M} a_i^k (\Phi(x_i) * \Phi(x)) \tag{22}$$

Supposed that Equation (15) is not valid. Then, the $K$ in Equation (21) is replaced by $\widetilde{K}$.

$$\widetilde{K}_{ij} = K_{ij} - \frac{1}{M} \sum_{m=1}^{M} l_{im} K_{mj} - \frac{1}{M} \sum_{n=1}^{M} K_{in} l_{nj} + \frac{1}{M^2} \sum_{m,n=1}^{M} l_{im} K_{mn} l_{nj} \tag{23}$$

where $l_{ij} = 1$ (for all $i, j$).

### 3.3. Kernel Function Methods

At present, there are several forms of kernel functions that can be chosen, as follows:

1. Linear kernel function (special case):

$$K(x, x_i) = x * x_i \tag{24}$$

2.   *P*-order polynomial kernel function:

$$K(x, x_i) = [(x * x_i) + 1]^p \qquad (25)$$

3.   Radial basis function (RBF):

$$K(x, x_i) = exp(-\frac{||x - x_i||}{\delta^2}) \qquad (26)$$

4.   Multilayer perceptual (MLP) kernel function:

$$K(x, x_i) = tanh[v(x * x_i) + c] \qquad (27)$$

The *P*-order polynomial kernel function, radial basis function and multilayer perceptual kernel function are used in the model of this paper.

## 4. EEG Data Processing Method Based on Sample Entropy and Principal Component Analysis/Kernel Principal Component Analysis

### 4.1. EEG Data Processing Method Based on Sample Entropy and Principal Component Analysis

Based on the above discussion about the method proposed in this paper, the algorithm that combined sample entropy and principal component analysis (SE_PCA) can be divided into the following three steps:

1.   Collect the EEG signal, and preprocess it; then, extract the sample entropy characteristics of the data by the formula Equations (1) to (8), and obtain a matrix $X_{m*n}$;
2.   Take the matrix $X_{m*n}$ into the formula Equations (9) to (14), then calculate its principal component;
3.   Construct a model, and use SVM to classify.

### 4.2. EEG Data Processing Method Based on Sample Entropy and Kernel Principal Component Analysis

Based on the above discussion about the method proposed in this paper, the SE_KPCA algorithm can be divided into the following four steps:

1.   Collect the EEG signal, and preprocess the EEG signal; then, extract the sample entropy characteristics of the data by the formula Equations (1) to (8), and obtain a matrix $X_{m*n}$;
2.   Select the kernel function $K(x, x_i)$, the matrix $X_{m*n}$ as an input of KPCA, and centralize it in high dimensional space; then, calculate matrix $\widetilde{K}$ according to Equation (23);
3.   Calculate the eigenvalues and eigenvectors of the matrix $\widetilde{K}$, as well as its nonlinear principal component;
4.   Construct a model, and use SVM to classify.

## 5. Method Testing and Result Analysis

### 5.1. Test Environment and Test Data

Test environment: The platform environment used in the experiment includes a static simulator (Beijing-China Joint Teaching Equipment Co., Ltd., ZY-31D vehicle driving simulator, Beijing, China), and this includes three 24-inch monitors and a software teaching system for driving simulations (ZM-601 V9.2). A 32-electrode EEG collecting cap, the computer system (windows 10 × 64), EEG collecting and preprocessing software (Neuroscan 3.2) and EEG analysis software (MATLAB R2014b) were used.

Test data description: The EEG signal data analyzed in this paper come from the EEG study, which simulated car driving training. Twenty five normal subjects were tested for the current fatigue level during the training, such as the sleep quality on the previous night, the diet on the day, etc., then two sets of experiment data were recorded by every subject, namely fatigue state and non-fatigue state. According to the precious experience in the fatigue-related experiment, each subject was asked to drive for 40 min without a break, then they were asked to take a questionnaire to check their states [21]. The EEG data are 32-electrode, 600 s time series at a sampling rate of 1000 Hz, which consisted of 300 s of rest (non-fatigue) and 300 s of fatigue. After collecting a person's EEG signal data, filtering and processing them (artifact removal, removal of eye movement interference, signal correction, etc. [21]) were conducted. This paper conducted two sets of experiments. The first one was taking some data from 10 individuals and 60 s for each person (the first 30 s in the non-fatigue state and the other 30 s in the fatigue state), which constructed a 600 * 30 data matrix (for which 600 is 600 s, 30 is the 30 electrodes), as shown in Figure 1. The other one was taking some data from 15 individuals and 60 s of each person, which constructed a 900 * 30 data matrix, as shown in Figure 2. At present, this paper merely compares the experimental results of 30 s, due to less data being able to reduce the time of experiment and the amount of data in 30 s being enough. However, the subsequent experiments would enlarge the selection of different time bands for testing.

$$
\textit{electrode} \quad \textit{number}
$$

$$
\begin{array}{c}
\quad\quad 1 \quad\quad 2 \quad \cdots \quad\quad 30 \\
\begin{array}{c} 1 \\ \textit{time } 2 \\ \cdots \\ 600 \end{array}
\begin{bmatrix}
x_{1,1} & x_{1,2} & \cdots & x_{1,30} \\
x_{2,1} & x_{2,2} & \cdots & x_{2,30} \\
\cdots & \cdots & \cdots & \cdots \\
x_{600,1} & x_{600,2} & \cdots & x_{600,30}
\end{bmatrix} = X
\end{array}
$$

**Figure 1.** Sample entropy data matrix (10 individuals).

$$
\textit{electrode} \quad \textit{number}
$$

$$
\begin{array}{c}
\quad\quad 1 \quad\quad 2 \quad \cdots \quad\quad 30 \\
\begin{array}{c} 1 \\ \textit{time } 2 \\ \cdots \\ 900 \end{array}
\begin{bmatrix}
x_{1,1} & x_{1,2} & \cdots & x_{1,30} \\
x_{2,1} & x_{2,2} & \cdots & x_{2,30} \\
\cdots & \cdots & \cdots & \cdots \\
x_{900,1} & x_{900,2} & \cdots & x_{900,30}
\end{bmatrix} = X
\end{array}
$$

**Figure 2.** Sample entropy data matrix (15 individuals).

*5.2. Driving Fatigue State Recognition Test Based on SE_PCA*

Firstly, the SE_PCA method was used to analyze the contribution ratio of the 30 electrode principal components. According to formula $\sum_{q=1}^{n} \lambda_q$ mentioned in Section 3.1, the contribution rate of the 30 electrode principal components was calculated as shown in Table 1, in which $i$ represents the principal components (or principal elements) and $C_i$ represents the contribution rate. From Table 1, each principal element corresponds to two sets of contribution rate, and the former data are from the first experiment and the latter from the second experiment. In the first experiment, the cumulative contribution rate of the top 10 principal components reached 90.63%. As a result, the amount of principal components was reduced from 30 to 10. Similarly, the cumulative contribution rate of the top 14 principal components reached 95.08%, and the cumulative contribution rate of the top 23 principal components reached 99.10%. In the second experiment, the cumulative contribution rate of the top eight principal components reached 90.14%. As a result, the amount of principal components was reduced from 30 to eight. Similarly, the cumulative contribution rate of the top 13 principal components reached 95.17%, and the cumulative contribution rate of the top 23 principal components reached 99.12%.

Secondly, we selected the main component through Equation (13) >90%. This article mainly tests three kinds of situations where the contribution rates reach 90%, 95% and 99%, respectively. The corresponding characteristic variables in the three cases were 10, 14 and 23 in the first experiment; the corresponding characteristic variables in the three cases were 8, 13 and 23 in the second experiment.

**Table 1.** Contribution rates of each principal component.

| $i$ | $C_i$ | $i$ | $C_i$ | $i$ | $C_i$ | $i$ | $C_i$ | $i$ | $C_i$ |
|---|---|---|---|---|---|---|---|---|---|
| 1 | 0.5664 0.6286 | 7 | 0.0275 0.0197 | 13 | 0.0097 0.0068 | 19 | 0.0043 0.0033 | 25 | 0.0016 0.0017 |
| 2 | 0.0721 0.0960 | 8 | 0.0209 0.0163 | 14 | 0.0086 0.0060 | 20 | 0.0038 0.0030 | 26 | 0.0013 0.0014 |
| 3 | 0.0633 0.0470 | 9 | 0.0191 0.0145 | 15 | 0.0079 0.0056 | 21 | 0.0031 0.0028 | 27 | 0.0012 0.0012 |
| 4 | 0.0486 0.0350 | 10 | 0.0154 0.0107 | 16 | 0.0065 0.0050 | 22 | 0.0027 0.0027 | 28 | 0.0011 0.0009 |
| 5 | 0.0425 0.0322 | 11 | 0.0138 0.0104 | 17 | 0.0051 0.0046 | 23 | 0.0023 0.0024 | 29 | 0.0011 0.0008 |
| 6 | 0.0306 0.0266 | 12 | 0.0125 0.0080 | 18 | 0.0044 0.0040 | 24 | 0.0018 0.0019 | 30 | 0.0008 0.0007 |

Finally, according to the three contribution rates, the accuracy of the recognition in driving fatigue was tested by the SVM classifier. This paper used a method based on $k$-fold cross-validation in which $k = 3$. Seventy percent of the data were used as a training set, then the other thirty percent of the data were used as a test set. The test results are shown in Tables 2 and 3. Compared with the driving fatigue recognition accuracy rates, which only used the sample entropy, when the contribution rate reached 0.99, the SE_KPCA method improved the recognition accuracy of the driving fatigue state compared with the SE method, and the time performance had also been reduced.

**Table 2.** Comparison of the sample entropy principal component analysis (SE_PCA) and SE methods (10 individuals).

| Contribution Rate | SE_PCA | | SE | |
|---|---|---|---|---|
| | Acc | Time | Acc | Time |
| 0.90 | 80.50% | 7.68 s | | |
| 0.95 | 81.83% | 8.25 s | 86.60% | 14.87 s |
| 0.99 | 88.00% | 10.34 s | | |

**Table 3.** Comparison of the SE_PCA and SE methods (15 individuals).

| Contribution Rate | SE_PCA | | SE | |
|---|---|---|---|---|
| | Acc | Time | Acc | Time |
| 0.90 | 59.00% | 20.18 s | | |
| 0.95 | 66.33% | 27.64 s | 71.44% | 39.66 s |
| 0.99 | 73.78% | 35.95 s | | |

### 5.3. Driving Fatigue State Recognition Test Based on SE_KPCA

First of all, we analyzed the principal components contribution rates by the SE_KPCA method. For example, when the kernel function chose a $P$-order polynomial kernel function and the parameter was set as $P = 2$, the calculation results were as shown in Table 4. As we can see, each principal element corresponds to two contribution rates; the former data were from the first experiment, and the latter data were from the second experiment, the same for Tables 5 and 6.

Then, we selected the main component, and the cumulative contribution rate calculation was consistent with the PCA method. For example, the kernel function chose a $P$-order polynomial kernel function testing for three cases with contribution rates of 90%, 95% and 99%. In the first experiment, Table 4 shows the result when parameter $P = 2$, and the characteristic variables in the three cases were: 8, 12, 26. Table 5 shows the result when parameter $P = 1$, and the characteristic variables in the three cases were: 9, 14, 23. Table 6 shows the result when parameter $P = 0.5$, and the characteristic variables in the three cases were: 10, 14, 22. For the second experiment, Table 4 shows the result when parameter $P = 2$, and the characteristic variables in the three cases were: 7, 12, 25. Table 5 shows the result when parameter $P = 1$, and the characteristic variables in the three cases were: 8, 13, 23. Table 6 shows the result when parameter $P = 0.5$, and the characteristic variables in the three cases were: 9, 13, 21.

**Table 4.** Contribution rates of each principal component of the $P$-order ($P = 2$) polynomial kernel function.

| $i$ | $C_i$ | $i$ | $C_i$ | $i$ | $C_i$ | $i$ | $C_i$ | $i$ | $C_i$ |
|---|---|---|---|---|---|---|---|---|---|
| 1 | 0.7026 0.7171 | 7 | 0.0190 0.0156 | 13 | 0.0059 0.0052 | 19 | 0.0027 0.0024 | 25 | 0.0013 0.0013 |
| 2 | 0.0534 0.0679 | 8 | 0.0157 0.0131 | 14 | 0.0057 0.0042 | 20 | 0.0023 0.0023 | 26 | 0.0013 0.0011 |
| 3 | 0.0412 0.0359 | 9 | 0.0141 0.0118 | 15 | 0.0048 0.0037 | 21 | 0.0019 0.0020 | 27 | 0.0011 0.0009 |
| 4 | 0.0318 0.0313 | 10 | 0.0093 0.0089 | 16 | 0.0037 0.0034 | 22 | 0.0018 0.0019 | 28 | 0.0010 0.0009 |
| 5 | 0.0269 0.0214 | 11 | 0.0077 0.0070 | 17 | 0.0034 0.0030 | 23 | 0.0016 0.0017 | 29 | 0.0009 0.0008 |
| 6 | 0.0213 0.0183 | 12 | 0.0073 0.0064 | 18 | 0.0030 0.0028 | 24 | 0.0015 0.0016 | 30 | 0.0008 0.0006 |

**Table 5.** Contribution rates of each principal component of the $P$-order ($P = 1$) polynomial kernel function.

| $i$ | $C_i$ | $i$ | $C_i$ | $i$ | $C_i$ | $i$ | $C_i$ | $i$ | $C_i$ |
|---|---|---|---|---|---|---|---|---|---|
| 1 | 0.5977 0.6255 | 7 | 0.0257 0.0205 | 13 | 0.0087 0.0066 | 19 | 0.0034 0.0033 | 25 | 0.0017 0.0016 |
| 2 | 0.0809 0.1035 | 8 | 0.0208 0.0177 | 14 | 0.0079 0.0055 | 20 | 0.0030 0.0031 | 26 | 0.0016 0.0013 |
| 3 | 0.0557 0.0456 | 9 | 0.0197 0.0139 | 15 | 0.0065 0.0046 | 21 | 0.0026 0.0026 | 27 | 0.0013 0.0012 |
| 4 | 0.0434 0.0390 | 10 | 0.0127 0.0112 | 16 | 0.0051 0.0046 | 22 | 0.0022 0.0025 | 28 | 0.0011 0.0009 |
| 5 | 0.0355 0.0288 | 11 | 0.0104 0.0091 | 17 | 0.0044 0.0042 | 23 | 0.0021 0.0023 | 29 | 0.0010 0.0008 |
| 6 | 0.0284 0.0255 | 12 | 0.0097 0.0076 | 18 | 0.0041 0.0042 | 24 | 0.0019 0.0021 | 30 | 0.0009 0.0006 |

**Table 6.** Contribution rates of each principal component of the $P$-order ($P = 0.5$) polynomial kernel function.

| $i$ | $C_i$ | $i$ | $C_i$ | $i$ | $C_i$ | $i$ | $C_i$ | $i$ | $C_i$ |
|---|---|---|---|---|---|---|---|---|---|
| 1 | 0.5055 0.5343 | 7 | 0.0314 0.0249 | 13 | 0.0107 0.0081 | 19 | 0.0041 0.0041 | 25 | 0.0020 0.0020 |
| 2 | 0.1051 0.1374 | 8 | 0.0250 0.0224 | 14 | 0.0097 0.0069 | 20 | 0.0036 0.0039 | 26 | 0.0019 0.0016 |
| 3 | 0.0685 0.0562 | 9 | 0.0238 0.0159 | 15 | 0.0079 0.0061 | 21 | 0.0032 0.0033 | 27 | 0.0015 0.0014 |
| 4 | 0.0552 0.0476 | 10 | 0.0157 0.0135 | 16 | 0.0062 0.0058 | 22 | 0.0027 0.0030 | 28 | 0.0013 0.0011 |
| 5 | 0.0424 0.0366 | 11 | 0.0128 0.0112 | 17 | 0.0054 0.0052 | 23 | 0.0025 0.0028 | 29 | 0.0012 0.0009 |
| 6 | 0.0353 0.0330 | 12 | 0.0128 0.0089 | 18 | 0.0051 0.0051 | 24 | 0.0023 0.0026 | 30 | 0.0010 0.0007 |

Last but not least, the three selected principal components for driving fatigue recognition accuracy were tested by the SVM classifier. The specific test was performed under three different principal component contribution rates by using the KPCA method of the $P$-order polynomial kernel function, radial basis function and multilayer perceptual kernel function, and every optimal parameters was obtained through multiple experiments. The test results are shown in Tables 7–9. The data test results of 15 people are shown in Tables 10–12. These tables also include the accuracy of the driving fatigue state recognition based on SE_PCA under the same contribution rates.

**Table 7.** Comparison between the sample entropy kernel principal component analysis (SE_KPCA) ($P$-order) method and the SE_PCA method (10 individuals).

| Contribution Rate | | SE_KPCA | | | SE_PCA |
|---|---|---|---|---|---|
| | Parameter | $P = 2$ | $P = 1$ | $P = 0.5$ | |
| 0.90 | Acc | 74.50% | 73.83% | 73.33% | 80.50% |
| | Time | 66.08 s | 14.74 s | 6.41 s | 7.68 s |
| 0.95 | Acc | 82.5% | 82.33% | 75.83% | 81.83% |
| | Time | 83.25 s | 15.70 s | 7.40 s | 8.25 s |
| 0.99 | Acc | 93.17% | 85.83% | 75.83% | 88.00% |
| | Time | 99.63 s | 17.77 s | 10.36 s | 10.34 s |

**Table 8.** Comparison between the SE_KPCA (RBF) method and the SE_PCA method (10 individuals).

| Contribution Rate | | SE_KPCA | | | | SE_PCA |
|---|---|---|---|---|---|---|
| | Parameter | $\sigma = 0.1$ | $\sigma = 0.2$ | $\sigma = 0.7$ | $\sigma = 1$ | |
| 0.90 | Acc | 93.80% | 98.33% | 89.50% | 81.80% | 80.50% |
| | Time | 91.60 s | 36.46 s | 7.58 s | 6.47 s | 7.68 s |
| 0.95 | Acc | 93.80% | 98.33% | 92.60% | 85.80% | 81.83% |
| | Time | 116.19 s | 45.00 s | 13.05 s | 8.57 s | 8.25 s |
| 0.99 | Acc | 93.80% | 98.33% | 92.80% | 86.30% | 88.00% |
| | Time | 138.92 s | 54.01 s | 28.50 s | 21.51 s | 10.34 s |

The test results from Tables 7–12 show that the SE_KPCA method was better than the SE_PCA method at identifying and classifying driving fatigue. In particular, when KPCA's kernel functions chose a radial basis function with a parameter of 0.2 and a contribution rate of 0.9, the classification accuracy of the SE_KPCA method reached 98.33%, and the time performance was good. The subsequent experiments in this paper all used the radial basis function, and the parameter $\sigma$ was 0.2, while the contribution rate was set as 0.9.

**Table 9.** Comparison between the SE_KPCA (MPL) method and the SE_PCA method (10 individuals).

| Contribution Rate | Parameter | SE_KPCA | | | | SE_PCA |
|---|---|---|---|---|---|---|
| | | $c = 0.1$ $v = 0.001$ | $c = 0.2$ $v = 0.01$ | $c = 0.7$ $v = 0.001$ | $c = 1$ $v = 0.01$ | |
| 0.90 | ACC | 70.60% | 70.30% | 70.67% | 70.50% | 80.50% |
| | Time | 4.03 s | 3.83 s | 3.97 s | 3.81 s | 7.68 s |
| 0.95 | ACC | 79.10% | 79.60% | 79.10% | 73.80% | 81.83% |
| | Time | 14.15 s | 13.82 s | 14.05 s | 13.48 s | 8.25 s |
| 0.99 | ACC | 88.00% | 86.33% | 88.00% | 85.83% | 88.00% |
| | Time | 7.09 s | 7.10 s | 7.07 s | 6.80 s | 10.34 s |

**Table 10.** Comparison between the SE_KPCA (P-order) method and the SE_PCA method (15 individuals).

| Contribution Rate | Parameter | SE_KPCA | | | SE_PCA |
|---|---|---|---|---|---|
| | | $P = 2$ | $P = 1$ | $P = 0.5$ | |
| 0.90 | Acc | 58.56% | 56.89% | 57.89% | 59.00% |
| | Time | 235.78 s | 40.49 s | 19.09 s | 20.18 s |
| 0.95 | Acc | 66.78% | 66.22% | 57.223% | 66.33% |
| | Time | 268.56 s | 43.84 | 22.24 s | 27.64 s |
| 0.99 | Acc | 75.56% | 72.11% | 58.56% | 73.78% |
| | Time | 325.98 s | 65.78 s | 35.64 s | 35.95 s |

**Table 11.** Comparison between the SE_KPCA (RBF) method and the SE_PCA method (15 individuals).

| Contribution Rate | Parameter | SE_KPCA | | | | SE_PCA |
|---|---|---|---|---|---|---|
| | | $\sigma = 0.1$ | $\sigma = 0.2$ | $\sigma = 0.7$ | $\sigma = 1$ | |
| 0.90 | Acc | 91.44% | 91.78% | 80.89% | 66.11% | 59.00% |
| | Time | 411.60 s | 259.83 s | 50.80 s | 41.20 s | 20.18 s |
| 0.95 | Acc | 91.44% | 91.67% | 83.89% | 74.33% | 66.33% |
| | Time | 416.19 s | 361.75 s | 72.98 s | 43.39 s | 27.64 s |
| 0.99 | Acc | 91.56% | 91.67% | 84.44% | 75.89% | 73.78% |
| | Time | 458.92 s | 395.39 s | 239.60 s | 99.86 s | 35.95 s |

**Table 12.** Comparison between the SE_KPCA (MPL) method and the SE_PCA method (15 individuals).

| Contribution Rate | Parameter | SE_KPCA | | | | SE_PCA |
|---|---|---|---|---|---|---|
| | | $c = 0.1$ $v = 0.001$ | $c = 0.2$ $v = 0.01$ | $c = 0.7$ $v = 0.001$ | $c = 1$ $v = 0.01$ | |
| 0.90 | ACC | 41.11% | 40.78% | 41.11% | 40.33% | 59.00% |
| | Time | 12.52 s | 11.91 s | 11.22 s | 11.94 s | 20.18 s |
| 0.95 | ACC | 62.78% | 61.78% | 62.78% | 61.33% | 66.33% |
| | Time | 14.15 s | 13.82 s | 14.05 s | 13.48 s | 27.64 s |
| 0.99 | ACC | 73.11% | 71.78% | 73.11% | 71.56% | 73.78% |
| | Time | 22.12 s | 21.26 s | 22.13 s | 21.19 s | 35.95 s |

## 5.4. Comparison Test between SE_KPCA and the Driving Fatigue Recognition Method Based on Fuzzy Entropy/Combination Entropy

In order to verify the classification effect of the SE_KPCA method further, traditional methods of feature extraction were used to compare the sample entropy, fuzzy entropy (FE) [13,22] and combination entropy (CE) [16]. Take the samples of 10 and 15 individuals' EEG signals as an example; fuzzy entropy and combination entropy were used for feature extraction, and then SVM was applied to identify the driving fatigue state; the test results are shown from Figures 3–6. After the comparison and analysis of the figure, the conclusion was draw that SE_KPCA had significantly improved the classification recognition rate compared with the traditional sample entropy, fuzzy entropy and combination entropy, and the time performance was good.

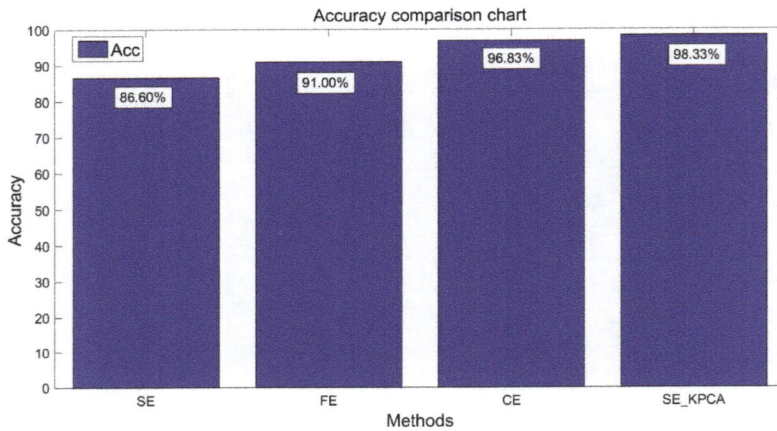

**Figure 3.** Comparison of the recognition accuracy rates among the sample entropy, fuzzy entropy, combination entropy and SE_KPCA methods (10 individuals).

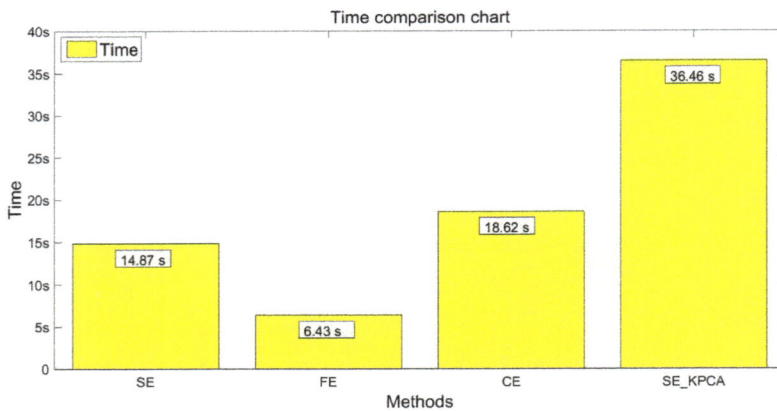

**Figure 4.** Comparison of the time among the sample entropy, fuzzy entropy, combination entropy and SE_KPCA methods (10 individuals).

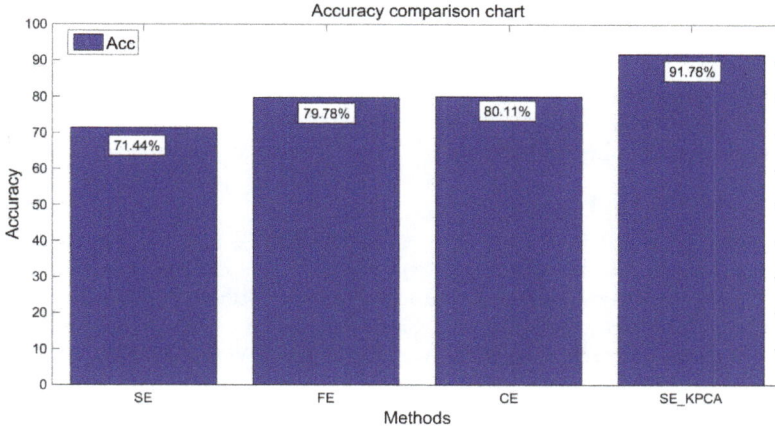

**Figure 5.** Comparison of the recognition accuracy rates among the sample entropy, fuzzy entropy, combination entropy and SE_KPCA methods (15 individuals).

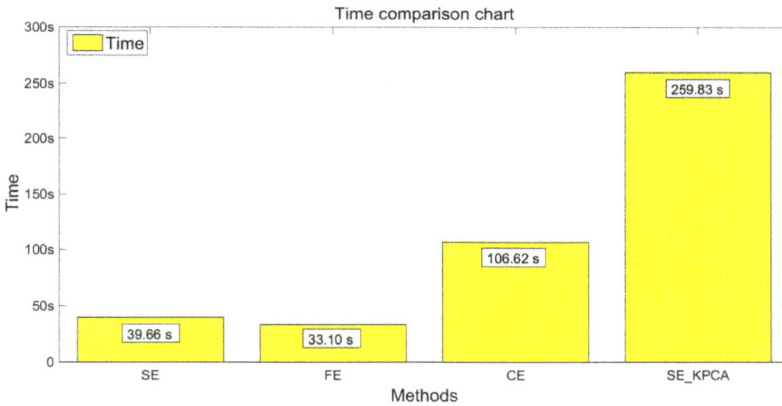

**Figure 6.** Comparison of the time among the sample entropy, fuzzy entropy, combination entropy and SE_KPCA methods (15 individuals).

*5.5. Comparison of the SE_KPCA Method Based on KPCA and Fuzzy Entropy/Combination Entropy for Driving Fatigue Identification*

5.5.1. Data Description

(1) The EEG data processing method based on KPCA and fuzzy entropy (FE_KPCA):

- Extract the features of fuzzy entropy from the collected EEG signals according to the fuzzy entropy formula in the literature [22];
- Select the kernel function $K(x, x_i)$; centralize the fuzzy entropy data in the high dimensional space, and then, calculate the matrix according to Equation (23);
- Calculate the eigenvalues and eigenvectors of the matrix $\tilde{K}$;
- Calculate its nonlinear principal component.

(2) The EEG data processing method based on the KPCA and combination entropy (CE_KPCA):

- Extract the features of the combination entropy from the collected EEG signals according to the combination entropy formula in the literature [21];
- Select the kernel function $K(x, x_i)$; centralize the combination entropy data in the high dimensional space, and then, calculate the matrix according to Equation (23);
- Calculate the eigenvalues and eigenvectors of the matrix $\tilde{K}$;
- Calculate its nonlinear principal component.

5.5.2. Experimental Results

After verifying the validity of the SE_KPCA method in Section 5.4, this paper compares KPCA combined with fuzzy entropy (FE_KPCA) with KPCA combined with combination entropy (CE_KPCA). As shown from Figures 7–10, through the same method, the SVM was adopted for classification and identification. After comparing and analyzing all the figures, our conclusion is that the classification recognition rate of SE_KPCA was obviously higher than FE_KPCA and CE_KPCA, and the temporal performance was lower.

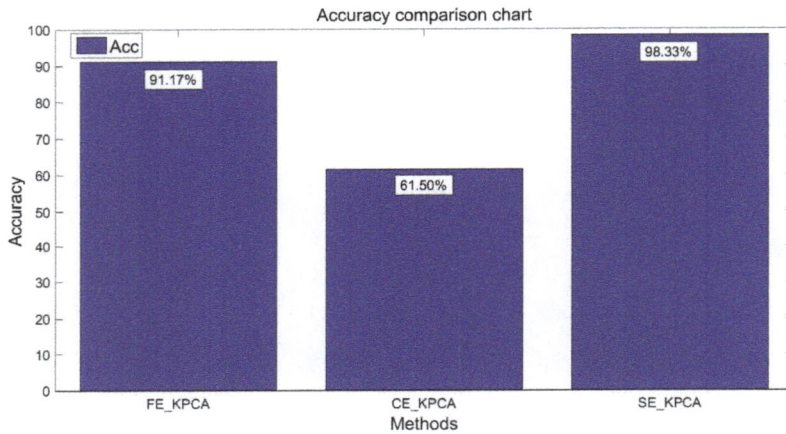

**Figure 7.** Comparison of the recognition accuracy rates among the fuzzy entropy KPCA (FE_KPCA), combination entropy KPCA (CE_KPCA) and SE_KPCA methods (10 individuals).

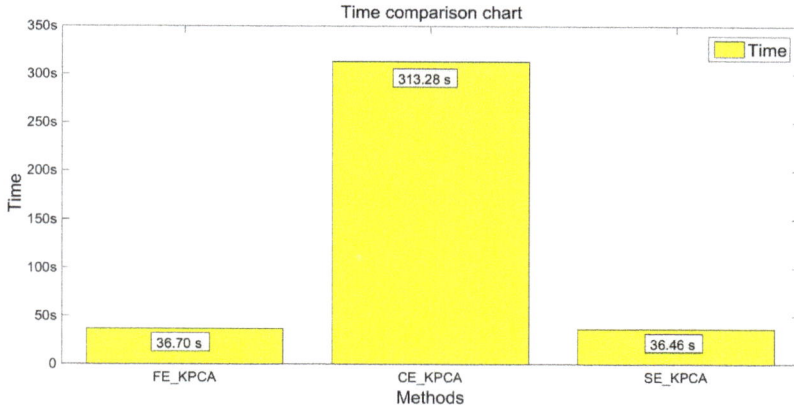

**Figure 8.** Comparison of the time among the FE_KPCA, CE_KPCA and SE_KPCA methods (10 individuals).

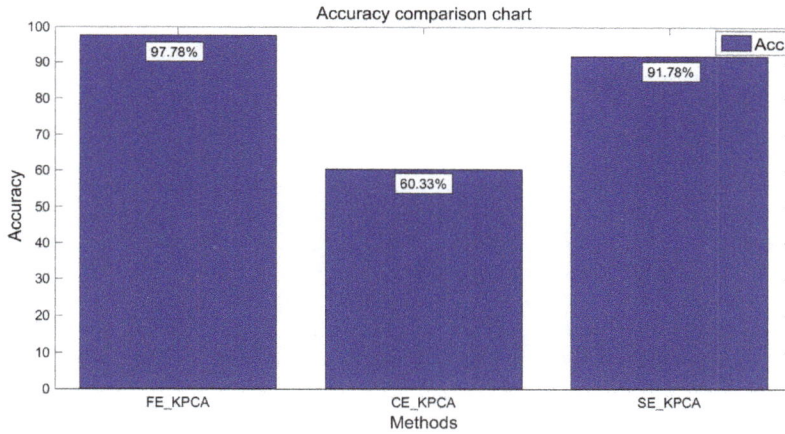

**Figure 9.** Comparison of the recognition accuracy rates among the FE_KPCA, CE_KPCA and SE_KPCA methods (15 individuals).

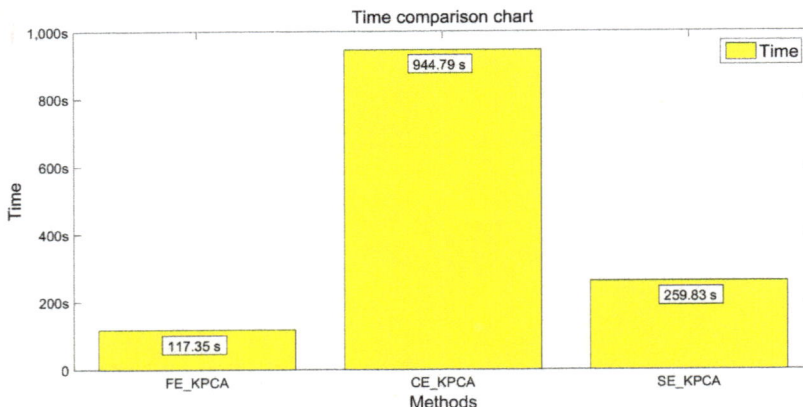

**Figure 10.** Comparison of the time among the FE_KPCA, CE_KPCA and SE_KPCA methods (15 individuals).

As shown from Figures 3–10, it can also be seen that FE_KPCA and CE_KPCA have no higher classification accuracy than traditional FE and CE, and the time performance is worse.

## 6. Conclusions

This paper studies the characteristics of EEG signals in two groups (fatigue state and non-fatigue state). Firstly, feature extraction of the EEG signal was conducted by applying sample entropy, then further feature extraction was made by using kernel principal component analysis, and the SVM classifier was used to classify and identify the two states of fatigue and non-fatigue. Through analysis and comparison of the experiment results that indicate when the kernel functions select the radial basis function, the classification recognition rate performs excellently. Besides, when compared with the traditional methods, the classification recognition rate is also significantly improved. This paper mainly researched the experiment results of the entropy, PCA and KPCA, but the subsequent experiments will introduce more methods for testing.

**Author Contributions:** T.Q. conceived of and designed the experiments. B.Y. performed the experiments. X.B. and P.L. analyzed the data. All the authors have read and approved the final manuscript.

**Funding:** This work is supported by the National Natural Science Foundation of China (Nos. 61070139, 81460769 and 61762045).

**Acknowledgments:** Thanks to Jianfeng Hu's team for providing EEG experiment data.

**Conflicts of Interest:** The authors declare no conflict of interest.

## References

1. Pei, Y.L.; Ma, Y.L. Effects of fatigue on drivers' perceptual judgment and operating characteristics. *J. Jilin Univ.* **2009**, *39*, 1151–1156. (In Chinese)
2. Li, X.F.; Ma, J.F. Theoretical classification and influencing factors of driving fatigue. *J. Decis. Mak.* **2017**, *8*, 87. (In Chinese)
3. Niu, L.B. Research on the Method of Driving Fatigue Based on ECG. Master's Thesis, Southwest Jiaotong University, Chengdu, China, 2017. Available online: http://www.wanfangdata.com.cn/details/detail.do?_type=degree&id=Y3206287 (accessed on 7 July 2018). (In Chinese)
4. Li, F.Q. Research on Driving Fatigue Recognition Algorithm Based on Eye Movement and Pulse Information Fusion. Master's Thesis, Shandong University, Jinan, China, 2015. Available online: http://www.wanfangdata.com.cn/details/detail.do?_type=degree&id=Y2792188 (accessed on 8 July 2018). (In Chinese)

5.  Zhao, C.; Zhao, M.; Yang, Y.; Gao, J.; Rao, N.; Lin, P. The reorganization of human brain networks modulated by driving mental fatigue. *IEEE J. Biomed. Health Inf.* **2017**, *21*, 743–755. [CrossRef] [PubMed]
6.  Zhao, X.P.; Meng, C.M.; Feng, M.K.; Chang, S.J. Fatigue detection based on cascade convolutional neural network. *J. Optoelectron. Laser* **2017**, *28*, 497–502. (In Chinese)
7.  Zhang, N.N.; Wang, H.; Fu, R.R. Feature extraction of driving fatigue EEG based on wavelet entropy. *J. Autom. Eng.* **2013**, *35*, 1139–1142. (In Chinese)
8.  Chai, R.; Naik, G.R.; Nguyen, T.N.; Ling, S.H.; Tran, Y.; Craig, A.; Nguyen, H.T. Driver fatigue classification with independent component by entropy rate bound minimization analysis in an EEG-based system. *IEEE J. Biomed. Health Inform.* **2017**, *21*, 715–727. [CrossRef] [PubMed]
9.  Zeng, H.; Yang, C.; Dai, G.; Qin, F.; Zhang, J.; Kong, W. EEG classification of driver mental states by deep learning. *Cognit. Neurodyn.* **2018**, *10*, 1–10. [CrossRef]
10. Chai, R.; Ling, S.H.; San, P.P.; Naik, G.R.; Nguyen, T.N.; Tran, Y.; Craig, A.; Nguyen, H.T. Improving eeg-based driver fatigue classification using sparse-deep belief networks. *Front. Neurosci.* **2017**, *11*, 103. [CrossRef] [PubMed]
11. Hu, J.; Min, J. Automated detection of driver fatigue based on EEG signals using gradient boosting decision tree model. *Cognit. Neurodyn.* **2018**, *12*, 1–10. [CrossRef] [PubMed]
12. Peng, J.Q.; Wu, P.D. Exploring EEG characteristics of driving fatigue. *J. Beijing Inst. Technol.* **2007**, *27*, 585–589. (In Chinese)
13. Ding, X.H.; Ma, Y.L. Research on Feature Extraction and Classification of EEG Signals in Sports Imaging. Master's Thesis, Hangzhou Dianzi University, Hangzhou, China, 2016. Available online: http://www.wanfangdata.com.cn/details/detail.do?_type=degree&id=D824056 (accessed on 6 July 2018). (In Chinese)
14. Guan, J.Q.; Yang, B.H.; Ma, S.W.; Yuan, L. Research on brain recognition of motor images based on PCA and SVM. *J. Beijing Biomed. Eng.* **2010**, *29*, 261–265. (In Chinese)
15. Zhang, Y.; Luo, M.W.; Luo, Y. Wavelet transform and sample entropy feature extraction method for EEG signals. *J. Intell. Syst.* **2012**, *7*, 339–344. (In Chinese)
16. Gao, X.W. Kernel PCA Feature Extraction Method and Its Application. Master's Thesis, Nanjing University of Aeronautics and Astronautics, Nanjing, China, 2009. Available online: http://www.wanfangdata.com.cn/details/detail.do?_type=degree&id=D077304 (accessed on 5 July 2018). (In Chinese)
17. Liu, C.; Zhao, H.B. Motion imagined EEG signal classification based on CSP and SVM algorithms. *J. Northeast. Univ.* **2010**, *31*, 1098–1101. (In Chinese)
18. Pincus, S.M. Approximate entropy as a measure of system complexity. *Proc. Natl. Acad. Sci. USA* **1991**, *88*, 2297-2301. [CrossRef]
19. Sun, Y.; Ye, N.; Xu, X.H. Feature extraction of EEG signals based on PCA and wavelet transform. In Proceedings of the 2007 Annual Conference of Chinese Control and Decision Science, Wuxi, China, 4–6 July 2007; Northeastern University Press: Shenyang, China, 2007; pp. 669–673.
20. Li, D.M. Classification of epileptic EEG signals based on PCA and the positioning of epileptogenic focus. *J. Biomed. Eng. Res.* **2017**, *36*, 218–223. (In Chinese)
21. Mu, Z.; Hu, J.; Min, J. Driver fatigue detection system using electroencephalography signals based on combined entropy features. *Appl. Sci.* **2017**, *7*, 150. [CrossRef]
22. Tian, J.; Luo, Z.Z. Feature extraction of motor-imagined EEG signals based on fuzzy entropy. *J. Huazhong Univ. Sci. Technol.* **2013**, *41*, 92–95. (In Chinese)

entropy

MDPI

*Article*

# Entropy-Based Structural Health Monitoring System for Damage Detection in Multi-Bay Three-Dimensional Structures

**Tzu-Kang Lin * and Ana Gabriela Laínez**

Department of Civil Engineering, National Chiao Tung University, Hsinchu 30010, Taiwan;
anagab93@hotmail.com
* Correspondence: tklin@nctu.edu.tw; Tel.: +886-03-571-2121-54919

Received: 28 November 2017; Accepted: 5 January 2018; Published: 11 January 2018

**Abstract:** In this paper, a structural health monitoring (SHM) system based on multi-scale cross-sample entropy (MSCE) is proposed for detecting damage locations in multi-bay three-dimensional structures. The location of damage is evaluated for each bay through MSCE analysis by examining the degree of dissimilarity between the response signals of vertically-adjacent floors. Subsequently, the results are quantified using the damage index (DI). The performance of the proposed SHM system was determined in this study by performing a finite element analysis of a multi-bay seven-story structure. The derived results revealed that the SHM system successfully detected the damaged floors and their respective directions for several cases. The proposed system provides a preliminary assessment of which bay has been more severely affected. Thus, the effectiveness and high potential of the SHM system for locating damage in large and complex structures rapidly and at low cost are demonstrated.

**Keywords:** multi-bay; three-dimensional; structural health monitoring; multi-scale; cross-sample entropy

## 1. Introduction

Structural health monitoring (SHM) has attracted considerable attention among engineers, because structures are inevitably subject to internal or external factors that affect their service lives. SHM enables early detection of any damage caused by factors such as the environment or faulty construction, thus facilitating timely maintenance or repair jobs. Over the past two decades, signal-processing techniques have mainly been applied in SHM methods to analyze the measured displacement, velocity or acceleration signals of structures to obtain dynamic characteristics such as basic vibration frequency (natural frequency) and damping. These characteristics contribute to the diagnosis of damage, as well as its condition and possible location [1,2].

In 1999, Wahab and De Roeck [3] used the change in dynamic parameters between the undamaged and damaged conditions of simply-supported and continuous beams to localize damage. In 2000, Maeck et al. [4] identified the degree and location of damage in reinforced concrete beams according to their modal characteristics for dynamic stiffness analysis. Subsequently, vibration-based SHM algorithms and their application limitations were examined and summarized by Chang et al. [5]. The feasibility of using measured modal parameters for damage detection of steel towers through Bayesian probability theory was examined by Lam and Yang [6]. Furthermore, Chen et al. [7] identified bridge bearing damage by using beam modal information as the input to a neural network.

German physicist Clausius [8] first introduced the concept of entropy in 1865 to evaluate the uncertainty of events in a thermodynamic system. Decades later, Shannon [9] proposed the influential Shannon entropy in the field of information theory. Moreover, Kolmogorov [10] defined the notion of entropy for a new class of dynamical systems; Sinai [11] then introduced a definition of entropy that can be applied to all dynamical systems. Jointly, the entropy defined in these two studies was called KS entropy, which was used to measure the complexity of measured time series in D-dimensional dynamic systems. Thereafter, research demonstrated that although KS entropy can be effectively applied to low-dimensional chaotic systems, it cannot be applied to experimental data. The calculation results are affected by various levels of noise that may be involved [12].

In 1991, Pincus [13] developed an analytical method called "approximate entropy" (ApEn) by modifying the KS entropy formula. This method permitted the quantification of regularity in a time series as a single number. The performance of ApEn was tested using clinical data, and the test results demonstrated the predictive and diagnostic capabilities of ApEn. An and Ou [14] proposed the mean curvature difference method based on the ApEn theory and successfully located the damage in the shear frame structures. A modification of ApEn, called "sample entropy" (SampEn), was proposed by Richman and Moorman [15] in 2000. The advantage of SampEn is that the entropy value obtained is not affected by the length of the time series. Moreover, higher relative consistency can be achieved under different parameters such as the threshold *r*, sample length *m* and signal length *N*. Richman and Moorman also argued that comparing a dataset with itself is meaningless; therefore, SampEn does not count self-matches and is free of any bias caused in the entropy estimation.

In 2002, multi-scale entropy (MSE) was proposed by Costa et al. [16,17] to measure the entropy in physiological time series. Traditional entropy-based algorithms that use only a single time scale for analysis may yield misleading results; therefore, a coarse-graining procedure was proposed to obtain more accurate results during entropy calculation. The MSE method was verified by applying it to heartbeat time series of healthy subjects, subjects with congestive heart failure and subjects with atrial fibrillation. The results showed that on multiple scales, healthy heartbeats had the highest entropy values; thus, diagnosing patients through the MSE method is feasible. MSE analysis has been used not only in the medical field, but also in the area of mechanical engineering. The damage condition of roller bearings has been successfully diagnosed through the MSE analysis of vibration signals [18,19].

Similar to ApEn, Cross-ApEn was introduced by Pincus and Singer [20] to analyze the degree of asynchrony of two related time series. Subsequently, Richman and Moorman [15] proposed Cross-SampEn as a superior alternative for the same purpose, because unmatched templates result in undefined probabilities in Cross-ApEn. The performance of both methods was tested using cardiovascular time series, demonstrating that the results obtained using Cross-SampEn had a higher level of relative consistency than those obtained using Cross-ApEn.

In 2013, Fabris et al. [21] applied the SampEn and Cross-SampEn algorithms to electroglottogram and microphone signals. The healthy patients and those with throat or vocal disorders can be identified by quantifying the degree of asynchrony between time series. Subsequently, SHM systems based on the MSE and MSCE algorithms introduced in [15–17] in the field of physiology have been proposed to identify damage locations and directions on a single-bay structure [22,23]. The vertical MSCE analysis was performed to identify the damaged floor, and planar MSCE analysis was performed to identify the damage directions. The resulting MSCE curves indicated that a higher degree of synchronicity between two signals yields lower entropy values. Furthermore, time series with high complexity have high entropy values, indicating damage. Considering the higher entropy values obtained for damaged floors, vertical and planar damage index (DI) values were proposed for efficiently quantifying damage. A comparison of healthy and damaged signals revealed positive DI values for damaged floors, whereas negative DI values were observed for healthy floors. Moreover, the parameters of sample entropy and wavelet transformation were optimized to detect the possible crack on a cantilever beam [24], and a cross-entropy optimization technique was also applied to identify the damage of the shear structure component [25].

The aim of the present study was to implement the vertical MSCE analysis coupled with the vertical DI on a large and complex three-bay bi-axial numerical model to identify the damaged floors and damaged bays of the structure. The remainder of this paper is organized as follows: The proposed SHM system is described in Section 2. In Section 3, a numerical evaluation of a three-bay, seven-story steel structure is presented. On the basis of the numerical results, the performance of the MSCE and DI analyses is discussed in Section 4. Finally, Section 5 provides the conclusions.

## 2. SHM Algorithm

### 2.1. SampEn

In this section, SampEn is first introduced to understand more clearly the methods used for the SHM system [13]. As a statistical method for analyzing time series, SampEn estimates the entropy value of a measured time series to quantify the complexity of a system. The results of SampEn, an unbiased refinement of ApEn, are not affected by the time series length or calculation parameters.

For a time series $\{X_i\} = \{x_1, \ldots, x_i, x_N\}$ with length $N$, a vector of $m$ data points $u_m(i) = \{x_i, x_{i+1}, \ldots, x_{i+m-1}\}, 1 \leq i \leq N - m + 1$ can be defined as the template. The combination of all templates with length $m$ is represented by the template space $T$ of the signal; for example, $[x_i, x_{i+1}, \ldots, x_{i+m-1}]$ represents the $i$-th template of the time series. Various $N - m + 1$ templates may constitute the time series. The template space $T$, which is the combination of all $N - m + 1$ templates with length $m$, is expressed as follows:

$$
T = \begin{bmatrix}
x_1 & x_2 & \cdots & x_m \\
x_2 & x_3 & \cdots & x_{m-1} \\
\vdots & \vdots & \ddots & \vdots \\
x_{N-m+1} & x_{N-m+2} & \cdots & x_N
\end{bmatrix} \tag{1}
$$

Let $d_{ij}$ be the maximum distance between two templates $i$ and $j$ and $r$ be a predetermined threshold.

$$
d_{ij} = \max\{|x(i+k) - x(j+k)| : 0 \leq k \leq m - 1\} \tag{2}
$$

Next, the number of similarities $n_i^m(r)$ between templates $u_m(i)$ and $u_m(j)$ can be calculated as follows:

$$
n_i^m(r) = \sum_{j=1}^{N-m} d[u_m(i), u_m(j)] \tag{3}
$$

where similarity is defined as follows:

$$
d[u_m(i), u_m(j)] = \begin{cases} 1 & d_{ij} \leq r \\ 0 & d_{ij} > r \end{cases} \tag{4}
$$

The distance between templates is calculated through Equation (2) and then substituted into Equation (4) to define the similarity between the two. The two templates are determined to be similar when the distance $d_{ij}$ does not exceed the threshold $r$. By contrast, the two templates are dissimilar when $d_{ij}$ exceeds $r$. Different pattern templates can be substituted for comparisons with template $i$. The degree of sample similarity $U_i^m(r)$ can then be calculated as follows:

$$
U_i^m(r) = \frac{n_i^m(r)}{(N - m + 1)} \tag{5}
$$

The average degree of sample similarity can be calculated as follows after obtaining the degree of sample similarity:

$$
U^m(r) = \frac{1}{N-m} \sum_{i=1}^{N-m} U_i^m(r) \tag{6}
$$

Here, $U^m(r)$ represents the average degree of similarity between all templates of length $m$ in the template space $T$. Subsequently, a new template space is created by assembling templates with length

$m + 1$. The average degree of similarity $U^{m+1}(r)$ of the new template space is calculated by repeating the aforementioned steps. Consequently, the SampEn values of the time series with parameters $m, r$ and $N$ can be obtained as follows:

$$S_E(m, r, N) = -\ln \frac{U^{m+1}(r)}{U^m(r)} \tag{7}$$

## 2.2. MSE

MSE can extract much more information of a time series, compared with single scale-based entropy methods. Briefly, a time series is subjected to a coarse-graining procedure to construct multiple time series at different time scales [15,16]. The procedure is described as follows: A discrete time series $x_1, x_2, \ldots, x_N$ with length $N$ is segmented into multiple time series with length $\tau$, where $\tau(\tau = 1, 2, \ldots, N)$ is the scale factor. Subsequently, a new time series $\{y_j^{(\tau)}\}$ is constructed by deriving the arithmetic mean of each set of data values according to the following equation:

$$y_j^{(\tau)} = \frac{1}{\tau} \sum_{i=(j-1)\tau+1}^{j\tau} x_i, 1 \leq j \leq \frac{N}{\tau} \tag{8}$$

The length of each coarse-grained time series is $N/\tau$, meaning that at scale 1, the coarse-grained time series is the original time series. Moreover, the length of the coarse-grained time-series decreases as $\tau$ increases. After the process is completed, SampEn is used to calculate the entropy values for each coarse-grained time series $\{y_j^{(\tau)}\}$. The obtained sample entropy values are the MSE of the time series and are plotted as a function of the scale factor $(f(\tau) = S_E)$.

## 2.3. Cross-SampEn

With a similar procedure to SampEn, Cross-SampEn is used to evaluate the degree of dissimilarity between two time series derived from the same system [20]. The estimation of Cross-SampEn can be summarized as follows. Consider two individual time series $\{X_i\} = \{x_1, \ldots, x_i, \ldots, x_N\}$ and $\{Y_j\} = \{y_1, \ldots, y_j, \ldots, y_N\}$ with length $N$. The signals are segmented into the following templates of length $m$: $u_m(i) = \{x_i, x_{i+1}, \ldots, x_{i+m-1}\}, 1 \leq i \leq N - m + 1$ and $v_m(j) = \{y_j, y_{j+1}, \ldots, y_{j+m-1}\}, 1 \leq j \leq N - m + 1$. The template space $T_x$ is presented as follows:

$$T_x = \begin{bmatrix} x_1 & x_2 & \cdots & x_m \\ x_2 & x_3 & \cdots & x_{m-1} \\ \vdots & \vdots & \ddots & \vdots \\ x_{N-m+1} & x_{N-m+2} & \cdots & x_N \end{bmatrix} \tag{9}$$

Similarly, template space $T_y$ is expressed as follows:

$$T_y = \begin{bmatrix} y_1 & y_2 & \cdots & y_m \\ y_2 & y_3 & \cdots & y_{m-1} \\ \vdots & \vdots & \ddots & \vdots \\ y_{N-m+1} & y_{N-m+2} & \cdots & y_N \end{bmatrix} \tag{10}$$

The number of similarities between $u_m(i)$ and $v_m(j)$, defined as $n_i^m(r)$, is calculated under the following criterion:

$$d[u_m(i), v_m(j)] \leq r, 1 \leq j \leq N - m \tag{11}$$

The similarity probability of the templates is evaluated using the following equations:

$$U_i^m(r)(v\|u) = \frac{n^m(r)}{(N-m)} \tag{12}$$

Then, the average probability of similarity of length $m$ is calculated as follows:

$$U^m(r)(v\|u) = \frac{1}{(N-m)} \sum_{i=1}^{N-m} U_i^m(r)(v\|u) \tag{13}$$

where $U^m(r)(v\|u)$ is the degree of dissimilarity between the two time series when $m$ points are segmented. Subsequently, new template spaces $T_x$ and $T_y$ are created by assembling templates with length $m+1$. The steps above are repeated to obtain the average similarity probability $U^{m+1}(r)(v\|u)$, and the Cross-SampEn values can be then derived as follows:

$$CS_E(m,r,N) = -\ln\frac{U^{m+1}(r)(v\|u)}{U^m(r)(v\|u)} \tag{14}$$

### 2.4. MSCE and DI

The SHM system proposed in this study relies on MSCE analysis for identifying structural damage. Because an extended three-bay model was analyzed in this study, it was also of interest to identify the damaged bay. However, diagnosing the location of damage through a simple observation of the obtained MSCE curves is typically difficult. Therefore, a DI based on [22,23] was proposed for rapidly and efficiently diagnosing the damaged floor, axis and bay in the structure.

For the three-bay structure described in the following section, the SHM process is applied to each bay separately to detect the possible damage, while the interaction between each bay has been included in the measured signals. Thus, the following procedure is repeated three times. Two groups of curves representing the healthy and damaged conditions of the structure are analyzed. For a biaxial structure with $N$ floors, the MSCE curves for the $x$- and $y$-axes under the healthy condition can be expressed as matrices:

$$MSCE_{undamaged} = \begin{Bmatrix} H_1^x \\ H_2^x \\ \vdots \\ H_N^x \end{Bmatrix} \quad MSCE_{undamaged} \begin{Bmatrix} H_1^y \\ H_2^y \\ \vdots \\ H_N^y \end{Bmatrix} \tag{15}$$

Similarly, the MSCE curves for the $x$- and $y$-axes under the damaged condition are expressed as follows:

$$MSCE_{damaged} = \begin{Bmatrix} D_1^x \\ D_2^x \\ \vdots \\ D_N^x \end{Bmatrix} \quad MSCE_{damaged} \begin{Bmatrix} D_1^y \\ D_2^y \\ \vdots \\ D_N^y \end{Bmatrix} \tag{16}$$

$H$ and $D$ represent the MSCE curves for the healthy and damaged conditions, respectively. The superscripts $x$ and $y$ represent the analyzed axes, and the subscript depicts the analyzed floor. For example, $H_N^x$ is the $x$-axis MSCE of the signal in the analyzed $N$-th floor and the signal of the floor beneath it under healthy conditions. This can be further expressed as follows:

$$H_N^{axis} = \{CS_E{}^1_{HNaxis}, CS_E{}^2_{HNaxis}, CS_E{}^3_{HNaxis}, \cdots, CS_E{}^\tau_{HNaxis}\} \tag{17}$$

Similarly, $D_N^{axis}$ can be arranged as follows:

$$D_N^{axis} = \{CS_E{}^1_{DNaxis}, CS_E{}^2_{DNaxis}, CS_E{}^3_{DNaxis}, \ldots, CS_E{}^\tau_{DNaxis}\} \tag{18}$$

where $CS_E$ denotes the Cross-SampEn value in each element, the superscript denotes the scale factor $\tau$, the first superscript denotes the health condition, the second subscript denotes the analyzed floor and the third subscript denotes the analyzed axis. Subsequently, the following formulae can be used to calculate the dual-axis DI per floor on a specific bay:

$$DI_{Nx} = \sum_{k=1}^{\tau}(CS_E{}^q_{DNx} - CS_E{}^q_{HNx}) \qquad DI_{Ny} = \sum_{k=1}^{\tau}(CS_E{}^q_{DNy} - CS_E{}^q_{HNy}) \tag{19}$$

The DI is evaluated by calculating the difference between the MSCE values of the damaged and healthy structures. Each bay of the structure has two DI values per floor: one on the *x*-axis and the other on the *y*-axis. For a specific floor, a positive DI indicates that the floor has sustained damage, whereas a negative DI indicates no damage on the floor, where the structure is under a more stable condition compared to the original healthy state. As verified in previous research [22], the proposed method can sustain the velocity changes for an approximately 10–20% noise level.

## 3. Numerical Simulation

To verify the feasibility of the SHM system proposed in this study, SAP2000 software (SAP2000 v9) was used to construct and analyze the three-bay, seven-story model, which is an extension of the areal benchmark structure commonly used at the National Center for Research on Earthquake Engineering (NCREE), for numerical simulation. The dimensions and characteristics of the numerical model are outlined as follows: The model was a steel structure comprising seven stories and three bays on the *x*-axis and a single bay on the *y*-axis. The height of each story was 1.06 m, and the widths of the bays on the *x*- and *y*-axes were 1.32 and 0.92 m, respectively. The columns were steel plates measuring $75 \times 50$ mm$^2$. The beams were steel plates of $70 \times 100$ mm$^2$. All sides of the structure were fitted with steel bracing chosen as L-shaped steel angles measuring $65 \times 65 \times 6$ mm. Apart from the self-weight of the structure, an additional 500-kg mass was added per bay on each story to simulate the actual characteristics of a structure.

Time history and modal analyses were performed on the model. To perform the time history analysis, white noise signals were first generated with a power intensity of 1 MW and selected as the input accelerations to excite the numerical model as a simulation of ambient vibrations. The direct integration time-history analysis was applied to solve the structural response for 150 s. The response signals of the undamaged scenario under ambient vibrations shown in Figure 1 were used as the reference of the SHM database for damage detection. Detailed setting of the finite element model is listed in Table 1.

**Table 1.** Numerical setting of the finite element model.

| Parameter | Setting |
|---|---|
| Constitutive equation | Linear elastic |
| Material | A36 steel |
| DOF of the foundation | Fixed in 6 DOF |
| Finite element for column, beam and bracing | Beam |
| Integration method | Direct-integration time-history analysis |
| Duration | 150 s |
| Ambient vibration | White noise of 1 mW power |

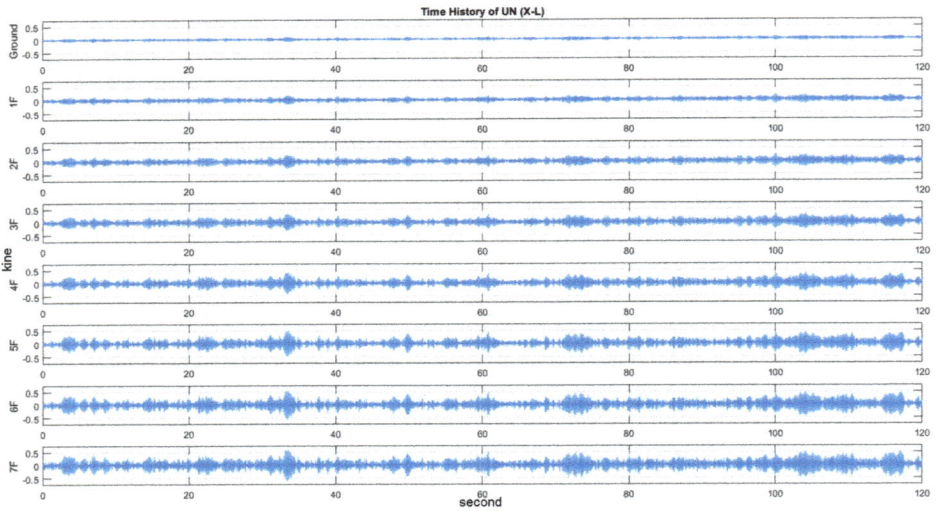

**Figure 1.** The time history of the velocity response of the undamaged scenario.

*Damage Database*

The numerical model of the steel structure was fitted with diagonal braces on all directions of every floor, implying that for the three-bay model, each floor had six braces on the strong (X) axis and four braces on the weak (Y) axis. Structural damage was simulated by removing the installed braces symmetrically; that is, two braces were removed on the *x*-axis to simulate damage on the strong axis, and two bracings were removed on the *y*-axis to simulate damage on the weak axis.

After the SHM database was constructed by performing the time history and modal analyses on the numerical model for each damage condition, biaxial velocity response data were extracted from the center of each floor per bay. Figure 2 shows the data extraction points of the structure. The various damage conditions along with their respective cases are presented in Table 2. The cases comprised various combinations of single-story, two-story or multistory damage, paired with single- or multi-bay and single- or multi-direction damage. The scenarios for the damage database were selected with a broad spectrum of diverse damage locations. In total, the damage conditions were classified into 12 categories and 26 cases. Numbers in the case names indicate damaged floors; X and Y represent the damaged axes; and L, C and R (left, center and right) denote the damaged bay. For example, Case 9 (6X-L & 6Y-R) represents a case involving damage on the sixth floor, *x*-axis, left bay and sixth floor, *y*-axis, right bay. Figure 3 illustrates the numerical model, with the dotted lines representing the braces removed to exemplify the damaged bracing for Case 9. A list of the modal analysis for all damage cases is shown in Table 3 to reflect the change of the global behavior of the numerical model. As depicted, the fundamental frequencies of the first and second modes drop accompanying the increase of damage level in both the X and Y directions. Significant changes are caused by damages located in the lower floors.

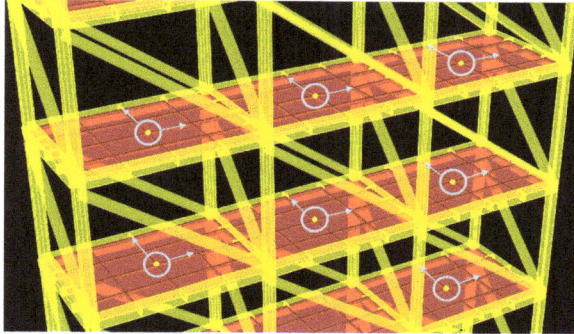

**Figure 2.** Signal extraction points.

**Table 2.** Damage cases for analysis.

| Case Number | Damage Group | Damaged Floor, Direction and Bay |
|---|---|---|
| 1 | | 5X-L |
| 2 | Single-story, single-bay, single-direction | 3Y-C |
| 3 | | 7Y-R |
| 4 | Single-story, single-bay, multidirectional | 4XY-L |
| 5 | | 6XY-C |
| 6 | Single-story, multi-bay, single-direction | 2X-L & 2X-C |
| 7 | | 5Y-L & 5Y-C & 5Y-R |
| 8 | Single-story, multi-bay, multidirectional | 3X-R & 3Y-C |
| 9 | | 6X-L & 6Y-R |
| 10 | Two-story, single-bay, single-direction | 3X-L & 6X-L |
| 11 | | 1Y-R & 5Y-R |
| 12 | Two-story, single-bay, multidirectional | 4X-C & 7Y-C |
| 13 | | 2XY-R & 3XY-R |
| 14 | | 5X-R & 7X-L |
| 15 | Two-story, multi-bay, single-direction | 2Y-C & 4Y-R |
| 16 | | 2X-L & C, 6X-C & R |
| 17 | Two-story, multi-bay, multidirectional | 4X-R & 2Y-L |
| 18 | | 6XY-R & 7XY-L |
| 19 | Multistory, single-bay, single-direction | 3X-L & 4X-L & 6X-L |
| 20 | | 1Y-R & 4Y-R & 7Y-R |
| 21 | Multistory, single-bay, multidirectional | 4X-L & 5Y-L & 6Y-L |
| 22 | | 1XY-C & 3XY-C & 5XY-C |
| 23 | Multistory, multi-bay, single-direction | 3X-L & 4X-C & 5X-R |
| 24 | | 6Y-L & 2Y-C & 7Y-R |
| 25 | Multistory, multi-bay, multidirectional | 1X-R & 2X-R & 1Y-L |
| 26 | | 7XY-R & 4Y-L & 6Y-C |

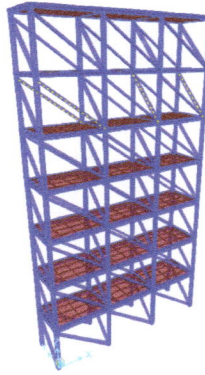

**Figure 3.** Three-dimensional view of the numerical model. The dotted braces represent the damaged bracing for Case 9 (6X-L & 6Y-R).

**Table 3.** Modal analysis of the numerical model.

| Case No. | Damage Case | Frequency (Hz) | | | |
|---|---|---|---|---|---|
| | | Mode 1(X) | Mode 2(X) | Mode 1(Y) | Mode 2(Y) |
| | Undamaged (UN) | 6.5 | 19.52 | 2.69 | 11.56 |
| 1 | 5X-L | 6.41 | 18.84 | 2.69 | 11.56 |
| 2 | 3Y-C | 6.50 | 19.53 | 2.65 | 11.55 |
| 3 | 7Y-R | 6.50 | 17.97 | 2.68 | 11.02 |
| 4 | 4XY-L | 6.12 | 18.69 | 2.60 | 11.32 |
| 5 | 6XY-C | 6.44 | 18.58 | 2.68 | 11.15 |
| 6 | 2X-L & 2X-C | 5.78 | 18.93 | 2.69 | 11.57 |
| 7 | 5Y-L & 5Y-C & 5Y-R | 6.27 | 17.41 | 2.44 | 8.88 |
| 8 | 3X-R & 3Y-C | 6.32 | 19.52 | 2.65 | 11.55 |
| 9 | 6X-L & 6Y-R | 6.43 | 17.02 | 2.66 | 10.48 |
| 10 | 3X-L & 6X-L | 6.29 | 18.81 | 2.69 | 11.57 |
| 11 | 1Y-R & 5Y-R | 5.88 | 17.55 | 2.49 | 9.21 |
| 12 | 4X-C & 7Y-C | 6.32 | 19.09 | 2.69 | 11.36 |
| 13 | 2XY-R & 3XY-R | 5.41 | 18.23 | 2.39 | 10.86 |
| 14 | 5X-R & 7X-L | 6.42 | 18.55 | 2.69 | 11.56 |
| 15 | 2Y-C & 4Y-R | 6.3 | 19.16 | 2.55 | 11.02 |
| 16 | 2X-L & C, 6X-C & R | 5.67 | 16.21 | 2.69 | 11.57 |
| 17 | 4X-R & 2Y-L | 5.91 | 18.66 | 2.57 | 11.00 |
| 18 | 6XY-R & 7XY-L | 6.43 | 15.50 | 2.66 | 10.19 |
| 19 | 3X-L & 4X-L & 6X-L | 6.13 | 18.52 | 2.69 | 11.57 |
| 20 | 1Y-R & 4Y-R & 7Y-R | 5.78 | 16.13 | 2.45 | 9.33 |
| 21 | 4X-L & 5Y-L & 6Y-L | 6.29 | 17.56 | 2.59 | 9.96 |
| 22 | 1XY-C & 3XY-C & 5XY-C | 5.92 | 18.07 | 2.57 | 10.54 |
| 23 | 3X-L & 4X-C & 5X-R | 6.12 | 18.46 | 2.70 | 11.57 |
| 24 | 6Y-L & 2Y-C & 7Y-R | 6.49 | 16.44 | 2.60 | 9.94 |
| 25 | 1X-R & 2X-R & 1Y-L | 5.67 | 18.09 | 2.56 | 10.40 |
| 26 | 7XY-R & 4Y-L & 6Y-C | 6.27 | 15.94 | 2.59 | 10.52 |

## 4. Results and Discussion

For the undamaged case and every damage condition, the velocity signals for the long (X) and short (Y) axes were extracted per bay from the center of each floor. Subsequently, the signals of two vertically-adjacent floors under the same damage condition were processed through Cross-SampEn at multiple scales (MSCE) to evaluate the dissimilarity between floors. After a series of optimization searches [23], where different combinations of parameters were considered for the best performance on damage detection accuracy, the parameters required for the calculation of Cross-SampEn, such as the template length $m$, threshold $r$ and signal length $N$, were selected as 4, 0.10× the standard deviation (SD) of the time series and 30,000, respectively. In addition, Cross-SampEn was calculated across 25 scales ($\tau = 25$).

After analyzing the undamaged case and the 26 damage cases through the MSCE method, the results were compared and quantified per bay using the DI. When a specific floor sustains damage, the structure experiences a loss of stiffness, which causes a rise in the complexity of the extracted response signals for that specific floor. Time series with high complexity have high entropy values, indicating damage. Thus, analysis of the changes in complexity or entropy of healthy and damaged signals results in positive DI values for damaged floors and negative DI values for healthy floors. Small positive values can be excluded by a predetermined threshold value of one, which was selected by experimentation. Thus, the damaged floor and direction can be easily detected.

Figure 4a–f presents the MSCE diagrams obtained for the healthy condition in the X- and Y-directions for the left, center and right bays; the diagrams of the X-direction are shown on the left side, and those of the Y-direction are on the right side. In these figures, G-1F denotes the curve for the first floor, 1F-2F denotes the curve for the second floor, 2F-3F denotes the curve for the third floor, and so forth. The Cross-SampEn for Channel G-1F was calculated from the velocity signals of the ground and the first floor.

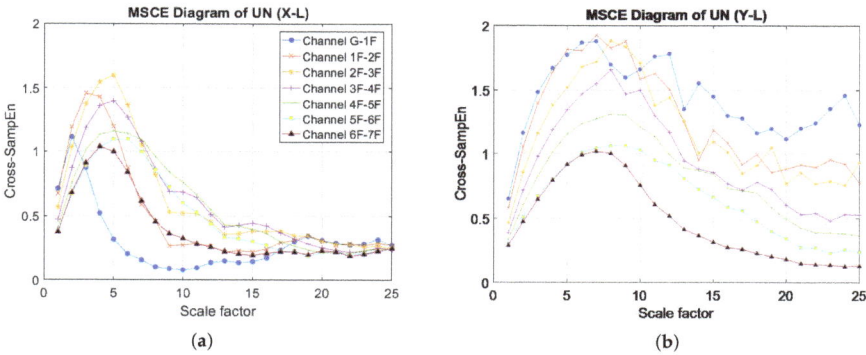

Figure 4. *Cont.*

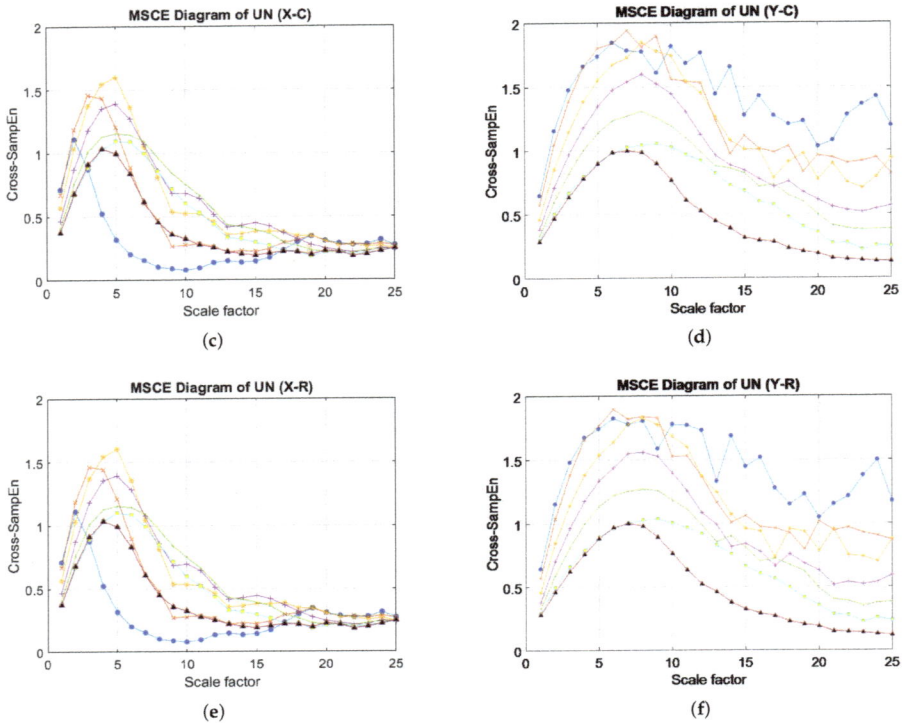

**Figure 4.** MSCE diagrams for undamaged condition: (**a**) left bay, X-direction; (**b**) left bay, Y-direction; (**c**) center bay, X-direction; (**d**) center bay, Y-direction; (**e**) right bay, X-direction; (**f**) right bay, Y-direction.

### 4.1. Single-Story, Single-Bay, Multidirectional Damage: Case 4 (4XY-L)

Figure 5 illustrates the MSCE diagrams for Case 4. In the X-direction, the rise of the curve for the fourth floor is evident in the MSCE diagrams of all bays, whereas the remaining curves are in similar positions as in the healthy condition. In the Y-direction, the rise of the fourth-floor curve and the dropping of the remaining curves are also noticeable in the diagrams of all bays. Furthermore, Figure 5b shows a large gap between the fourth-floor curve and the remaining curves at scale factors ranging from 8–25. These results indicate damage on the fourth floor in both directions, with a higher degree of damage possibly occurring on the left bay.

The DI analysis results shown in Figure 6 indicate that the fourth floor sustained damage in the X- and Y-directions. Figure 6c,d enables a closer evaluation of the DI values of the fourth floor. In both directions, the bar representing the left bay is higher than the remaining bars, indicating that this bay could have sustained more damage. In the Y-direction, the DI values are shown to be higher, and the differences among bays are shown to be more significant than those in the X-direction. Because the analyzed model involved three bays in the X-direction and a single bay in the Y-direction, the removal of bracings in the short (Y) axis could foster a larger impact on its stiffness; therefore, more significant differences could be observed.

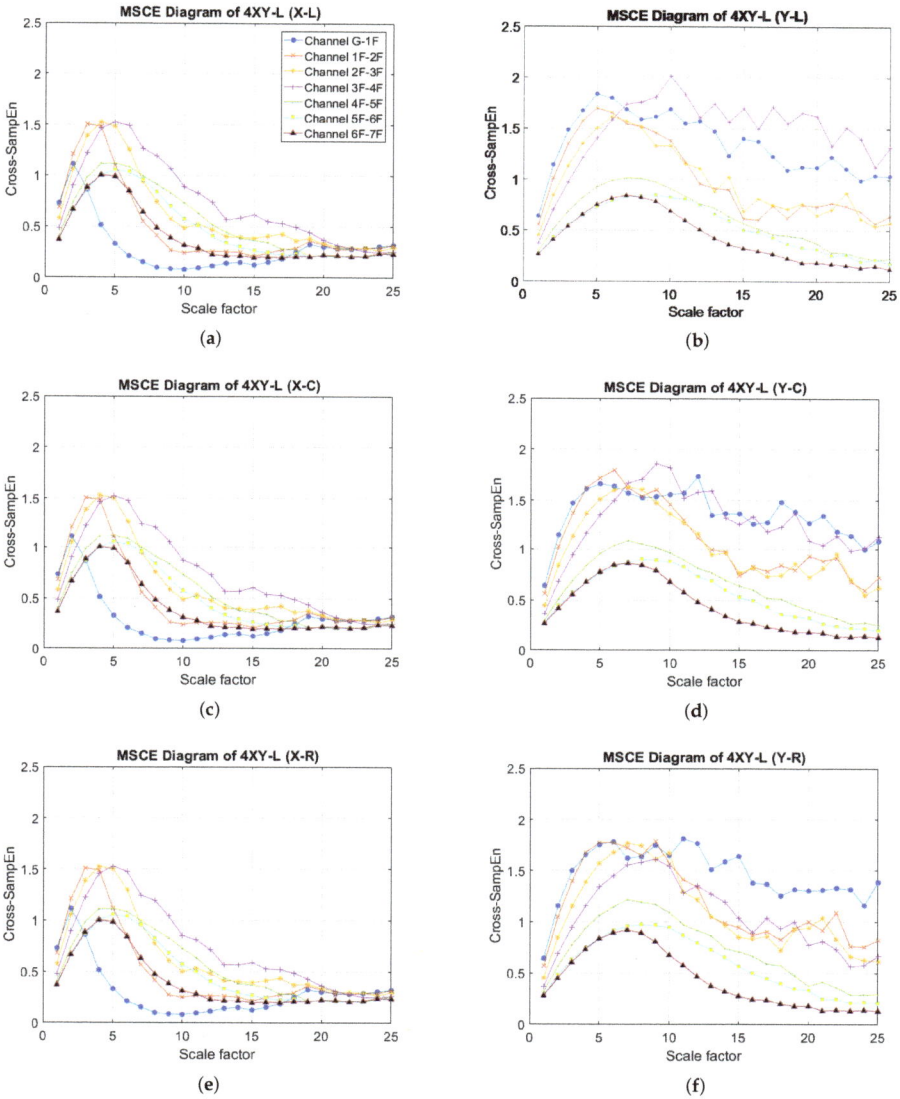

**Figure 5.** MSCE diagrams for Case 4: damage on fourth floor, left bay, X- and Y-directions: (**a**) left bay, X-direction; (**b**) left bay, Y-direction; (**c**) center bay, X-direction; (**d**) center bay, Y-direction; (**e**) right bay, X-direction; (**f**) right bay, Y-direction.

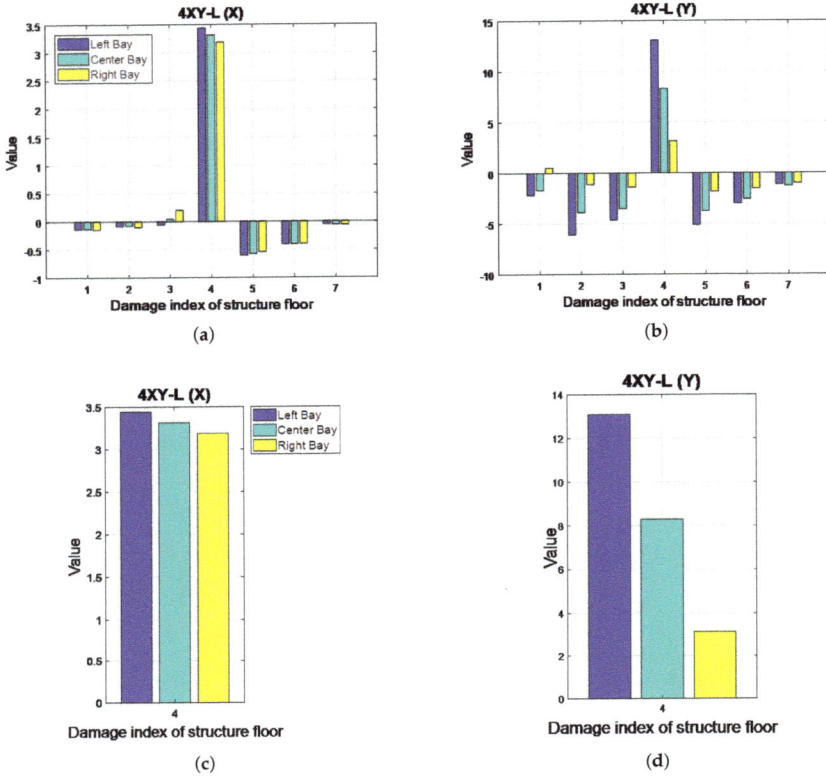

**Figure 6.** Damage index (DI) diagrams for Case 4: (**a**) X-direction; (**b**) Y-direction; (**c**) close up of DI diagrams for the fourth floor in Case 4: X direction; (**d**) Y-direction.

### 4.2. Single-Story, Multi-Bay, Multidirectional Damage: Case 8 (3X-R & 3Y-C)

The results for the MSCE analysis of Case 8 are shown in Figure 7. In the X-direction, the MSCE diagrams for the three bays are very similar to one another. The figure reveals that the curves for the third floor are in higher positions than they are in the diagram for the healthy condition. Furthermore, at scales ranging from 4–24, the third-floor curves are shown to maintain significantly higher positions than the remaining curves. In the Y-direction, the curves for the third floor are also shown to rise relative to their original positions in the diagram for the undamaged condition, with the difference being slightly higher for the MSCE diagram for the center bay (Figure 7d). Case 8 was therefore classified as involving damage in both directions of the third floor.

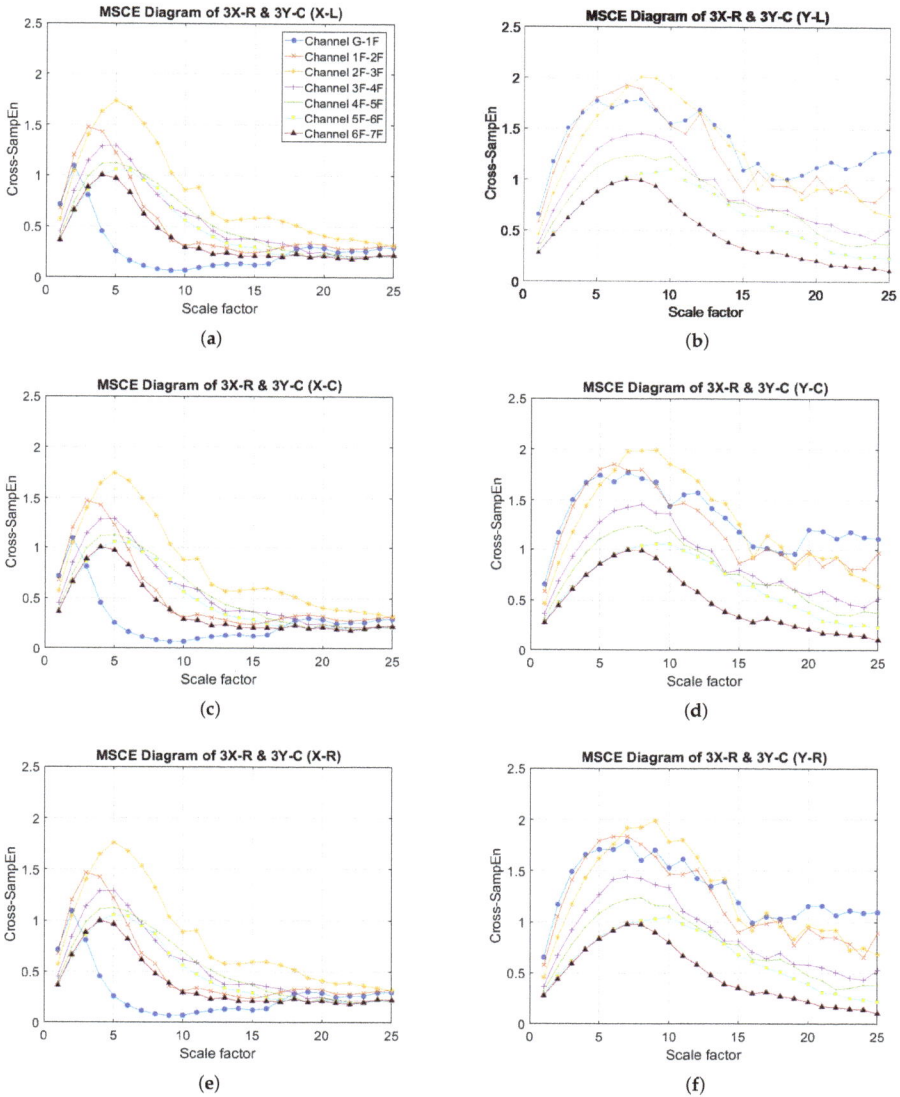

**Figure 7.** MSCE diagrams for Case 8: damage on the third floor, right bay, X-direction and third floor, center bay, Y-direction: (**a**) left bay, X-direction; (**b**) left bay, Y-direction; (**c**) center bay, X-direction; (**d**) center bay, Y-direction; (**e**) right bay, X-direction; (**f**) right bay, Y-direction.

The DI diagrams for Case 8 are illustrated in Figure 8. In the X-direction, a positive DI value is shown for the third floor, with the right bay bar revealing the highest DI. Although the second floor also had a positive DI, it was excluded because the value fell under the threshold. Because damage happened on the third floor, an effect on the adjacent second floor is natural due to the production of a rigid body response. The DI diagram for the Y-direction also reveals a positive DI for the third floor. In this case, the center bay bar is shown to be the highest. These results, which can be examined more closely in Figure 8c,d, confirm the MSCE analysis findings, with a further indication of the possible damaged bay.

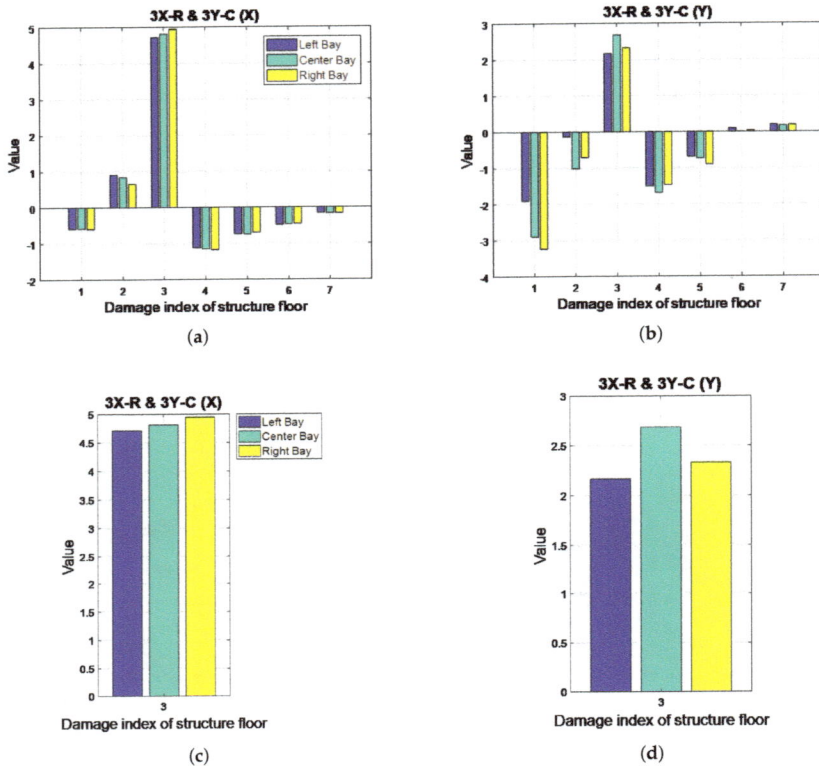

**Figure 8.** DI diagrams for Case 8: (**a**) X-direction; (**b**) Y-direction; close-up of DI diagrams for the third floor in Case 8: (**c**) X-direction; (**d**) Y-direction.

### 4.3. Two-Story, Single-Bay, Single-Direction Damage: Case 11 (1Y-R & 5Y-R)

The results obtained from the MSCE analysis of Case 11 are presented in Figure 9. In the X-direction, the MSCE curves for all bays remain almost identical to those of the undamaged condition; therefore, no damage occurred in the X-direction. Figure 9b illustrates the MSCE curves for the left bay in the Y-direction, revealing that that the curve for the first floor is at a higher position than the remaining curves at most scales. Furthermore, at scales of 5–15, a rise in complexity can be observed for the fifth floor. The changes in the curves for the first and fifth floors are even more evident in the MSCE diagrams for the center and right bays (Figure 9d,f). Therefore, the first and fifth floors were damaged in the Y-direction. Additionally, the curve for the fifth floor in the right bay displays higher entropy values than in the center bay for scales ranging from 10–25, suggesting a higher degree of damage to the right bay.

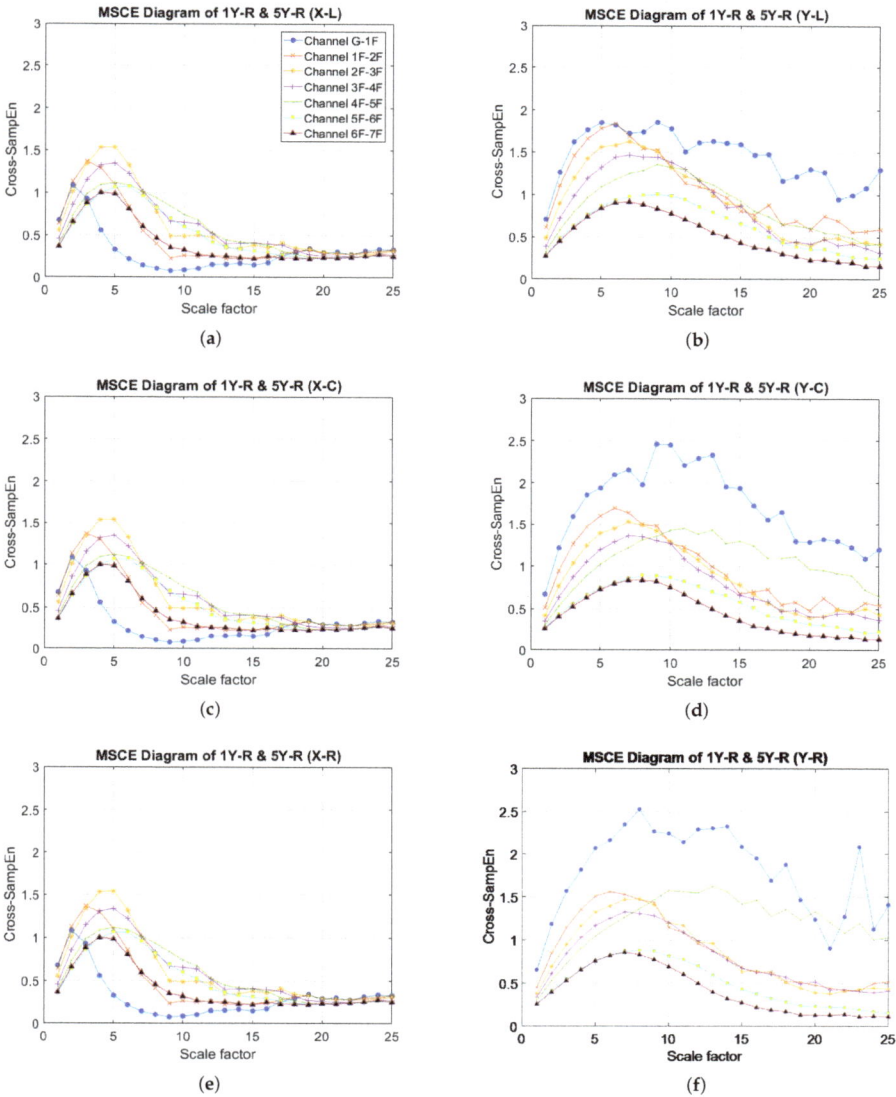

**Figure 9.** MSCE diagrams for Case 11: damage on first floor, right bay, Y-direction and fifth floor, right bay, Y-direction: (**a**) left bay, X-direction; (**b**) left bay, Y-direction; (**c**) center bay, X-direction; (**d**) center bay, Y-direction; (**e**) right bay, X-direction; (**f**) right bay, Y-direction.

The DI analysis results are shown in Figure 10, indicating no damage in the X-direction. Moreover, in the Y-direction, the first and fifth floors are demonstrated to have positive DI values. Figure 10c shows a close-up of the DI values obtained for the first and fifth floors in the Y-direction; the bars representing the right bay exhibit the highest values. Therefore, the first and fifth floors were damaged in the Y-direction, with the right bay sustaining a higher degree of damage. Hence, the MSCE analysis results are confirmed for this case.

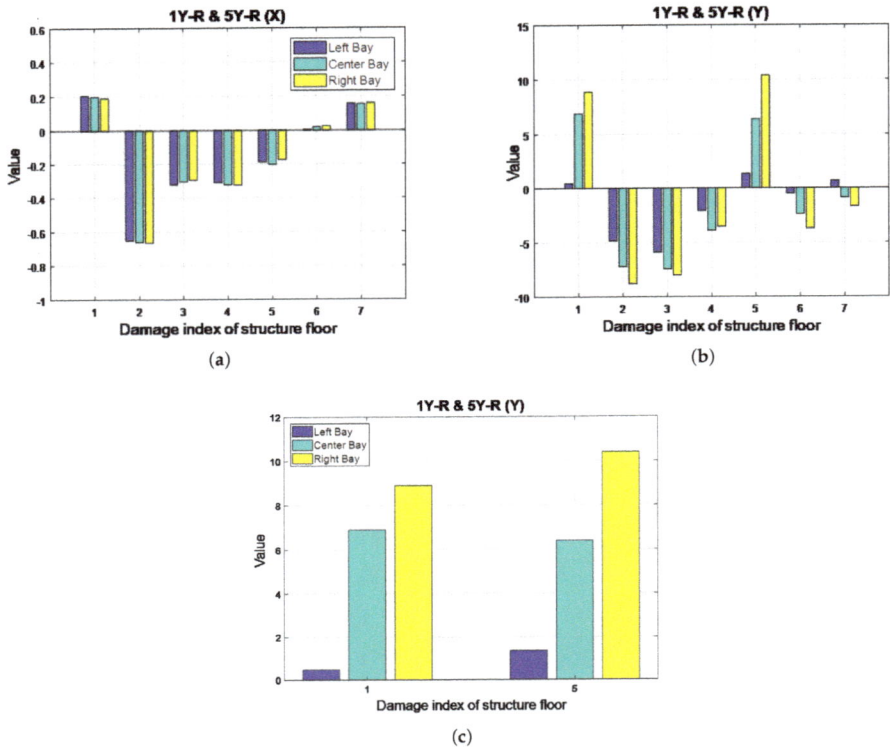

**Figure 10.** DI diagrams for Case 11: (**a**) X-direction; (**b**) Y-direction; (**c**) close-up of DI diagrams for the first and fifth floors in Case 11 (Y-direction).

### 4.4. Two-Story, Multi-Bay, Single-Direction Damage: Case 16 (2X-L & C, 6X-C & R)

Figure 11 illustrates the MSCE curves obtained from the analysis of Case 16. The curves for the second and sixth floors are shown to clearly separate and maintain a wide gap from the remaining curves in the MSCE diagrams obtained from analysis of the three bays in the X-direction. The curves for the second and sixth-floor reach their peaks at scales 5 and 7, respectively. In the Y-direction, no evident changes are shown in the diagrams for any bay; therefore, damage was determined to have occurred on only the second and sixth floor in the X-direction.

The DI diagrams for Case 16 are shown in Figure 12. In the X-direction, positive DI values are shown for the second and sixth floors, thus resulting in these floors being classified as damaged. The DI value of the second floor is higher than that in the sixth floor because the damage on the second floor can relatively affect the global behavior of the structure compared with similar damage conditions. In this particular case, the second floor was damaged on the left and center bays, whereas the sixth floor was damaged on the center and right bays. The damaged bay could not be identified by the DI during the analysis of damage to a single floor in adjacent bays; however, the results could still indicate that the second and sixth floors received more damage in the left and right bays, respectively. Figure 12c enables a closer observation of these results. Finally, the DI diagram for the Y-direction indicates no damage on this axis.

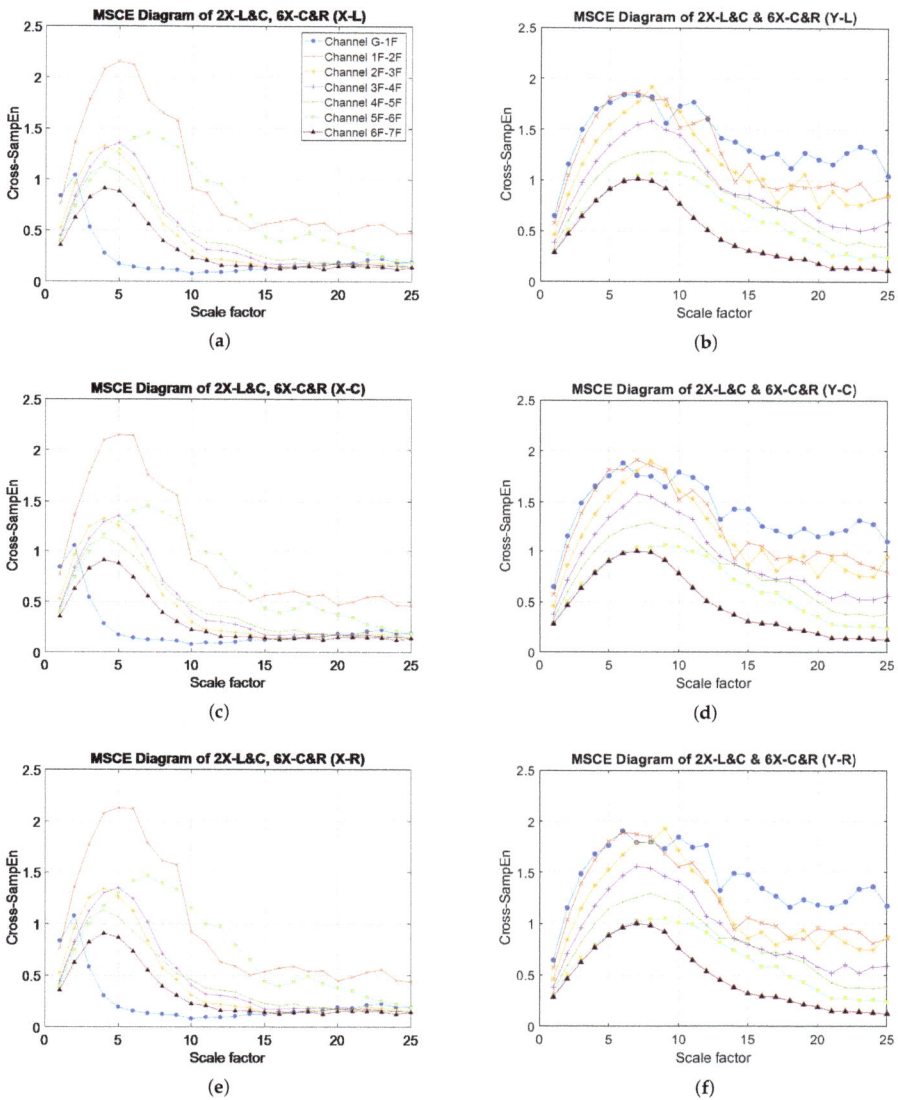

**Figure 11.** MSCE diagrams for Case 16: damage on second floor, left and center bays, X-direction and sixth floor, center and right bays, X-direction: (**a**) left bay, X-direction; (**b**) left bay, Y-direction; (**c**) center bay, X-direction; (**d**) center bay, Y-direction; (**e**) right bay, X-direction; (**f**) right bay, Y-direction.

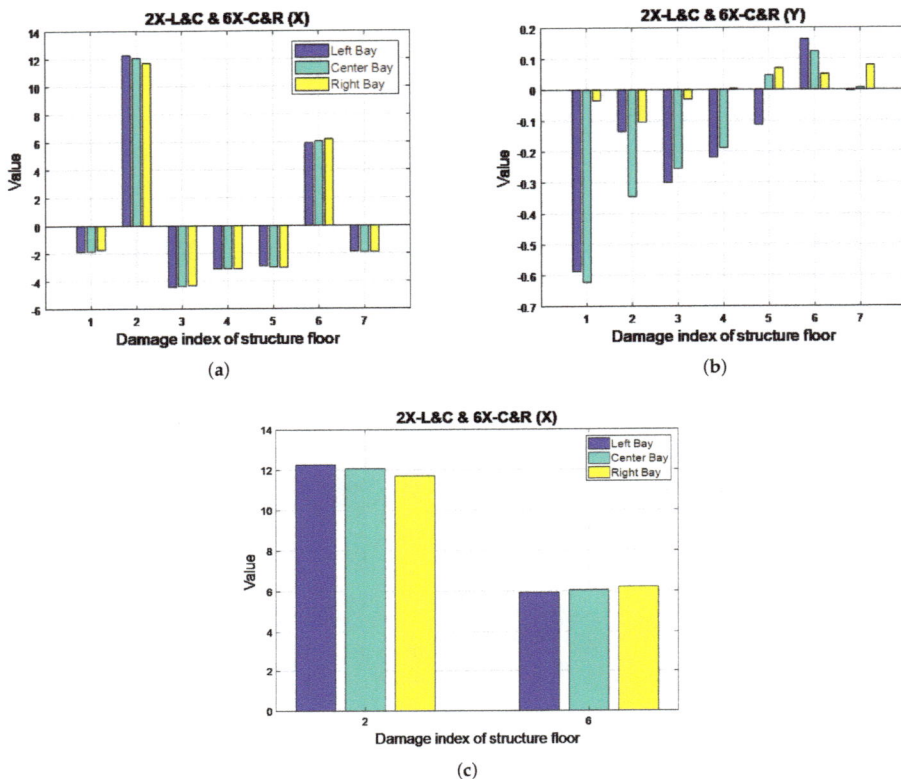

**Figure 12.** DI diagrams for Case 16: (**a**) X-direction; (**b**) Y-direction; (**c**) close-up of DI diagrams for the second and sixth floors in Case 16 (X-direction).

### 4.5. Multistory, Multi-Bay, Single-Direction Damage: Case 23 (3X-L & 4X-C & 5X-R)

The MSCE diagrams for Case 23 are illustrated in Figure 13. For the three bays in the X-direction, an increase in complexity is shown for the third and fourth floors. The fifth-floor curve is shown to remain in an almost identical position as in the diagram for the healthy condition, whereas the remaining curves are illustrated to decrease relative to their positions under the healthy condition. Although the fifth floor was incorrectly classified as undamaged, the third and fourth floors were determined as damaged in the X-direction. To determine the damaged bays, further examination was required through DI analysis. No evident changes were observed in the Y-direction.

The DI diagrams for this case are presented in Figure 14, revealing positive DI values for the third and fourth floors in the X-direction. The resulting diagram, a close-up of which is provided in Figure 14c, suggests that the third floor sustained more damage in the left bay, whereas the bar is slightly higher for the center bay on the fourth floor. The fifth floor was not identified as damaged. In the Y-direction, small positive DI values are shown in the figure; however, these values were determined to be negligible because they fell under the predetermined threshold of one, thereby confirming the MSCE analysis results.

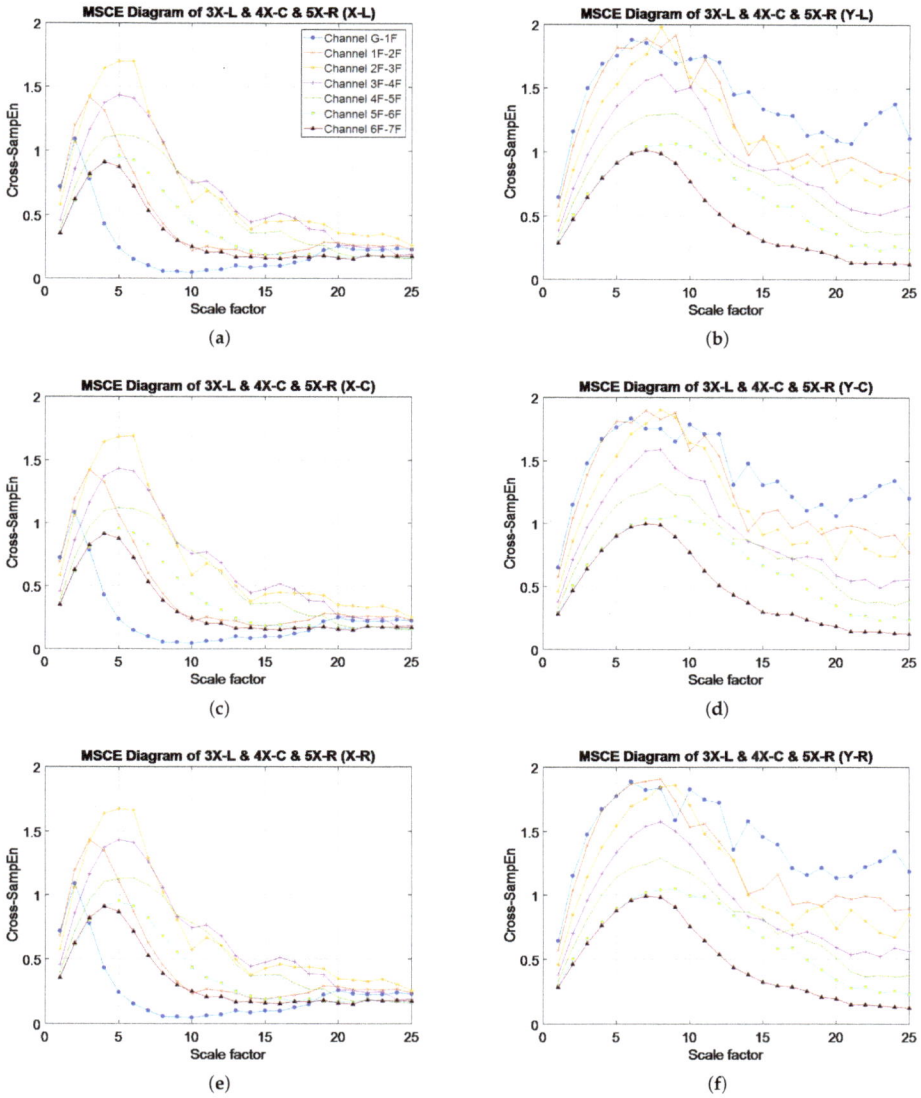

**Figure 13.** MSCE diagrams for Case 23: damage on the third floor, left bay, X-direction; fourth floor, center bay, X-direction; fifth floor, right bay, X-direction: (**a**) left bay, X-direction; (**b**) left bay, Y-direction; (**c**) center bay, X-direction; (**d**) center bay, Y-direction; (**e**) right bay, X-direction; (**f**) right bay, Y-direction.

(a)

(b)

(c)

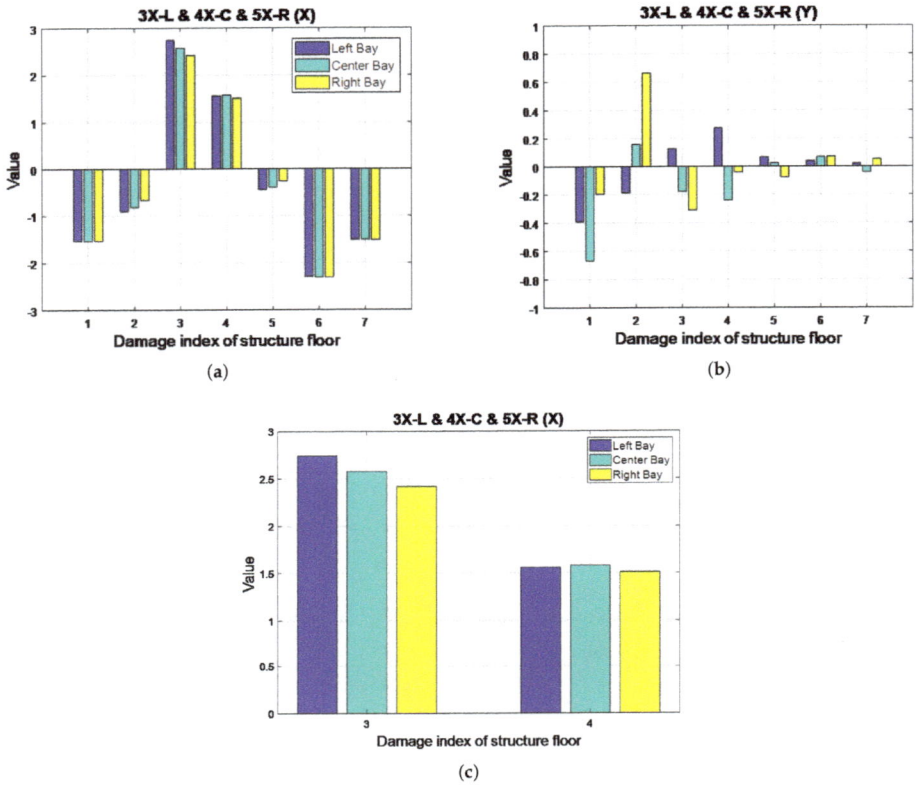

**Figure 14.** DI diagrams for Case 23: (**a**) X-direction; (**b**) Y-direction; (**c**) close-up of DI diagrams for the third and fourth floors in Case 23 (X-direction).

*4.6. Multistory, Multi-Bay, Multidirectional Damage: Case 26 (7XY-R & 4Y-L & 6Y-C)*

Figure 15 illustrates the MSCE curves for Case 26. In the X-direction, the curves for the seventh floor for all bays are shown to increase and reach their peaks at scale 5. However, the curves for the remaining floors are shown to drop or maintain their positions as in the diagrams for the healthy condition. Therefore, the seventh floor was determined to have sustained damage in the X-direction. In the Y-direction, the MSCE diagram for the left bay (Figure 15b), reveals that the curve for the fourth-floor increases dramatically. The trend for the first floor also increases, albeit slightly erratically. A slight increase in the curves for the sixth and seventh floors can be observed in the MSCE diagram for the center bay (Figure 15d). Moreover, a change is observed in the fourth-floor curve, but to a lower degree than that in the left bay MSCE diagram for the left bay. Regarding the MSCE diagram for the right bay (Figure 15f), the curves for the first, sixth and seventh floors increase significantly. On the basis of these results, the first, fourth, sixth and seventh floors sustained damage; however, this was false for the first floor.

Figure 16 illustrates the DI diagrams obtained for this case, with a close-up presented in Figure 16c,d. Positive DI values can be observed for the seventh floor in the X-direction, with the bar representing the right bay having the highest DI value. The first, fourth, sixth and seventh floors were determined to have sustained damage in the Y-direction. However, slightly erratic curves are shown in the figure for the first floor, which may have caused its misclassification. The DI diagrams suggest that the fourth and seventh floors received more damage in the left and right bays, respectively. For the sixth floor, the bar representing the right bay indicates the highest value, possibly because the damage to the center bay also affected the right bay because the bracing was shared by both. The results obtained in the DI analysis for this case demonstrate the accuracy of the proposed system.

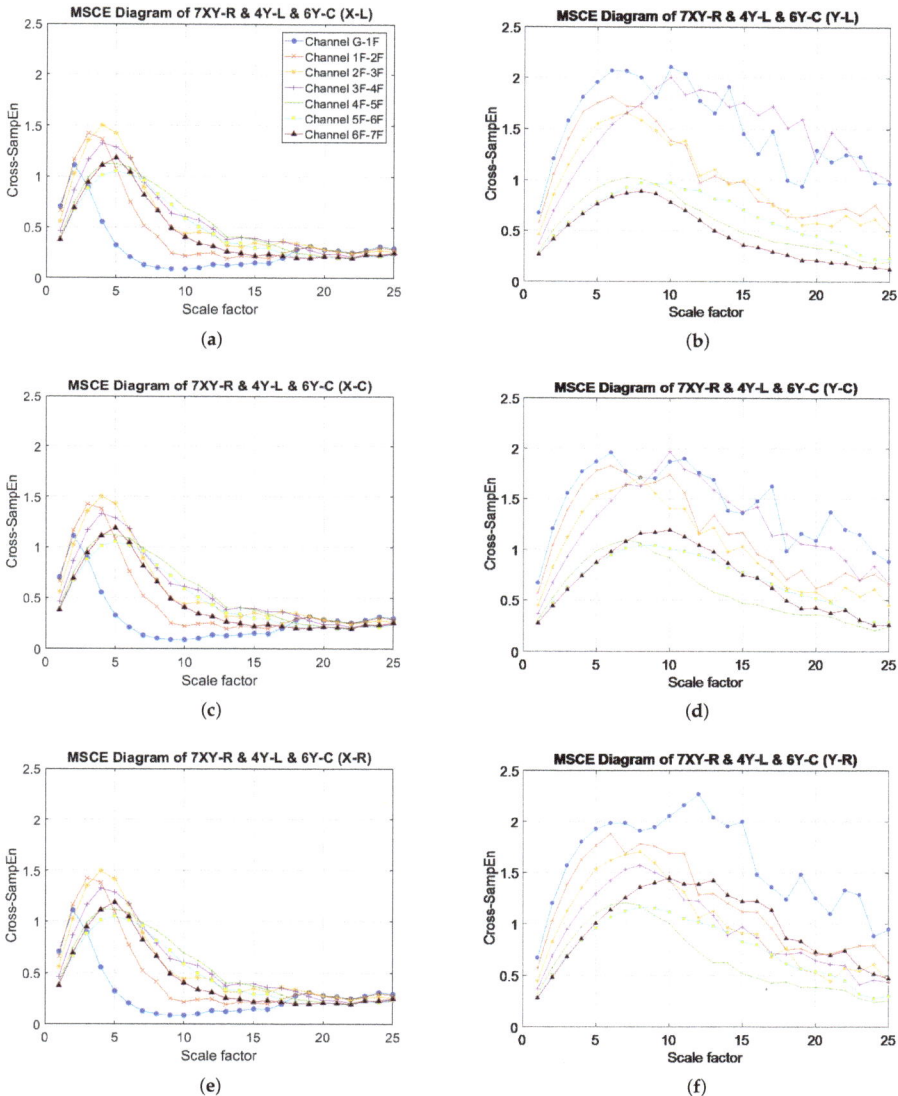

**Figure 15.** MSCE diagrams for Case 26: damage on seventh floor, right bay, X- and Y-directions; fourth floor, left bay, Y-direction; sixth floor, center bay, Y-direction: (**a**) left bay, X-direction; (**b**) left bay, Y-direction; (**c**) center bay, X-direction; (**d**) center bay, Y-direction; (**e**) right bay, X-direction; (**f**) right bay, Y-direction.

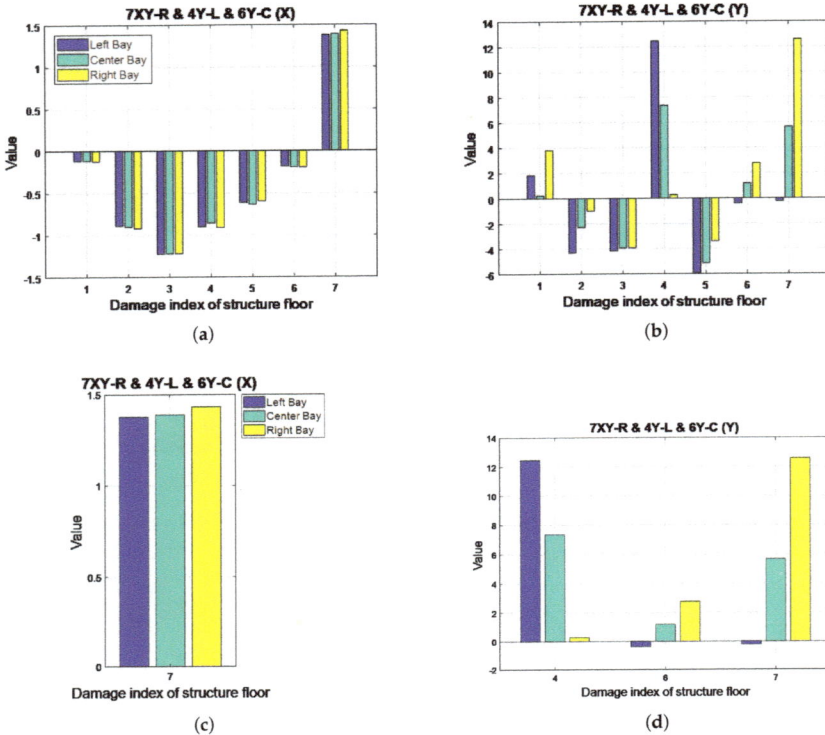

**Figure 16.** DI diagrams for Case 26: (**a**) X-direction; (**b**) Y-direction; (**c**) seventh floor, X-direction; (**d**) fourth, sixth and seventh floor, Y-direction.

### 4.7. General Discussion

In this study, 26 damage cases classified into 12 categories were examined to verify the feasibility of MSCE and DI analyses for damage detection in a complex three-bay structure. The complete results are summarized in Table 4. The DI values are lower when there are more damaged floors as the structural complexity is redistributed after different damage conditions. These were further analyzed in two stages. First, the DI results obtained for the identification of the damaged floors and damage directions were quantified through a precision and recall analysis. For the precision and recall analyses, the DI results for the X- and Y-directions were first classified into four categories: true positives (TP), representing damaged floors that have been correctly identified; false positives (FP), representing floors that have been misclassified as damaged; true negatives (TN), representing undamaged floors that have been correctly classified; and false negatives (FN), representing damaged floors that have been misclassified as undamaged. Precision and recall can then be calculated as follows:

$$\text{Precision} = \frac{TP}{TP + FP} \qquad \text{Recall} = \frac{TP}{TP + FN} \tag{20}$$

**Table 4.** Classification results of DI analysis.

| Case Number | Damage Group | Damage Case | Damage Index (X-Direction) | Damage Index (Y-Direction) |
|:---:|:---:|:---:|:---:|:---:|
| 1 | Single-story, | 5X-L | OK | OK |
| 2 | single-bay, | 3Y-C | OK | OK |
| 3 | single-direction | 7Y-R | OK | 1F [1] |
| 4 | Single-story, single- | 4XY-L | OK | OK |
| 5 | bay, multi-directional | 6XY-C | 6F [2] | OK |
| 6 | Single-story, multi- | 2X-L & 2X-C | 2F [2] | OK |
| 7 | bay, single-direction | 5Y-L & 5Y-C & 5Y-R | OK | 5F [2] |
| 8 | Single-story, multi- | 3X-R & 3Y-C | OK | OK |
| 9 | bay, multidirectional | 6X-L & 6Y-R | OK | 1F [1] |
| 10 | Two-story, single-bay, | 3X-L & 6X-L | OK | OK |
| 11 | single-direction | 1Y-R & 5Y-R | OK | OK |
| 12 | Two-story, single-bay, | 4X-C & 7Y-C | OK | OK |
| 13 | multidirectional | 2XY-R & 3XY-R | OK | $1F^1$ |
| 14 | Two-story, | 5X-R & 7X-L | OK | OK |
| 15 | multi-bay, | 2Y-C & 4Y-R | OK | 1F [1] & 2F [2] |
| 16 | single-direction | 2X-L & C, 6X-C & R | 2F [2] & 6F [2] | OK |
| 17 | Two-story, multi- | 4X-R & 2Y-L | OK | 1F [1] |
| 18 | bay, multidirectional | 6XY-R & 7XY-L | OK | 1F [1] & 2F [1] |
| 19 | Multistory, single- | 3X-L & 4X-L & 6X-L | 6F [3] | OK |
| 20 | bay, single-direction | 1Y-R & 4Y-R & 7Y-R | OK | OK |
| 21 | Multistory, single- | 4X-L & 5Y-L & 6Y-L | OK | 1F [1] & 2F [1] |
| 22 | bay, multidirectional | 1XY-C & 3XY-C & 5XY-C | 1F [3] & 5F [3] | 1F [3] & 3F [2] |
| 23 | Multistory, multi-bay, | 3X-L & 4X-C & 5X-R | 5F [3] | OK |
| 24 | single-direction | 6Y-L & 2Y-C & 7Y-R | OK | 1F [1] & 2F [3] |
| 25 | Multistory, multi-bay, | 1X-R & 2X-R & 1Y-L | 1F [3] & 2F [3] | OK |
| 26 | multidirectional | 7XY-R & 4Y-L & 6Y-C | OK | 1F [1] & 6F [2] |

[1] Indicates that the floor has been misclassified as damaged; [2] indicates that the damaged bay has not been successfully identified; [3] indicates that the damaged floor has not been detected.

## 4.7.1. Damage Location

High precision denotes few false positives, which represents the percentage of detecting the damage location correctly, and a high recall indicates few false negatives, which indicates the reliability of not missing the possible damage. The combined results for both directions are summarized in Table 5. The derived precision and recall were 83% and 87%, respectively. These results indicate that 83% of the floors and their respective directions classified by the DI as damaged were true positive. As most of the damages were simulated along the X-direction, better performance was achieved in the X-direction for 100% than the Y-direction, which is 73% in the study. Moreover, 87% of all actual damaged floors and their respective directions were correctly classified as damaged. The high-percentage of accuracy proves that the system can reliably detect any level of damage with only a small probability of misdiagnoses, which are false negatives. The results have demonstrated the capacity of the proposed SHM system to locate the damaged floor and direction of a large, three-bay numerical model.

**Table 5.** Precision and recall analysis results (X- and Y-directions).

| Direction | True Positives | False Positives | True Negatives | False Negatives | Precision | Recall |
|---|---|---|---|---|---|---|
| X | 25 | 0 | 151 | 6 | 100% | 81% |
| Y | 30 | 11 | 139 | 2 | 73% | 94% |
| Total | 55 | 11 | 290 | 8 | 83% | 87% |

#### 4.7.2. Damaged Bay Identification

A separate analysis of the identification accuracy of the damaged bays was then performed; therefore, the results pertaining to the damaged bays were not considered for the precision and recall analyses. Because the removal of bracings on a specific bay would inevitably affect those adjacent to it, the calculated DI values would be fairly close. The combined results for both directions are shown in Table 6, indicating an average identification accuracy of 75%. Better performance was achieved in the Y-direction, as it only has one bay in the Y-direction. For the complicated damage combinations in the X-direction, the accuracy dropped to 71%. Furthermore, the proposed system is limited when more than one bay is damaged. Therefore, these results are merely a suggestion as to which bay might have been affected more severely. With the support of the damaged bay identification, the proposed method can be easily extended to any complex structure.

**Table 6.** Identification accuracy of damaged bays (X- and Y-directions).

| Direction | Damage Instances | Correctly Identified Bay | Accuracy |
|---|---|---|---|
| X | 34 | 24 | 71% |
| Y | 34 | 27 | 79% |
| Total | 68 | 51 | 75% |

### 5. Conclusions

In this study, the feasibility of detecting damage in a complex three-bay, seven-story numerical model by SHM methods was examined. Through the proposed SHM system, damage locations can be rapidly and effectively detected by only measuring the velocity response data of the model. The ambient vibrations from the center of each floor before and after damage occurs were first recorded, and subsequently, the complexity of the signals can be analyzed by the MSCE method. The reliability and viability of the proposed SHM system were examined through the numerical analysis of 26 damage cases in 12 categories representing several degrees of damage severity. The results of the analyses were examined in two stages. First, the results pertaining to the damaged floor and direction indicate that 83% of the floors and their respective directions were truly damaged. Furthermore, 87% of all the actually damaged floors and their respective directions were correctly classified as damaged by the DI. Subsequently, the identification of the damaged bays was analyzed, and an identification accuracy of 75% was obtained. Identification of the damaged bays through the proposed SHM system is limited, especially when multi-bay damage exists on a single floor.

The obtained results verify the feasibility and further potential of the proposed SHM system for the detection and localization of damage in large and complex structures.

**Acknowledgments:** This work is supported by the Ministry of Science and Technology (MOST) (105-2221-E-009-056-) at Taipei, Taiwan.

**Author Contributions:** Tzu-Kang Lin conceived of and put forward the research ideas. Ana Gabriela Laínez carried out the numerical analysis and wrote the paper.

**Conflicts of Interest:** The authors declare no conflict of interest.

## References

1. Friswell, M.; Penny, J.; Garvey, S. Parameter subset selection in damage location. *Inverse Probl. Eng.* **1997**, *5*, 189–215.
2. Doebling, S.W.; Farrar, C.R.; Prime, M.B. A summary review of vibration-based damage identification methods. *Shock Vib. Dig.* **1998**, *30*, 91–105.
3. Wahab, M.A.; De Roeck, G. Damage detection in bridges using modal curvatures: Application to a real damage scenario. *J. Sound Vib.* **1999**, *226*, 217–235.
4. Maeck, J.; Wahab, M.A.; Peeters, B.; De Roeck, G.; De Visscher, J.; De Wilde, W.; Ndambi, J.M.; Vantomme, J. Damage identification in reinforced concrete structures by dynamic stiffness determination. *Eng. Struct.* **2000**, *22*, 1339–1349.
5. Chang, P.C.; Flatau, A.; Liu, S. Health monitoring of civil infrastructure. *Struct. Health Monit.* **2003**, *2*, 257–267.
6. Lam, H.F.; Yang, J. Bayesian structural damage detection of steel towers using measured modal parameters. *Earthq. Struct.* **2015**, *8*, 935–956.
7. Chen, Z.; Fang, H.; Ke, X.; Zeng, Y. A new method to identify bridge bearing damage based on Radial Basis Function Neural Network. *Earthq. Struct.* **2016**, *11*, 841–859.
8. Clausius R. On several convenient forms of the fundamental equations of the mechanical theory of heat. *Ann. Phys.* **1865**, *201*, 352–400.
9. Shannon, C.E. A mathematical theory of communication, Part I, Part II. *Bell Syst. Tech. J.* **1948**, *27*, 623–656.
10. Kolmogorov, A.N. New metric invariant of transitive dynamical systems and endomorphisms of Lebesgue spaces. *Dokl. Russ. Acad. Sci.* **1958**, *119*, 861–864.
11. Sinai, Y.G. On the notion of entropy of a dynamical system. *Dokl. Akad. Nauk. SSSR* **1959**, *124*, 768–771.
12. Pincus, S.M.; Gladstone, I.M.; Ehrenkranz, R.A. A regularity statistic for medical data analysis. *J. Clin. Monit. Comput.* **1991**, *7*, 335–345.
13. Pincus, S.M. Approximate entropy as a measure of system complexity. *Proc. Natl. Acad. Sci. USA* **1991**, *88*, 2297–2301.
14. An, Y.-H.; Ou, J.-P. Structural damage localisation for a frame structure from changes in curvature of approximate entropy feature vectors. *Nondestruct. Test. Eval.* **2014**, *29*, 80–97.
15. Richman, J.S.; Moorman, J.R. Physiological time-series analysis using approximate entropy and sample entropy. *Am. J. Physiol.-Heart Circ. Physiol.* **2000**, *278*, H2039–H2049.
16. Costa, M.; Goldberger, A.L.; Peng, C.K. Multiscale entropy analysis of complex physiologic time series. *Phys. Rev. Lett.* **2002**, *89*, 068102.
17. Costa, M.; Goldberger, A.L.; Peng, C.K. Multiscale entropy analysis of biological signals. *Phys. Rev. E* **2005**, *71*, 021906.
18. Zhang, L.; Xiong, G.; Liu, H.; Zou, H.; Guo, W. Bearing fault diagnosis using multi-scale entropy and adaptive neuro-fuzzy inference. *Expert Syst. Appl.* **2010**, *37*, 6077–6085.
19. Liu, H.; Han, M. A fault diagnosis method based on local mean decomposition and multi-scale entropy for roller bearings. *Mech. Mach. Theory* **2014**, *75*, 67–78.
20. Pincus, S.; Singer, B.H. Randomness and degrees of irregularity. *Proc. Natl. Acad. Sci. USA* **1996**, *93*, 2083–2088.
21. Fabris, C.; De Colle, W.; Sparacino, G. Voice disorders assessed by (cross-) sample entropy of electroglottogram and microphone signals. *Biomed. Signal Process. Control* **2013**, *8*, 920–926.
22. Lin, T.K.; Liang, J.C. Application of multi-scale (cross-) sample entropy for structural health monitoring. *Smart Mater. Struct.* **2015**, *24*, 085003.
23. Lin, T.K.; Tseng, T.C.; Lainez, A.G. Three-dimensional structural health monitoring based on multiscale cross-sample entropy. *Earthq. Struct.* **2017**, *12*, 673–687.
24. Wimarshana, B.; Wu, N.; Wu, C. Crack identification with parametric optimization of entropy & wavelet transformation. *Struct. Monit. Maint.* **2017**, *4*, 33–52.
25. Guan, X.; Wang, Y.; He, J. A probabilistic damage identification method for shear structure components based on Cross-Entropy optimizations. *Entropy* **2017**, *19*, 27.

entropy

MDPI

*Article*

# Complexity Analysis of Carbon Market Using the Modified Multi-Scale Entropy

Jiuli Yin [1,†], Cui Su [1,†], Yongfen Zhang [2,†] and Xinghua Fan [1,†,*]

[1]  Center for Energy Development and Environmental Protection Strategy Research, Jiangsu University, Zhenjiang 212013, China; yjl@ujs.edu.cn (J.Y.); 2211702020@stmail.ujs.edu.cn (C.S.)
[2]  College of Business, Shanghai University of Finance & Economics, Shanghai 200433, China; zhangyf@sou.edu.cn
*  Correspondence: fan131@ujs.edu.cn; Tel.: +86-511-8878-0164
†  These authors contributed equally to this work.

Received: 28 April 2018; Accepted: 1 June 2018; Published: 5 June 2018

**Abstract:** Carbon markets provide a market-based way to reduce climate pollution. Subject to general market regulations, the major existing emission trading markets present complex characteristics. This paper analyzes the complexity of carbon market by using the multi-scale entropy. Pilot carbon markets in China are taken as the example. Moving average is adopted to extract the scales due to the short length of the data set. Results show a low-level complexity inferring that China's pilot carbon markets are quite immature in lack of market efficiency. However, the complexity varies in different time scales. China's carbon markets (except for the Chongqing pilot) are more complex in the short period than in the long term. Furthermore, complexity level in most pilot markets increases as the markets developed, showing an improvement in market efficiency. All these results demonstrate that an effective carbon market is required for the full function of emission trading.

**Keywords:** complexity; entropy; carbon market; multi-scale entropy

---

## 1. Introduction

The carbon market is a market in which carbon emission allowances are traded. The price of carbon emission allowances determined by demand and supply in the market is the carbon price. The first carbon emissions trading scheme (ETS) was initiated by the European Union (EU) in 2005. There were 19 ETSs by the end of 2017, which was more than three times of its number in 2012. They covered over 15% of global carbon emissions which account for more than seven billion tons of greenhouse gas emissions equivalent. The coverage would double as China introduced its national carbon trading system in 2018 after more than four years of pilot work [1].

How well the carbon market performs is particularly important for traders, investors as well as policymakers [2]. In a well-performed market, prices at any point in time can "fully reflect" available information. This is the essence of the Efficient Market Hypothesis (EMH) given by Fama [3]. The EMH has been extensively tested in various markets, such as stock markets, various commodity futures markets, off-the-counter markets, bond markets, options markets [4]. However, the EMH has not been well tested in carbon markets.

Research emphasis on carbon markets has been paid to the EU ETS. Newell et al. [5] sum up the lessons of the carbon market and look ahead to global policy, pointing out that policymaking is essential to the carbon market. Fan et al. [6] study the complexity of the EU carbon market and concludes that the complexity of carbon market corresponds to the extreme socio-political events. Most existing studies confirm the weak efficiency of EU carbon markets. The EU spot carbon market is verified to be fully effective [7]. Yang et al. [8] believe the EU carbon trading market has been characterized by weak-form efficiency. Adopting the variance ratio method, Alberto et al. [9] find the weak efficiency of EUA carbon market in the second stage. However, some authors reject the weak market efficiency of three main exchanges under the EU ETS [10] and the Intercontinental Exchange between 2008 and 2011 [11]. Charles et al. [12] point out that the lack of the cost-of-carry relationship is the reason of inefficiency of main European carbon markets. However, the market efficiency is found to be improved over the period [13]. The EUA future carbon market is found to be not efficient by using event-study methodology [14] but is found to be efficient within one month [15]. Besides the spot and future markets, the efficiency of the EUA options market is also studied [2]. As to the China's ETS, Lo [16] believes that the implementation of carbon trading in China is of great significance. Zhao et al. [17] show that the market efficiency in China is not satisfactory although the country has made a preliminary achievement in system designs. In a later research, Zhao et al. [18] find signs of restoring market efficiency in four pilot carbon markets.

As known in the literature, there are many types of market efficiency, such as allocative efficiency, operational efficiency, informational efficiency [19]. A market is informational efficient if the current market price instantly and fully reflects all relevant available information. We limit the type of "market efficiency" to be "informational efficiency" in this study.

We apply complexity characteristics to measure the efficiency of carbon markets. The reasons lie in the following two aspects. Firstly, complexity characteristic of a nonlinear system not only embraces or is at least closely connected to all other data features [20] but also determines the characteristics of different internal factors and their relationships [21]. Secondly, the carbon market is regarded as a complex system in which traders have different strategic choices and act in complicated ways with mixed, and often intricate incentives [22].

Entropy, together with fractality and chaos, is generally taken as the measurement of complexity [22]. The concept of entropy was originally developed from the classic Shannon entropy [23]. There are various entropy measures such as approximate entropy [24], E-R entropy [25], Kolmogorov–Smirnov entropy [26] and multi-scale entropy(MSE) [27]. MSE outperforms the previous ones in that it considers the multi-scale property of the underlying system, thereby avoiding misguiding results for complexity multi-scale system. It is pointed out that fuzzy entropy provides improved evaluation of signal complexity [28]. However, this study still apply the sample entropy for the continence to compare the complexity between Chinese pilots markets and the EU carbon market [6].

The modified MSE (MMSE) [29] is an improvement of the MSE. The implementation of MSE consists of two steps: (1) Scale extraction and (2) entropy estimation. The coarse-grain procedure [26] is used to determine the scales of data. The sample entropy (SampEn) is generally employed as the entropy measure. It is pointed out that sample entropy estimation presents larger variance for greater scale factor because the coarse-grained time series becomes shorter [30]. The moving average algorithm [29] handles this issue in a good manner for short time series. Using this method, one can estimate the entropy with a better accuracy or get less undefined entropy.

Before the end of 2013, China had already launched seven ETS pilots in five biggest cities of Beijing, Chongqing, Tianjin, Shanghai, and Shenzhen, and two provinces of Hubei and Guangdong [1]. These seven pilots became fully functional before the end of 2015. Considering the relatively short time of the establishment of the market, this paper studies the complexity of carbon market by the modified MSE method. We describe the methods used in Section 2. Then we present the experiment results and discussion in Section 3. The last section provides the overall conclusions.

## 2. Methods

The modified MSE (MMSE) depends on the calculation of the sample entropy in a certain range of scales [29]. The essence of the modified multi-scale entropy method is trying to largely reserve the data length by using a moving-average process. In this process, one generates the new time series by moving a window with a length of the given scale point by point through the entire time series. The system dynamics is now presented by the newly generated time series on different scales. Then the sample entropy algorithm is applied to the generated time series and the MSE is obtained.

## 2.1. Moving-Average

For the return series of carbon price $P(i), i = 1, 2, \cdots, N$, the moving averaged time series $x(i, \tau)$ at the scale factor $\tau$ is calculated as

$$x(i, \tau) = \frac{1}{\tau} \sum_{j=i}^{i+\tau-1} P(j), \qquad 1 \leq i \leq N - \tau + 1. \tag{1}$$

## 2.2. Sample Entropy for Each Moving Average Time Series

Sample Entropy [31] is equal to the negative natural logarithm of an estimate of the conditional probability that subseries of length $m$ that match pointwise within some tolerance, $r$, will also match when their length is increased by one.

Given the moving averaged sequence $\{x(i), i = 1, \cdots, n = N - \tau + 1\}$ at the scale factor $\tau$ (for clarity, we drop the symbol $\tau$ in this subsection), we first define a subseries of length $m$ as

$$X_i = \{x(i), x(i+1), \ldots, x(i+m)\} \qquad i = 1, 2, \ldots, n - m. \tag{2}$$

Let the distance $d_m[X_i, X_j]$ between template vectors $X_i$ and $X_j$ with length $m$ be the largest absolute difference between their corresponding elements

$$d_m[X_i, X_j] = \max_{0 \leq k \leq m-1} |x(i+k) - x(j+k)|, 1 \leq i, j \leq n - m, i \neq j. \tag{3}$$

Then we denote $C^m(r)$ and $C^{m+1}(r)$ respectively the number of pairs of series of length $m$ and $m+1$ having distance smaller than $r$.

Finally, the sample entropy is calculated by

$$SampEn(m, r, N) = -\ln \frac{C^{m+1}(r)}{C^m(r)}. \tag{4}$$

It is clear that $SampEn(m, r, N)$ will be always non-negative. A smaller value of sample entropy indicates less level of complexity or more self-similarity in a data set.

Empirically, the value of entropy is not very dependent on the specific values of $m$ and $r$ [32]. For small $m$, especially $m = 2$, estimation of SampEn can be achieved with relatively few points [24]. Considering the length of the sample in this study, we select the case $m = 2$ and only this one. The literature uses $r$ values between 0.1 and 0.25 of the standard deviation [24,33]. We use $r = 0.15\sigma$, where $\sigma$ is the standard deviation of the data points. This study calculates sample entropy values for the scale factors from 1 to 60 ($\tau = 1, 5, 20, 60$ represent scale of one day, one week, one month, and one quarter respectively).

## 3. Experimental Results and Discussion

### 3.1. Data

We consider seven pilot carbon markets in China, namely, Beijing, Chongqing, Guangdong, Hubei, Shanghai, Shenzhen, and Tianjin market. These markets were established and functioned sequentially during 2013 and 2014 with the presumed purpose of providing for experiences to its future national scheme. The sample data are daily trading prices obtained from the Carbon Trading Network (http://k.tanjiaoyi.com/), covering the period from the first trading date of each market to the end of the year 2017. Because there are multiple vintage years of carbon allowances in the Shenzhen market, 2014 Shenzhen carbon emission allowances, known as SZA-2014, is selected as the representative carbon price for Shenzhen pilot. The return series $x_t = log(p_t) - log(p_{t-1})$ is adopted as the experiment data to calculate entropy value, where $p_t$ denote the carbon prices on day $t$.

Figure 1 presents a graphical representation of the sample data. The price in Beijing market is the highest in the most time and fluctuates around 50 Yuan per tonne. The carbon price in Guangdong fluctuates the valiant, ranging from 7.53 to 77 Yuan. Hubei market has a small volatility while markets in Chongqing and Shenzhen have many horizontal segments. The carbon price looks like a radome walk in Beijing, Hubei, and Shenzhen markets while those show a declining trend in Chongqing, Guangdong, and Tianjin markets. All these observations infer that those pilot markets develop at different levels, with Beijing (Hubei and Shenzhen ) market be the relatively high-level of complexity while others the low.

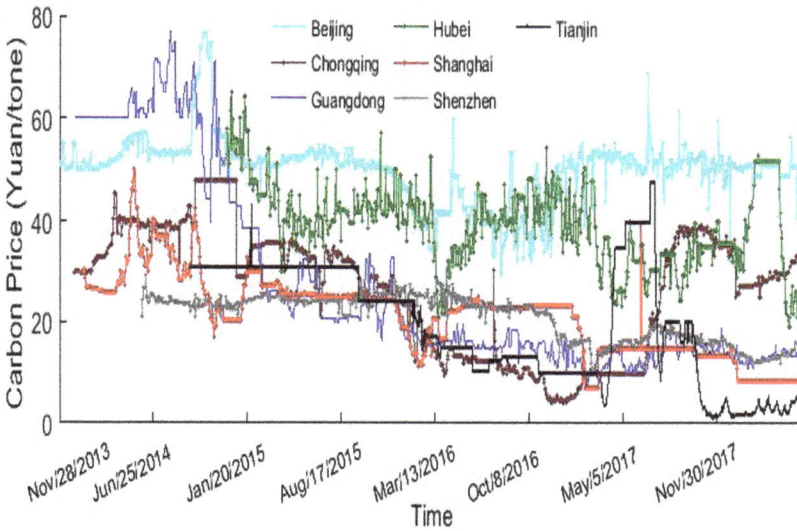

**Figure 1.** The daily carbon price in seven pilot markets

### 3.2. Complexity Analysis in Overall Time

We consider the complexity from the overall perspective, that is, using the classical sample entropy of the time series without any coarse-grain procedures. For comparison, we consider a white noise with the length to be the largest length of data in the seven pilots. Figure 2 shows the results. Entropy in the pilot markets are all far smaller than that of the white noise ( less than a half). Six entropies out of seven are smaller than one with the entropy of Chongqing close to zero and that of Hubei is close to one, meaning that all pilot carbon markets present a low-level complexity. The low-level complexity indicates that all the pilot markets are quite immature in lack of market efficiency. However, marked differences exist among individual markets. It is obvious that Hubei isolates from other markets with the highest entropy of 1.1674, while Chongqing market, accompanied with the market of Tianjin, has the lower entropy around 0.1. We divide the range of the MSE of the pilots equally into three intervals, namely, $(0, 0.4]$, $(0.4, 0.8]$ and $(0.8, 1.2]$ and refer to them as small, medium and large entropy interval respectively. The remaining markets show a medium value of complexity ranging from 0.4–0.8.

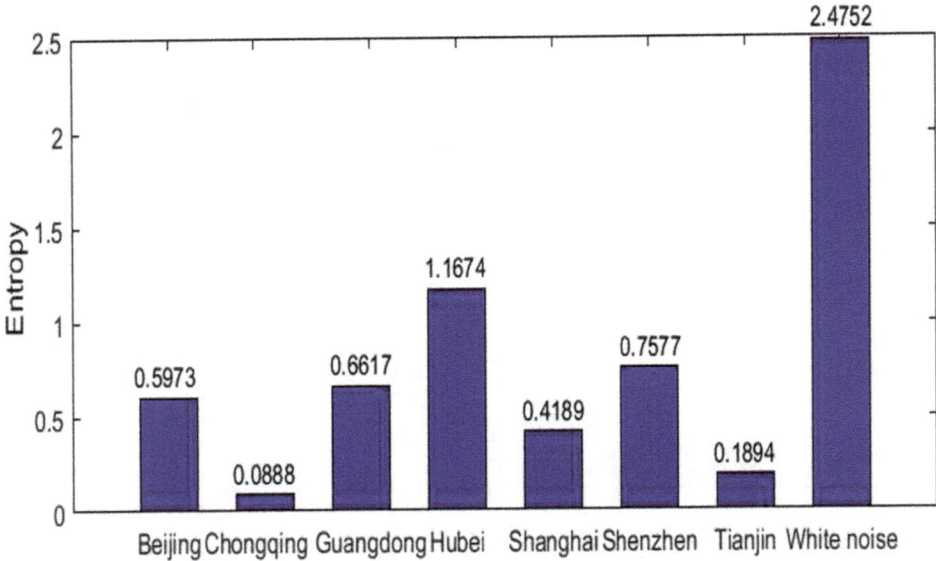

**Figure 2.** Sample entropy for whole data dynamics.

*3.3. Complexity Analysis in Multiple Scales*

Multiple-scale entropy reflects the complexity from different time scales. Figure 3 reports the results of the multi-scale entropy analysis. Compared with the multiple-scale entropy, the overall perspective is the special case when the time scale is one. This is verified by the coincidence of values in Figure 2 and those in Figure 3 with time scale $\tau = 1$.

First, the entropy for every pilot market is monotonically decreasing with the time scale, indicating a decline in complexity level. Second, the curves of the entropy almost level when the time scale is greater than 20. This critical time scale was also found in the European carbon market [6]. These two results suggest that there are different factors affecting the efficiency of the pilot carbon markets. Within the small time scale (shorter than one month) , the inner market features might present more irregular factors leading to the price fluctuations, while fluctuations in a larger time scale (longer than one month) are more related to certain regular (conventional and not occasional) factors or smooth trends.

Third, the rank of complexity changes for some pilot markets. Shenzhen market ranks the first in complexity level for most of short time scales (except for that Hubei ranks the first for the smallest three scales) but is surpassed by Gongdong market for long time scales. Chongqing and Tianjin markets always rank the last for both the short and long time scales. This result suggests that Shenzhen market is the best one in short-term fluctuation while Guangdong market is the best in long-term development, but both Chongqing and Tianjin markets perform the poorest, either in market fluctuation or development.

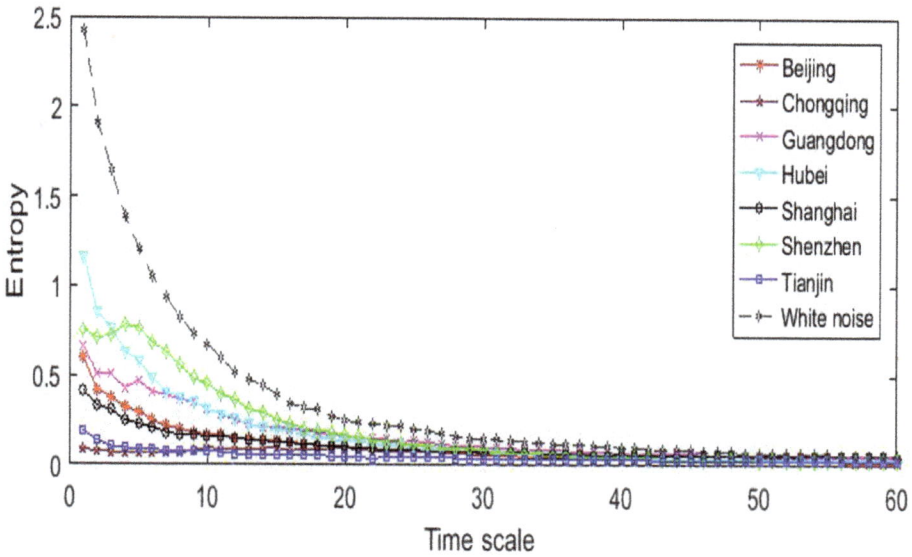

**Figure 3.** The behavior of multi-scale entropy.

### 3.4. Evolution of Complexity

To reflect the dynamics of local situations, rolling window technique is applied to analyze the time-variation of complexity. We use a fixed window width $N_w = 250$, which is about a year. The step length of the window is set as a single trading day. In detail, we compute the sample entropy of the first window, covering the series from the first data point to the 250th point. Then, the window slides forward by deleting the first point and adding the 251th point. Fixing a scale $\tau$ in the interval $[2, 60]$, we assign the entropy of the window to its middle point. The results are shown by color diagrams in the lower panel of Figure 4, which also displays the graphes for the corresponding price return series.

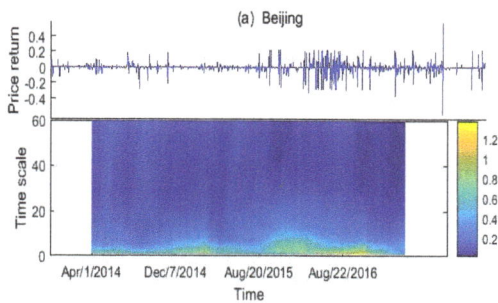

**Figure 4.** *Cont.*

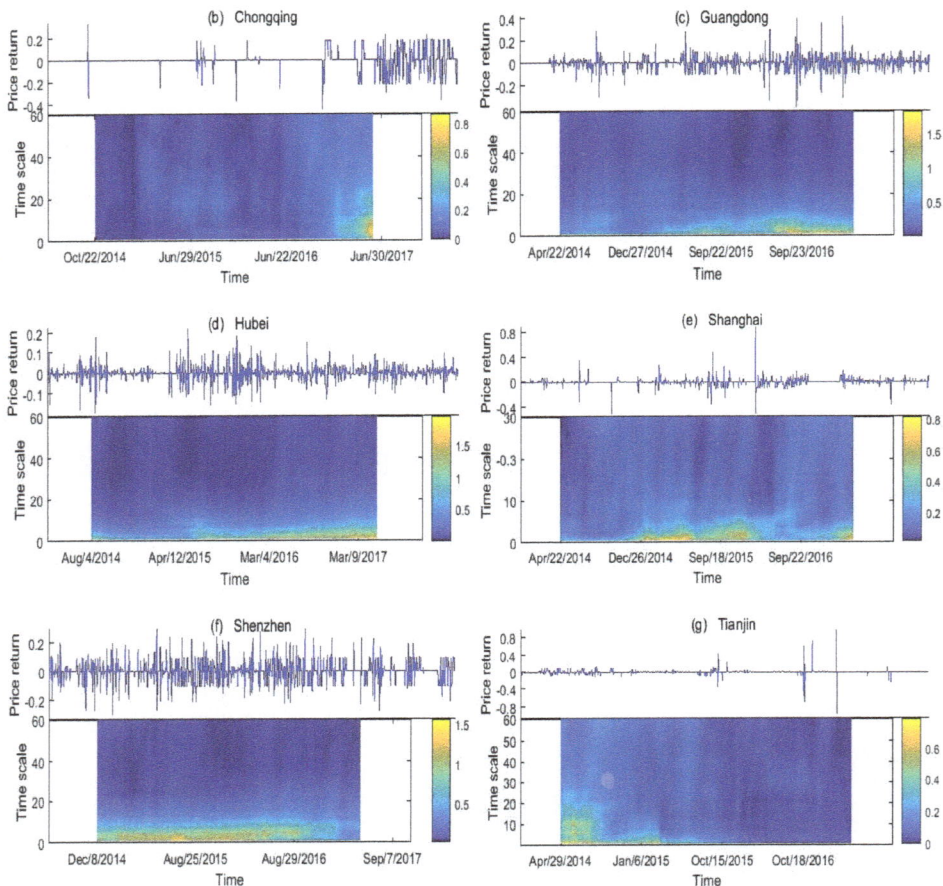

**Figure 4.** Time dependency of the return series (**upper panel**) and the dependency of the sample entropy to the time and scale (**lower panel**). (**a**) Beijing; (**b**) Chongqing; (**c**) Guangdong; (**d**) Hubei; (**e**) Shanghai; (**f**) Shenzhen; (**g**) Tianjin.

There are several common features for the entropy diagrams in Figure 4. First, the diagrams are primarily dominated by large part of small entropies (*entropy* < 0.4). This result confirms that all the markets present a low-level complexity, in lack of mark efficiency. Second, larger entropy locates in the lower part of the diagrams while smaller entropy the upper part. This confirms the finding in Section 3.2 that these markets have a higher complexity in small time scale than in large scale. Third, we note one or three narrow yellow peaks in each diagram. The peak values are higher than 0.8 for Beijing, Guangdong, Hubei, Shenzhen and around 0.7 in other markets. Near those yellow peaks, there are larger regions of other peaks, lower (typically 0.4) than the yellow ones but significantly higher than the other values. Except for Shenzhen and Tianjin markets, those yellow peaks appear at the right end of the time axis, showing global increases in the multi-scale sample entropy with the increase in Chongqing being the most obvious. Those increases indicate that the five markets have improved their market efficiency recently. Fourth, we note a relative large area of yellow peaks for Shenzhen and Hubei markets (Figure 4d,f), covering more than half of the time period at the lower part of the diagrams. This indicates the relatively higher level complexity of the two markets than others.

*Entropy* **2018**, *20*, 434

At last, entropy is positively correlated with fluctuation intensity rather than the amplitude of the return series. The more intensive fluctuation, the greater entropy. An apprarent large MSE is observed (Figure 4b) when the return series fluctuates highly after a certain date. Contrast to this, the entropy is close to zero when the return series is mainly flat. Shenzhen and Hubei markets (Figure 4d,f) have much higher entropy than Tianjin (Figure 4g) although amplitudes of their return series are only about one thirds of that in Tianjin.

## 4. Conclusions

We presented a detailed investigation of the correlation of market performance on different time scales. The modified multi-scale entropy by using moving average algorithm was applied to present the complexity of pilot carbon markets in China due to the short length of the sample data.

Our analysis indicates an overall low complexity in those carbon markets, far smaller than that of the Europe carbon market [6]. This inferior market efficiency to that in Europe may be caused by the economic discourse of climate change and excessive state intervention in China's carbon trading political economy. However, the complexity role was verified in this study by the fact that the complexity of the carbon market (except Chongqing) is higher in small time scales than in large scales. A same critical time scale with the European carbon market was found. Our results also show that complexity is improved as the pilot carbon markets developed while differences in complexity exist.

It is should be noticed that for the time series, the wavelet transform technique [34] may be more efficient in many applications than the moving average technique. One might obtained some different interesting results by using different method.

Results obtained in this study provide a robust basis for investment decisions and policy arrangements. It is also quite important to establish a more consistent quantitative scenario of the recent past.

**Author Contributions:** J.Y. and C.S. conceived and designed the experiments; X.F. and J.Y. performed the experiments; C.S. and Y.Z. analyzed the data; J.Y. and Y.Z. analyzed the experiment result. X.F. and C.S. wrote the paper. All authors have read and approved the final manuscript.

**Funding:** This research was funded by [National Natural Science Foundation of China] grant number [Nos. 71673116 and 71690242] and [Humanistic and Social Science Foundation from Ministry of Education of China] grant number [16YJAZH007].

**Conflicts of Interest:** The authors declare no conflict of interest. The founding sponsors had no role in the design of the study; in the collection, analysis, or interpretation of data; in the writing of the manuscript, and in the decision to publish the results.

## References

1. Pizer, W.A.; Zhang, X. *China's New National Carbon Market*; Working Paper of Nicholas Institute for Environmental Policy Solutions; Duke: Durham, NC, USA, 2018.
2. Krishnamurti, C.; Hoque, A. Efficiency of European emissions markets: Lessons and implications. *Energy Policy* **2011**, *39*, 6575–6582. [CrossRef]
3. Fama, E. Efficient capital markets: A review of theory and empirical work. *J. Financ.* **1970**, *25*, 383–417. [CrossRef]
4. Jensen, M.C. Some anomalous evidence regarding market efficiency. *J. Financ. Econ.* **1978**, *6*, 95–101. [CrossRef]
5. Newell, R.G.; Pizer, W.A.; Raimi, D. Carbon market lessons and global policy outlook. *Science* **2014**, *343*, 1316–1317. [CrossRef] [PubMed]
6. Fan, X.; Li, S.; Tian, L. Complexity of carbon market from multi-scale entropy analysis. *Physica A* **2016**, *452*, 79–85. [CrossRef]
7. Seifert, J.; Uhrig-Homburg, M.; Wagner, M. Dynamic behavior of $CO_2$ spot prices. *J. Environ. Econ. Manag.* **2008**, *56*, 180–194. [CrossRef]

8.  Yang, X.; Liao, H.; Feng, X.; Yao, X. Analysis and tests on weak-form efficiency of the EU carbon emission trading market. *Low Carbon Econ.* **2018**, *9*, 1–17. [CrossRef]
9.  Alberto, M.; Vries, D.; Frans, P. Carbon trading thickness and market efficiency. *Energy Econ.* **2010**, *32*, 1331–1336.
10. Daskalakis, G.; Markellos, R.N. Are the European carbon markets efficient. *Rev. Futures Mark.* **2008**, *17*, 103–128.
11. Daskalakis, G. On the efficiency of the European carbon market: New evidence from Phase II. *Energy Policy* **2013**, *54*, 369–375. [CrossRef]
12. Charles, A.; Darné, O.; Fouilloux, J. Market efficiency in the European carbon markets. *Energy Policy* **2013**, *60*, 785–792. [CrossRef]
13. Joyeux, R.; Milunovich, G. Testing market efficiency in the EU carbon futures market. *Appl. Financ. Econ.* **2010**, *20*, 803–809. [CrossRef]
14. Miclăuş, P.G.; Lupu, R.; Dumitrescu, S.A.; Bobircă, A. Testing the efficiency of the European carbon futures market using event-study methodology. *Int. J. Energy Environ.* **2008**, *2*, 121–128.
15. Tang, B.; Shen, C.; Gao, C. The efficiency analysis of the European $CO_2$ futures market. *Appl. Energy* **2013**, *112*, 1544–1547. [CrossRef]
16. Lo, A.Y. Carbon trading in a socialist market economy: Can China make a difference. *Ecol. Econ.* **2013**, *87*, 72–74. [CrossRef]
17. Zhao, X.; Jiang, G.; Nie, D.; Chen, H. How to improve the market efficiency of carbon trading: A perspective of China. *Renew. Sustain. Energy Rev.* **2016**, *59*, 1229–1245. [CrossRef]
18. Zhao, X.; Wu, L.; Li, A. Research on the efficiency of carbon trading market in China. *Renew. Sustain. Energy Rev.* **2017**, *79*, 1–8. [CrossRef]
19. Norden, L.; Weber, M. Informational efficiency of credit default swap and stock markets: The impact of credit rating announcements. *J. Bank. Financ.* **2004**, *28*, 2813–2843. [CrossRef]
20. Tang, L.; Yu, L.; Liu, F.; Xu, W. An integrated data characteristic testing scheme for complex time series data exploration. *Int. J. Inf. Technol. Decis. Making* **2013**, *12*, 491–521. [CrossRef]
21. Tang, L.; Yu, L.; He, K. A novel data-characteristic-driven modeling methodology for nuclear energy consumption forecasting. *Appl. Energy* **2014**, *128*, 1–14. [CrossRef]
22. Tang, L.; Lv, H.; Yu, L. An EEMD-based multi-scale fuzzy entropy approach for complexity analysis in clean energy markets. *Appl. Soft Comput.* **2017**, *56*, 124–133. [CrossRef]
23. Shannon, C.E. A mathematical theory of communication. *Bell Syst. Tech. J.* **1948**, *27*, 379–423. [CrossRef]
24. Pincus, S.M. Approximate entropy as a measure of system complexity. *Proc. Natl. Acad. Sci. USA* **1991**, *88*, 2297–2301. [CrossRef] [PubMed]
25. Eckmann, J.P.; Ruelle, D. Ergodic theory of chaos and strange attractors. *Rev. Modern Phys.* **1985**, *57*, 617–656. [CrossRef]
26. Grassberger, P.; Procaccia, I. Estimation of the Kolmogorov entropy from a chaotic signal. *Phys. Rev. A* **1983**, *28*, 2591–2593. [CrossRef]
27. Costa, M.; Goldberger, A.L.; Peng, C. Multiscale entropy analysis of complex physiologic time series. *Phys. Rev. Lett.* **2002**, *89*, 068102. [CrossRef] [PubMed]
28. Chen, W.; Zhuang, J.; Yu, W.; Wang, Z. Measuring complexity using FuzzyEn, ApEn, and SampEn. *Med. Eng. Phys.* **2009**, *31*, 61–68. [CrossRef] [PubMed]
29. Wu, S.; Wu, C.; Lee, K.; Lin, S. Modified multiscale entropy for short-term time series analysis. *Physica A* **2013**, *392*, 5865–5873. [CrossRef]
30. Heurtier, A.H. The multiscale entropy algorithm and its variants : A review. *Entropy* **2015**, *17*, 3110–3123. [CrossRef]
31. Richman, J.S.; Moorman, J. Physiological time-series analysis using approximate entropy and sample entropy. *Am. J. Physiol. Heart Circ. Physiol.* **2000**, *278*, 2039–2049. [CrossRef] [PubMed]
32. Costa, M.; Peng, C.K.; Goldberger, A.L.; Hausdorff, J.M. Multiscale entropy analysis of human gait dynamics. *Physica A* **2003**, *330*, 53–60. [CrossRef]

*Entropy* **2018**, *20*, 434

33. Pincus, S.M. Assessing Serial Irregularity and Its Implications for Health. *Ann. N. Y. Acad. Sci.* **2001**, *954*, 245–267. [CrossRef] [PubMed]

34. Chou, C.M. Wavelet-based multi-scale entropy analysis of complex rainfall time series. *Entropy* **2011**, *13*, 241–253. [CrossRef]

*Article*

# Low Computational Cost for Sample Entropy

**George Manis [1,*], Md Aktaruzzaman [2] and Roberto Sassi [3]**

[1]  Department of Computer Science and Engineering, University of Ioannina, Ioannina 45110, Greece
[2]  Department of Computer Science and Engineering, Islamic University Kushtia, Kushtia 7003, Bangladesh; md.aktaruzzaman@cse.iu.ac.bd
[3]  Dipartimento di Informatica, Università degli Studi di Milano, Crema 26013, Italy; roberto.sassi@unimi.it
*   Correspondence: manis@cs.uoi.gr; Tel.: +30-2651-008-806

Received: 28 November 2017; Accepted: 9 January 2018; Published: 13 January 2018

**Abstract:** Sample Entropy is the most popular definition of entropy and is widely used as a measure of the regularity/complexity of a time series. On the other hand, it is a computationally expensive method which may require a large amount of time when used in long series or with a large number of signals. The computationally intensive part is the similarity check between points in $m$ dimensional space. In this paper, we propose new algorithms or extend already proposed ones, aiming to compute Sample Entropy quickly. All algorithms return exactly the same value for Sample Entropy, and no approximation techniques are used. We compare and evaluate them using cardiac inter-beat ($RR$) time series. We investigate three algorithms. The first one is an extension of the $kd$-trees algorithm, customized for Sample Entropy. The second one is an extension of an algorithm initially proposed for Approximate Entropy, again customized for Sample Entropy, but also improved to present even faster results. The last one is a completely new algorithm, presenting the fastest execution times for specific values of $m$, $r$, time series length, and signal characteristics. These algorithms are compared with the straightforward implementation, directly resulting from the definition of Sample Entropy, in order to give a clear image of the speedups achieved. All algorithms assume the classical approach to the metric, in which the maximum norm is used. The key idea of the two last suggested algorithms is to avoid unnecessary comparisons by detecting them early. We use the term *unnecessary* to refer to those comparisons for which we know a priori that they will fail at the similarity check. The number of avoided comparisons is proved to be very large, resulting in an analogous large reduction of execution time, making them the fastest algorithms available today for the computation of Sample Entropy.

**Keywords:** Sample Entropy; algorithm; fast computation; $kd$-trees; bucket-assisted algorithm

## 1. Introduction

The use of conditional entropy to measure the regularity (or complexity) of time series or signals has become quite popular. The two most commonly used measures of entropy are Approximate Entropy (*ApEn*) and Sample Entropy (*SampEn*), which have been used extensively in biological signals analysis over the last 20 years [1].

Approximate Entropy was first proposed by Pincus [2] as a measure of systems complexity. It quantifies the unpredictability of fluctuations in a time series; the *approximate* part of its name came from the fact that the index was derived from the estimate of Kolmogorov–Sinai [3,4] entropy—a theoretical metric employed in the context of nonlinear dynamical systems. Many potential applications [5–10] of this metric for biological signals analysis are found in the literature.

To date, hundreds of published papers have employed *ApEn*, first praising its quality but also, over the years, evidencing its limits. A related index, Sample Entropy (*SampEn*), was introduced by Richman and Moorman [11], and is actually a slightly different way to compute the metric. *SampEn* attempts to improve *ApEn*, being a less biased metric for the complexity of the system (at the price of a larger variance of the estimates). This is obtained by evaluating the conditional Rényi entropy of order 2, instead of the classical conditional entropy. Like *ApEn*, *SampEn* has also been used in various scientific fields, such as neonatal heart rate signals analysis [12], effects of mobile phones radiation on heart rate variability (HRV) [13], sleep apnea detection [14], epilepsy detection from electroencephalogram (EEG) signals [15], detection of atrial fibrillation [16], in the analysis of human postural data [17], etc.

The computation of each metric requires checking the similarity of small patterns (or templates of size *m*), constructed from the series. The number of similarity checking, which is the most computationally intensive part of their computations, increases quadratically when increasing the series length *N*. The proposed study provides powerful algorithms which might be helpful for the usage of *SampEn* in real-time applications from the computational point of view. Earlier stages of this work have been presented in [18] and [19], where Approximate Entropy was investigated. This paper focuses on Sample Entropy. The contribution of the paper can be summarized in the following points:

- an improvement to the *kd*-algorithm used by other researchers [20,21] for the fast computation of Sample Entropy is introduced
- an algorithm computing Sample Entropy quickly is proposed, which is an extension to the bucket-assisted algorithm [19] initially introduced for Approximate Entropy. This algorithm has been customized to compute Sample Entropy, and has also been extended to present even faster execution times by sorting the points inside the buckets and by tuning the size of the buckets
- a completely new algorithm is presented which is faster than any other algorithm when used for specific values of *m*, *r*, and signal lengths
- finally, a comparison of all algorithms is presented, based on experimental results collected using implementations of the algorithms in C programming language. The implementation in C allows the programmer to optimize the code in a relatively low level, without heavy software layers lying between the programmer and the hardware.

This paper assumes the classical definition of Sample Entropy, in which the maximum norm is used as a distance between the vectors. Algorithms 1 and 4 can be easily modified to support some other norms, instead of the maximum one. Algorithms 2 and 3 are not appropriate for other norms. However, we must note that in almost all applications of Sample Entropy, the maximum norm is used as the distance between two vectors.

## 2. Sample Entropy

Suppose a time series with $N$ points is given:

$$x = x_1, x_2, \cdots, x_N,\tag{1}$$

from which a new series $\vec{x}$ of vectors of size $m$ is constructed. Sometimes this series is also referred to in the literature as *pattern* or *template*:

$$\vec{x} = \vec{x}_1, \vec{x}_2, \cdots, \vec{x}_{N-m+1}, \quad \vec{x}_i = (x_i, x_{i+1}, \cdots, x_{i+m-1}).\tag{2}$$

The two vectors $\vec{x}_i$ and $\vec{x}_j$ are considered similar if the maximum distance between all of their corresponding elements is within a selected threshold $r$. This threshold is also termed the *tolerance of mismatch* between two vectors; i.e.,:

$$|x_{i+k} - x_{j+k}| \leq r, \quad \forall\{i,j\}, \ 0 \leq k \leq m-1.\tag{3}$$

In the following, the notation $||\vec{x}_i - \vec{x}_j||_m \leq r$ will be used to express the similarity of two vectors of size $m$. Given the distance $r$, the number of vectors of length $m$ similar to $\vec{x}_i$ are given by $n_i^m(r)$:

$$n_i^m(r) = \sum_{\substack{j=1 \\ j \neq i}}^{N-m} \Theta(i, j, m, r), \tag{4}$$

where:

$$\Theta(i, j, m, r) = \begin{cases} 1: & ||\vec{x}_i - \vec{x}_j||_m \leq r \\ 0: & otherwise. \end{cases} \tag{5}$$

Similarly, for vectors of length $m + 1$:

$$n_i^{m+1}(r) = \sum_{\substack{j=1 \\ j \neq i}}^{N-m} \Theta(i, j, m+1, r). \tag{6}$$

In Equations (4) and (6), please note that $j \neq i$, meaning that self-matches are excluded (comparison of a vector with itself).

The measures of similarity $B_i^m(r)$ and $A_i^m(r)$ between templates of length $m$ and $m + 1$, respectively, are:

$$B_i^m(r) = \frac{1}{N-m} n_i^m, \quad i = 1, 2, \cdots, N-m, \tag{7}$$

$$A_i^m(r) = \frac{1}{N-m} n_i^{m+1}, \quad i = 1, 2, \cdots, N-m. \tag{8}$$

The mean values of these measures of similarity are computed next:

$$B^m(r) = \frac{1}{N-m} \sum_{i=1}^{N-m} B_i^m(r), \tag{9}$$

$$A^m(r) = \frac{1}{N-m} \sum_{i=1}^{N-m} A_i^m(r). \tag{10}$$

Sample Entropy is given by the formula:

$$SampEn(m, r) \begin{cases} \to \infty, & when \ A = 0 \\ = \ln B/A, & otherwise. \end{cases} \tag{11}$$

## 3. The Straightforward Implementation

In the implementation of the definition presented above (Section 2), we need two variables $A$ and $B$ to count the total number of similar points and a nested loop to compare all vectors with each other. An algorithm computing Sample Entropy follows (Algorithm 1). This algorithm is based on the definitions, and some basic improvements have been introduced that made the implementation simpler and, at the same time, faster. wo

---

**Algorithm 1:** Straightforward

---

```
01:   A = B = 0              // initialize similarity counters
02:   for  i = 1...N−m:            // create all pairs of vectors
03:      for  j = i+1...N−m:
04:         for  k = 0...m−1:            // check vectors i and j in m-dimensional space
05:            if |x_{i+k} − x_{j+k}|>r  then: break
06:         if  k = m  then:            // if found to be similar
07:            B = B+1            // increase similarity counter
08:            if |x_{i+m} − x_{j+m}|<r  then:            // check for similarity in m+1
       dimensional space
09:               A = A+1            // increase similarity counter
10:   A = A/(N−m)² ;  B = B/(N−m)²            // counters become probabilities
11:   if  A = 0  then:  SampEn → ∞
12:   else:  SampEn = ln B/A            // SampEn is finally computed
```

---

The input time series is $x$, $m$ is the embedding dimension, and $r$ is the threshold distance. The counters $A$ and $B$ (line 01) are initialized to zero. Then, all possible pairs of vectors are checked for similarity (lines 02–03). The index $j$ of the second *for* ranges from $i+1$ to $N−m$ to avoid unnecessary double checks: it is not necessary to check pair $(\vec{x}_j, \vec{x}_i)$ when pair $(\vec{x}_i, \vec{x}_j)$ has already been checked. Additionally, self matching checks are avoided; i.e., vector $\vec{x}_i$ with vector $\vec{x}_i$. In lines 04–05, each pair of vectors in the $m$-dimensional space is checked for similarity. Please note that in the similarity check, not all $m$ comparisons between the elements are necessary. If one comparison fails, then the similarity test stops immediately, exiting the loop. If the vectors are found to be similar (line 06), the counter $B$ is increased (line 07). Then, the similarity check is performed for the corresponding vectors in the $(m+1)$-dimensional space. Since the similarity check for the $m$ first elements has already succeeded (lines 04–06), only the last elements of the two vectors need to be checked (line 08). In case of success, $A$ is increased (line 09). Next, the probability of two vectors being similar in the $m$ and $m+1$ dimensional space is computed (line 10), even though this is not necessary, since the two denominators will be simplified in division in the next step. Finally (lines 11–12), $SampEn$ is returned as the logarithm of the ratio of $A$ and $B$, when $A \neq 0$. Otherwise, it is infinite.

Some implementation details: It is important to note that the code was optimized after several tests. The use of *break* in C was proved to be the optimal solution, significantly affecting the overall execution time. The same technique was selected for all algorithms, when this was possible.

## 4. Computation Using *kd*-Trees

A *kd-tree* is a binary tree, each node of which is a vector. The tree is organized like a binary tree. However, when transversing it, we decide if we have to move towards the left or the right child by comparing the $k_{th}$ element of the vector we are looking for, with the $k_{th}$ element of the vector stored in the node. The value of $k$ is computed from the level of the visited node: $k = lv \mod m$, where $lv$ is the level of the visited node (the level of the root is considered as 0) and $m$ is the size of the vector.

In our problem, the purpose of transversing is not to locate a specific node, but all nodes which are similar to the given vector. We call this kind of searching *range search*. In range search it might be necessary to visit both children, according to the value of $r$. For example, if the vector we are looking for is (3,5,6), the vector in the node is (3,4,6), $r=2$ and $lv=1$, then we compare the second elements of the vectors (i.e., 5 and 6) and we decide to move towards the right child. However, since $r=2$, nodes with their second element equal to 3 are also candidates for being similar and are located under the left child. Thus, in this example we have to visit both left and right children.

The algorithms [20,21] have been proposed for fast computation of Sample Entropy using *kd*-trees. They first construct the *kd*-tree and then use range search for finding the similar vectors. Proposed here is an algorithm which searches for similar points, before the final *kd*-tree is constructed. This is in accordance with the definition of Sample Entropy which avoids self matches. It is also a trick to avoid the comparison between the pair of vectors $\vec{x}_j$ and $\vec{x}_i$, when the pair of vectors $\vec{x}_i$ and $\vec{x}_j$ has already been tested for similarity in a previous step. This improvement makes the algorithm two times faster, compared to the descriptions given in [20,21]. The pseudocode follows (Algorithm 2):

---

**Algorithm 2:** *kd*-Tree Based

---

```
01 :   A = B = 0              // initializations
02 :   kd = empty
03 :   for i = 1...N−m:       // for every vector
04 :       t_A, t_B = range_search(kd, i)    // count the similar vectors already in
       the tree
05 :       A = A+t_A ;  B = B+t_B    // update the similarity counters
06 :       insertkd(kd, i)           // and then insert the vector in the tree
07 :   if A = 0 then:  SampEn → ∞
08 :   else:  SampEn = ln B/A        // SampEn is finally computed
```

---

The algorithm is simple. Similarity counters $A$ and $B$ are initialized to zero (line 01) and the *kd*-tree to *empty* (line 02). Next, for every vector which is to be inserted in the *kd*-tree (line 03), we first perform range search to find the similar vectors already in the tree (line 04), we update the similarity counters $A$ and $B$ (line 05), and then we insert the vector in the tree (line 06). Sample Entropy is computed at lines 07 and 08.

Some implementation details: Recursive functions were not used in order to avoid function call delays. Instead, a stack was implemented, again without the use of functions for the stack operations. For the *kd*-trees representation, three integer arrays were used, each of size $N$. The first had the indexes of the vector in the time series, the second the indexes of the left children, and the third the indexes of the right children, avoiding delays due to structure complexity, pointer handling, and dynamic memory allocation for each tree node.

## 5. The Bucket-Assisted Algorithm

The bucket-assisted algorithm is an extension of the algorithm published in [19] for the computation of Approximate Entropy. The algorithm has been adapted to the definition of Sample Entropy and also improved to present even faster execution times. These two modifications speed up the algorithm remarkably, and will be described in this section.

In [19], we proposed a fast algorithm for computing $ApEn$. In that algorithm, we used a series of buckets, and we put the candidate points to be similar to each other in neighboring buckets.

The main idea was to integrate the given series $x$ and create a new series $X$ such that:

$$X = X_1, X_2, \cdots, X_{N-m+1} \tag{12}$$

where:

$$X_i = x_i + x_{i+1} + \cdots + x_{i+m-1}. \tag{13}$$

Consider a set of buckets:

$$B = \{B_1, B_2, \cdots, B_{h_N}\}, \tag{14}$$

which consists of $h_N$ buckets of equal size $r$, where

$$h_N = \lceil X_{max}/r \rceil. \tag{15}$$

Now, point $X_i$ is mapped into bucket $B_h$ when $h = \lceil X_i/r \rceil$. When a point $X_i$, which corresponds to the vector $\vec{x}_i$, is mapped into the bucket $B_h$, then all points similar to $X_i$ are mapped into one of the buckets: $B_{h-m}, B_{h-m+1}, \cdots, B_h, B_{h+1}, \cdots, B_{h+m}$. Please see [19] for the proof.

A graphical explanation of the main idea of the bucket-assisted algorithm is shown in Figure 1, where $m=2$ and the bucket size is 10 ms. Vectors in the bucket $BC$ (solid lines) can be similar only to the vectors between lines $A$ and $D$ (dashed lines). However, it is not necessary to examine both pairs $(\vec{x}_i, \vec{x}_j)$ and $(\vec{x}_j, \vec{x}_i)$ for similarity, as discussed above. Thus, the vectors in the bucket $BC$ are checked for similarity only with those vectors located between lines $A$ and $B$, and then between lines $B$ and $C$.

One of the main contributions of this work is an extension to the bucket-assisted algorithm, which further speeds up the execution time. The modifications are the following:

- points in the buckets are sorted according to the first element of the vector
- buckets are divided again into smaller buckets (of size smaller than $r$).

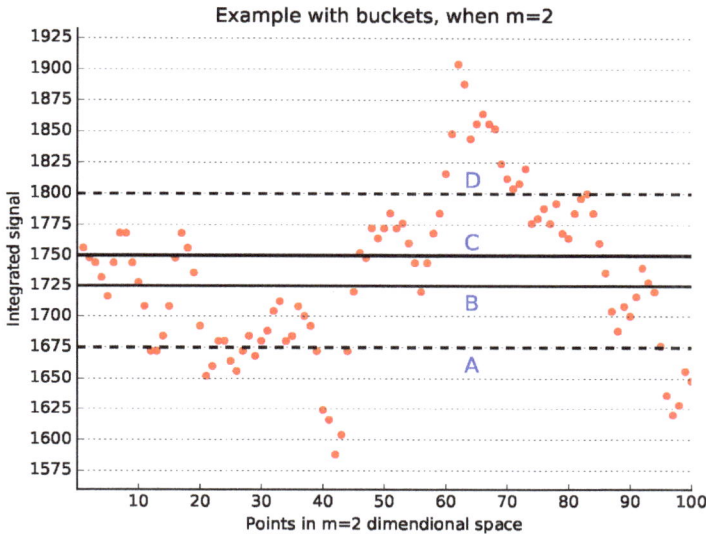

**Figure 1.** Example of the bucket-assisted algorithm. The integrated signal is depicted here. Points between the solid lines $B$ and $C$ can be similar only to those points laying between the dashed lines $A$ and $D$. However, it is sufficient to check for similarity only in those points located between lines $B-C$ and $A-B$.

These two modifications are enough for a significant speedup, as shown later in Section 7. Please remember that the similarity test fails if the absolute value of at least one of the differences between the corresponding elements of the examined vectors is larger than $r$; i.e.,:

$$||\vec{x}_i - \vec{x}_j||_m \leq r$$

$$\Leftrightarrow |x_i - x_j| \leq r, |x_{i+1} - x_{j+1}| \leq r, \cdots, |x_{i+m-1} - x_{j+m-1}| \leq r. \tag{16}$$

Thus, a reasonable approach would be to have the points in the buckets sorted according to the first element of the vector, perform the comparison $|x_i - x_j| \leq r$, and exclude the points that failed this test from the following comparisons. A low overhead binary search $O(n \log n)$ algorithm could be used. Then, the points could be sorted again based on the second element of the vector and perform the comparison $|x_{i+1} - x_{j+1}| \leq r$ until the last comparison $|x_{i+m-1} - x_{j+m-1}| \leq r$ is reached. The points that pass these tests can be considered as similar. However, this approach requires more sorting, since for every examined point we have to sort up to $m-1$ times. This is relatively expensive, even when we use a low overhead sorting algorithm such as quick sort $O(n \log n)$. Thus, the approach we selected was to sort the points in the buckets only once and excluded from further comparisons only those points that failed the first of the above tests; i.e., $|x_i - x_j| \leq r$. The algorithm continues by performing the rest of the comparisons of Equation (16) by testing the corresponding elements of each pair of vectors. It can be proved experimentally that the proposed modification adds a significant speedup to the execution time of the algorithm.

Some more speedup (also significant) can be achieved by dividing the large buckets into smaller ones. A finer-grained distribution of the points can be achieved by dividing large buckets into smaller ones, avoiding even more comparisons, as shown in Figure 2. When using the large buckets $B_1, B_2, B_3, B_4$ for $m = 3$, the point marked by a small circle belongs to bucket $B_4$, must be compared for similarity with every other point in buckets $B_1, B_2, B_3, B_4$. When using the smaller buckets $b_1, b_2, \cdots, b_{20}$, the same point belonging in bucket $b_{19}$ need to be compared only with points in buckets $b_4, b_5, \cdots, b_{19}$. With this refinement, we can exclude a considerable number of smaller buckets (4 out of 20 in our example) from the comparisons. The number by which a larger bucket is split into smaller ones is a parameter for the problem. We will call it the *split* factor and symbolize it as $r_{split}$. The number of smaller buckets that can be excluded from the comparisons is determined by $1/r_{split}$ of the total number of the smaller buckets.

**Example of splitting when split factor = 5 and m=3**

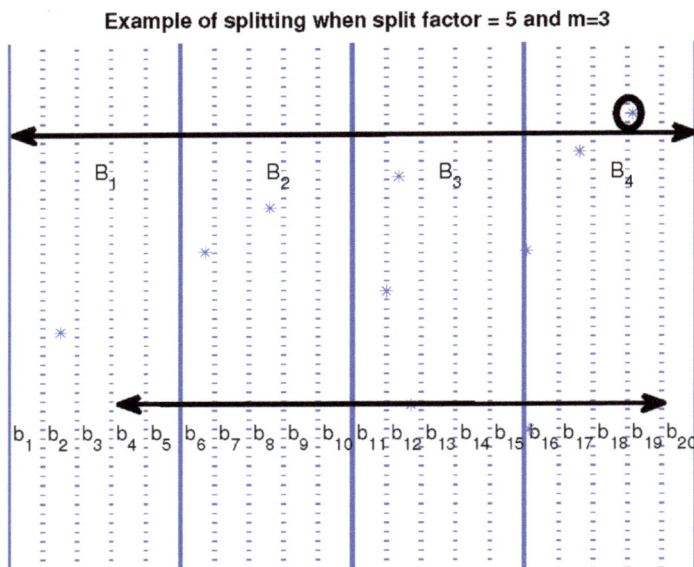

**Figure 2.** Splitting large buckets into smaller ones. Asterisks are points of the integrated signal. This splitting of the buckets into smaller ones can lead to an increased number of avoided comparisons. For example, for the point marked by the small circle belonging to the bucket $b19$, comparisons are reduced by approximately 20%.

The algorithm in a detailed description in pseudocode follows (Algorithm 3). Again, $x$ is the input signal, $m$ the embedding dimension, and $r$ the threshold distance.

---

**Algorithm 3:** Bucket-Assisted

---

01: **for** $i = 1 \ldots N - m$: $\quad X_i = \sum_{k=0}^{m-1} x_{i+k}$       // integrated signal

02: $X_{min} = \min(X_i)$

03: **for** $i = 1 \ldots N - m$: $\quad X_i = X_i - X_{min} + 1$       // normalization

04: $N_b = \lceil max(X_i)/r/r_{split} \rceil$       // number of buckets

05: **for** $i = 1 \ldots N - m$: $\quad bucket_b = $ **empty**

06: **for** $i = 1 \ldots N - m$:       // fill in the buckets

07: $\quad b = \lceil X_i/r/r_{split} \rceil$

08: $\quad bucket_b = bucket_b \cup \{ \vec{x}_i \}$

09: **for** $b = 1 \ldots N - m$:       // sort vectors according to first element

10: $\quad b_{ordered} = \{ \vec{x}_i \in bucket_b : x_i \leq x_{i+1} \}$

11: $\quad bucket_b = b_{ordered}$

12: $A = B = 0$

13: **for** $i_b = 1 \ldots N_b$:       // for every bucket

14: $\quad$ **for** $j_b = i_b - m \cdot r_{split} \ldots i_b - 1, \ j_b 0$:    // visit all buckets possibly containing
similar vectors

15: $\quad\quad$ **for** $\vec{x}_i, \ \vec{x}_i \in bucket_{i_b}$:

16: $\quad\quad\quad$ candidates $= \{ \vec{x}_j \in bucket_{i_b} : \ x_j - r \leq x_i \leq x_j + r, \ ij \}$ 17:
$\cup \{ \vec{x}_j \in bucket_{j_b} : \ x_j - r \leq x_i \leq x_j + r \}$ // exploit sorting to exclude some comparisons

18: $\quad\quad\quad$ **for** $\vec{x}_j, \ \vec{x}_j \in$ candidates:

19: $\quad\quad\quad\quad$ **if** $\| \vec{x}_i - \vec{x}_j \|_m \leq r$ **then:**       // similarity check

20: $\quad\quad\quad\quad\quad B = B + 1$

21: $\quad\quad\quad\quad\quad$ **if** $|x_{i+m} - x_{j+m}| \leq r$ **then:** $\quad A = A + 1$

22: **if** $A = 0$ **then:** $SampEn \to \infty$

23: **else:** $SampEn = \ln B/A$       // SampEn is finally computed

---

A less formal description of the algorithm follows. We first integrate the signal using a window of size $m$ (line 01). The integrated signal $X$ is normalized so that $min(X_i) = 1$ (lines 02–03). The number of buckets is equal to the maximum value of the integrated signal divided by the threshold distance $r$ and by the split factor $r_{split}$ (line 04). Next (line 05), we initialize the set of buckets *bucket* to the empty set. To fill the buckets, we select the appropriate bucket for each vector $\vec{x}_i$ (line 07) and we add it in this bucket (line 08). Next, we sort the vectors in each bucket according to their first element (lines 09–11).

For the similarity check, we need two counters. We use $B$ for the $m$ dimensional space and $A$ for the $m+1$ dimensional space. These two counters are initialized to zero (line 12). For every bucket $i_b$ (line 13), we check for similar points in all $j_b$ buckets in which similar points are possible to be found (line 14). For every point in the examined bucket $i_b$ (line 15), we find all points that are not excluded from similarity due to the distance of their first elements (lines 16–17). Since points are sorted according to their first element, this procedure is rapid with complexity only $O(\log n)$. In the next step, we check for similarity all pairs of candidate points (lines 18–21) with the same method as it was described in the simple algorithm. *SampEn* is computed at the two last lines of the pseudocode (lines 22–23).

## 6. A Lightweight Algorithm

Typically, values of the parameter $m$ which are used for *SampEn* estimations are $m = 1, \ldots, 3$ [12,14,22]. However, recently in [23,24] it was recommended that $m = 1$ in short time series keeps the variation smaller and improves the confidence of the estimates of entropy. Here, we propose an algorithm for fast computation of Sample Entropy which is straightforward when $m = 1$. However, the algorithm is also fast for small signal lengths and other values of $m$. Since it has a simple implementation, we will call it a *lightweight* algorithm.

The algorithm reduces the number of comparisons between points by sorting the original series $x_i$. For this purpose, a fast sorting algorithm is used with $O(n \log n)$ complexity. Then, we consider only those sequences for which the first elements are within the allowed tolerance: $\vec{x}_i$ and $\vec{x}_j$, i.e., $x_j \leq x_i + r$. Since the original series is sorted, it is not necessary to consider those cases for which $x_j \geq x_i - r$, as they were already included. The pseudocode follows (Algorithm 4):

---

**Algorithm 4:** Lightweight

```
01:  ord_x = { x_i : x_i ≤ x_{i+1} }              // sort x in ascending order
02:  pos_x = { i : ord_{x_i} = x_{pos_{x_i}} }     // remember original positions
03:  A = B = 0
04:  for i = 1 ... N−m:
05:      candidates_i = { ord_{x_j} : ord_{x_j} ≤ ord_{x_i} + r } // points of the ordered series
     matching other points within r
06:      a = pos_{x_i}
07:      for ord_{x_j} ∈ candidates_i:
08:          b = pos_{x_j}
09:          if ||x̄_a − x̄_b||_m ≤ r then:          // similarity check
10:              B = B+1
11:              if |x_{a+m} − x_{b+m}| ≤ r then:  A = A+1
12:  if A = 0 then:  SampEn → ∞
13:  else:  SampEn = ln B/A                         // SampEn is finally computed
```

---

In the lightweight algorithm, the series is sorted (line 01) and the original positions of the elements are stored for later reference (line 02). Similarity counters $A$ and $B$ are initialized to zero (line 03). Then, for any sample $x_i$ of the ordered series, starting from its beginning, all those other samples $x_j$ such that $x_j \leq x_i + r$ are included in the list of possible candidate matches (lines 04–05). The search for candidates is performed on the sorted series, with a binary search, which at worst is $O(\log n)$. The stored positions $pos_x$ are used to locate the original locations $a$ and $b$ for $x_i$ and any of the $x_j$ elements in the candidates list, respectively (lines 06–08). The vectors $\vec{x}_a$, $\vec{x}_b$ of length $m$ starting at $a$ and $b$ are checked, and if $||\vec{x}_a - \vec{x}_b||_m \leq r$, the counter $B$ is incremented. If the two further elements at positions $a + m$ and $b + m$ are closer than the threshold $r$, the counter $A$ is also incremented (lines 09–11). Sample Entropy is finally computed at lines 12 and 13.

## 7. Experimental Results

In order to evaluate/compare the four algorithms, we performed four experiments with two different datasets. Both datasets are publicly available from Physionet [25]. The first one consists of 24 h of recordings of healthy subjects in normal sinus rhythm (*nsr2* dataset). The second one consists of 24 h of recordings of congestive heart failure patients (*chf2* dataset). For both datasets, the original electrocardiogram (ECG) recordings were digitized at 128 samples per second, and the beat annotations were obtained by automated analysis with manual review and correction.

The two datasets present different signal characteristics. As expected, the mean value of the *chf2* dataset is lower than that of *nsr2*. Due to the large number of ectopic beats, the *chf2* dataset presents larger standard deviation. The existence of ectopic beats influences both the mean value and the standard deviation of the signals. Thus, we removed the ectopic beats (a common practice in HRV analysis) and created two more datasets with different characteristics. The resulting four datasets were the basis for our comparisons. We will refer to them as *nsr2*, *chf2*, *nsr2_f*, and *chf2_f*, where the index $f$ comes from the word *filtered*. Average values for the mean and the standard deviation of each dataset are presented in Table 1. It is not a surprise that the standard deviation of the *chf2_f* dataset is the lowest of all, since it reflects the reduction of the complexity of the heart as a system, due to the heart failure disease.

**Table 1.** Signal characteristics of the examined datasets.

|  | *nsr2* | *chf2* | *nsr2$_f$* | *chf2$_f$* |
|---|---|---|---|---|
| mean | 809 msec | 681 msec | 807 msec | 667 msec |
| standard deviation | 204 msec | 369 msec | 156 msec | 45 msec |

Experiments with all four algorithms were conducted on a 4-core desktop computer (3.6 GHz Intel Xeon E5-1620 processor; 16 GB of RAM; Linux OpenSuse 42.2 x86_64 OS). Code was optimized as much as possible for all algorithms, and the parameter $-O3$ was used in the GNU Compiler Collection (GCC 4.8.5).

Ten signals were randomly selected from each dataset. The mean execution time for each algorithm and each dataset was computed. Each experiment was performed 100 times, thus the reported execution time is the mean value of 1000 runs.

In order to exclude overheads from the computation time, we first read all input data and stored them into matrices. Then, inside the outer loop (which repeats the experiment 100 times), and before the inner loop (which computes Sample Entropy for the ten signals), we started the timer by using the C function call *clock_tclock(void)*. We used the same call after the inner loop and accumulated all time intervals of all 100 repetitions to estimate the total and then the execution time.

We will start with the experimental result collected from the *nsr2* dataset, and then we will discuss differences observed in the other datasets. Figure 3 shows execution times for all four algorithms and the typical values $m = 2$ and $r = 0.2$. The straightforward implementation is the slowest of all, becoming especially slow for large values of $N$. The improved version of *kd*-trees—as described earlier in this paper—is faster, but not as fast as the other two algorithms. In this figure, the bucket-assisted seems to present the best results, followed by the lightweight algorithm

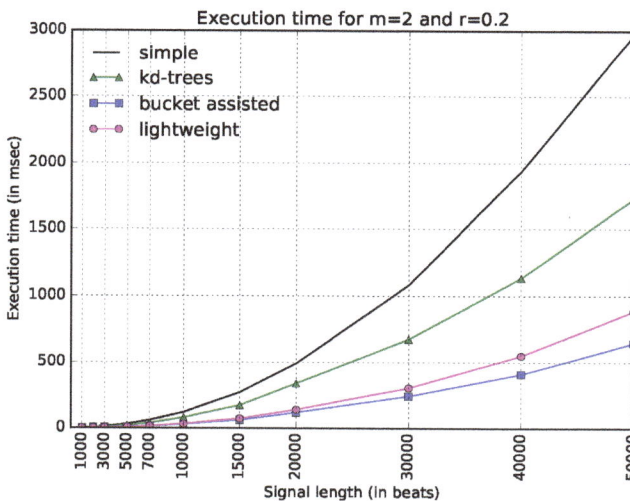

**Figure 3.** Execution time of all algorithms for the typical values of parameters $m$ and $r$ ($m = 2$ and $r = 0.2$) and various signal lengths ($N$).

The parameter $r_{split}$ for the bucket-assisted algorithm was selected to be equal to 5. We did not try to completely optimize it. We ran the code for several inputs and the typical parameters $m = 2$ and $r = 0.2$, and selected a good and easy-to-remember value of $r_{split}$ for them. Since we did not want to be less fair to the rest of the algorithms and not optimize the results of the bucket-assisted algorithm with an additional parameter, we kept the value $r_{split} = 5$ the same for the rest of our experiments. However, we also did some sensitivity analysis on the value of $r_{split}$, which will be discussed at the end of this section.

One can note that in Figure 3, it is difficult to see the behavior of the algorithms for low values of $N$. For this reason, we added another figure, which gives the same information in a different way. In Figure 4, the $x$-axis is in logarithmic scale. The values in the $y$-axis do not express execution time, but speedup, dividing the execution time of each algorithm with respect to the straightforward one, which we considered as a reference. Expressing the results in terms of a well-defined algorithm—also implemented in a standard programming language, which introduces minimal overhead—allows other researchers to compare their results easily with the ones given in this paper.

Thus, there are only three curves in this figure. The information is depicted in a clearer way. Here, one can see that the bucket-assisted algorithm outperforms the other algorithms for values of $N$ approximately larger than 3000 beats. For signals smaller than 3000 beats, the lightweight algorithm gives the lowest execution times.

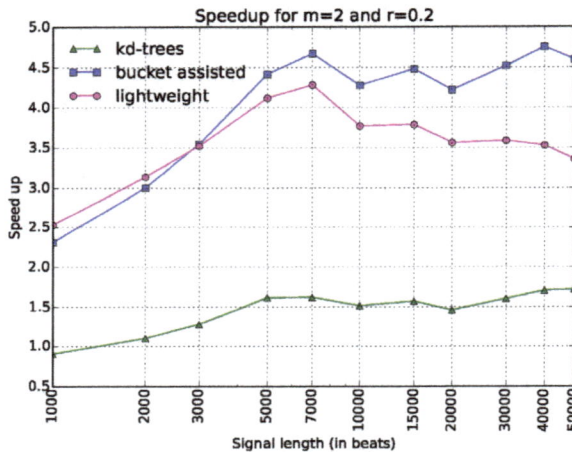

**Figure 4.** Execution time of all algorithms as a speedup gained from the simple one. Typical values of parameters $m$ and $r$ ($m = 2$ and $r = 0.2$) have been selected. The $x$-axis is in a logarithmic scale.

Since this kind of diagram seems more illustrative than the one in Figure 3, we will present the rest of the diagrams in the same way. In Figure 5, speedups for the parameters $m = 1$ and $r = 0.2$ are shown. One can note here that the lightweight algorithm is always faster. The *kd*-tree algorithm presents poor results. In Figure 6, the speedups when $m = 2$ and $r = 0.1$ are presented. Here, the bucket-assisted algorithm is always faster. The *kd*-tree algorithm again presents poorer results.

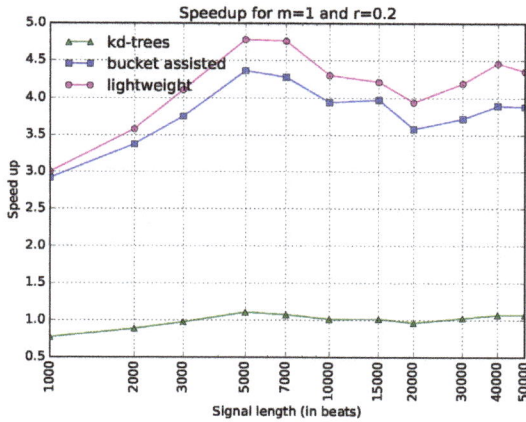

**Figure 5.** Execution time of all algorithms as a speedup gained from the simple one, when $m=1$ and $r=0.2$. The $x$-axis is in a logarithmic scale.

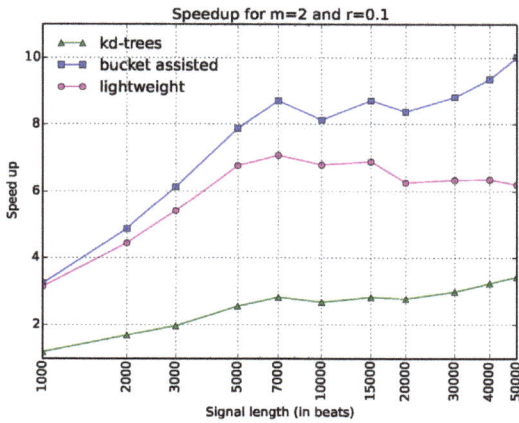

**Figure 6.** Execution time of all algorithms as a speedup gained from the simple one, when $m=2$ and $r=0.1$. The $x$-axis is in a logarithmic scale.

The experiments with the three other datasets gave similar but interesting results, since they helped us to make some additional conclusions. In all cases, the *kd*-tree algorithm was slower than both the bucket-assisted and the lightweight algorithm. We will continue the comparison between the two latter.

For large values of $N$ ($N > 10,000$) the bucket-assisted algorithm was always faster than the lightweight one. As the value of $N$ decreases, the lightweight algorithm presents lower execution times.

The removal of ectopic beats gave an advantage to the bucket-assisted algorithm. The lightweight algorithm gave its best execution times for the $chf2$ dataset and then for the $nsr2$ dataset—the two datasets with the higher variability. The removal of ectopic beats decreased this variability, and for the same values of $m$, $r$, and $N$, the bucket-assisted significantly improved its performance, almost always presenting better results with the $chf2_f$ dataset (the one with the smallest variability).

The relation of the input parameters, the characteristics of the input signals, and the performance of the algorithms is difficult to predict or model. Some general conclusions can be made, but it is certain that each algorithm has a different reason to be used. To give a more detailed image of the relation of the input parameters, the characteristics of the signals, and the execution times, we added a table presenting—for each value of $m$, $r$, and $N$—the number of datasets for which each algorithm performed better. In Table 2, the first number is the number of datasets for which the bucket-assisted algorithm was faster, while the second one is the number of datasets for which the lightweight algorithm was faster. The table depicts only values of $N \leq 10,000$.

**Table 2.** Comparison of bucket-assisted and lightweight algorithms.

| | $m = 1$ | | | $m = 2$ | | | $m = 3$ | | |
|---|---|---|---|---|---|---|---|---|---|
| | $r = 0.1$ | $r = 0.2$ | $r = 0.3$ | $r = 0.1$ | $r = 0.2$ | $r = 0.3$ | $r = 0.1$ | $r = 0.2$ | $r = 0.3$ |
| $N = 1000$ | 3/1 | 2/2 | 1/3 | 3/1 | 3/1 | 1/3 | 2/2 | 1/3 | 1/3 |
| $N = 2000$ | 3/1 | 1/3 | 1/3 | 3/1 | 3/1 | 1/3 | 2/2 | 1/3 | 1/3 |
| $N = 3000$ | 3/1 | 1/3 | 1/3 | 4/- | 3/1 | 2/2 | 4/- | 4/- | 1/3 |
| $N = 5000$ | 3/1 | 1/3 | 2/2 | 4/- | 3/1 | 2/2 | 4/- | 4/- | 4/- |
| $N = 7000$ | 3/1 | 1/3 | 2/2 | 4/- | 3/1 | 2/2 | 4/- | 4/- | 4/- |
| $N = 10,000$ | 3/1 | 2/2 | 3/1 | 4/- | 3/1 | 3/1 | 4/- | 4/- | 4/- |

A last issue to discuss is the question of which value should be selected for the $r_{split}$ factor. We chose 5 for the reasons we mentioned earlier; however, this value is not necessarily the optimal one. In order to perform a sensitivity analysis for the $r_{split}$ factor, we selected the typical values used for Sample Entropy for the parameters $m$ and $r$ ($m = 2$, $r = 0.2$). We ran the algorithm for different values of $r_{split}$ factor and for different values of $N$. The optimal value of the $r_{split}$ factor was selected for each $N$. We noticed that the larger the value of $N$, the larger the value of $r_{split}$ factor that gave the optimal results. For $N < 3000$, the best values ranged from 2 to 5. For $N20,000$, the optimal value was close to 15. Despite the small values of $N$, the selection of the $r_{split}$ factor was not crucial, since there was a plateau of values which presented similar execution times. If we try to explain this behavior, a large number of samples would lead to overcrowded buckets. By splitting the buckets into smaller ones, we can achieve a much better distribution, which leads to better execution times.

## 8. Discussion of Related Work

To the best of the authors' knowledge, the first algorithm for fast computation of *ApEn* was published in [26]. This algorithm is of $O(N^2)$ complexity, does not avoid comparisons, and also has a $O(N^2)$ spatial complexity, even though it can be implemented with a spatial complexity of $O(N)$.

Another algorithm for *SampEn* is available in [25]. The algorithm builds up templates matching within the tolerance $r$ until no match is found, and keeps track of template matches in counters $A_k$ and $B_k$ for all lengths $k$ up to $m$. Once all the matches are counted, Sample Entropy is computed. This algorithm has been designed to compute Sample Entropy for all values of $m$ at once. Thus, a straight comparison with the proposed algorithms may not be fair, since the target of the algorithms is different. However, the core of the algorithm is similar to the straightforward implementation we described. Additionally, the modification of the proposed algorithms to target the computation of all values of $m$ is possible, but is not an aim of this paper.

In [20], apart from the algorithm for *kd*-trees, they also presented an algorithm for computing Approximate and Sample Entropy for signals whose elements belong in a definite set of values. It is also based on *kd*-trees, and exploits the fact that the height of the tree can be limited, and that more than one vector can be stored in the tree node.

The authors of [21] made a theoretical study of the problem which leads to a lower complexity algorithm again based on the *kd*-trees, which might perform better in very long signals. However, as in [20], with the size of the signal we used, the overhead for constructing the *kd*-tree and the overhead introduced for a single comparison led to much larger execution times than those obtained with the bucket-assisted or the lightweight algorithms. The same conclusion was drawn in a paper which proposed a fast algorithm for fractal dimension estimation [27]—a similar problem with the one studied here. This paper proposed an algorithm checking for neighboring points in an $m$-dimensional space by separating the $m$-space into orthogonal subspaces and mapping $m$-dimensional points onto these subspaces. It also compared this approach with another one, published before, which used $kd$-trees for the same purpose [28] and concluded that the algorithm with the subspaces was faster. The approach with the buckets reduces the complexity of handling $m$-dimensional spaces, since handling $m$-dimensional subspaces requires a large amount of memory or alternatively significant overhead to map the $m$-dimensional subspaces onto simpler structures and then manage these structures.

## 9. Conclusions

In this paper, three Sample Entropy computation algorithms were compared with each other, and with an algorithm resulting directly from the definition of the method, in order to decide which one is the fastest (and for which input parameters). The first algorithm was a modified version of an existing one, based on $kd$-trees. The second one is an extension of another algorithm (the bucket-assisted one), initially proposed for Approximate Entropy, but customized for Sample Entropy and extended to provide even smaller execution times. The last one is a completely new algorithm, which we call *lightweight* since it is "light-weight" compared to the $kd$-tree-based and the bucket-assisted one. Despite the fact that it was improved, the $kd$-tree algorithm showed worse execution times than the bucket-assisted and the lightweight algorithms. The lightweight one gave better execution times for specific values of $m$ and $r$, and for smaller values of $N$. Thus, the bucket-assisted algorithm and the lightweight one act complementarily, and the one of choice must be selected according to the problem at hand.

**Acknowledgments:** We would like to thank Elias Kalivas, undergraduate student in the University of Ioannina, for helping us with the experimentations with the code.

**Author Contributions:** G.M. and R.S. have designed the algorithms. All authors worked in the implementation and performed the experiments. All authors have read and approved the final manuscript.

**Conflicts of Interest:** The authors declare no conflict of interest.

## References

1. Yentes, J.M.; Hunt, N.; Schmid, K.K.; Kaipust, J.P.; McGrath, D.; Stergiou, N. The appropriate use of approximate entropy and sample entropy with short data sets. *Ann. Biomed. Eng.* **2013**, *41*, 349–365.
2. Pincus, S.M. Approximate Entropy as a measure of system complexity. *Proc. Natl. Acad. Sci. USA* **1991**, *88*, 2297–2301.
3. Kolmogorov, A.N. Entropy per unit time as a metric invariant of automorphism. *Dokl. Russ. Acad. Sci.* **1959**, *124*, 754–755.
4. Sinai, Y.G. On the Notion of Entropy of a Dynamical System. *Dokl. Russ. Acad. Sci.* **1959**, *124*, 768–771.
5. Signorini, M.G.; Sassi, R.; Lombardi, F.; Cerutti, S. Regularity patterns in heart rate variability signal: The approximate entropy approach. In Proceedings of the 20th International Conference of the IEEE Engineering in Medicine and Biology Society, Hong Kong, China, 1 November 1998; pp. 306–309.
6. Beckers, F.; Ramaekers, D.; Aubert, A.E. Approximate Entropy of Heart Rate Variability: Validation of Methods and Application in Heart Failure. *Cardiovasc. Eng.* **2001**, *1*, 177–182.
7. Valenza, G.; Allegrini, P.; Lanatà, A.; Scilingo, E.P. Dominant Lyapunov exponent and approximate entropy in heart rate variability during emotional visual elicitation. *Front. Neuroeng.* **2012**, *5*, 3, doi:10.3389/fneng.2012.00003.
8. Srinivasan, V.; Eswaran, C.; Sriraam, N. Approximate Entropy-Based Epileptic EEG Detection Using Artificial Neural Networks. *Trans. Inf. Tech. Biomed.* **2007**, *11*, 288–295.
9. Ocak, H. Automatic detection of epileptic seizures in EEG using discrete wavelet transform and approximate entropy. *Expert Syst. Appl.* **2009**, *36*, 2027–2036.

10. Cerutti, S.; Corino, V.D.A.; Mainardi, L.T.; Lombardi, F.; Aktaruzzaman, M.; Sassi, R. Non-linear regularity of arterial blood pressure variability in patient with atrial fibrillation in tilt-test procedure. *Europace* **2014**, *16*, iv141–iv147.

11. Richman, J.S.; Moorman, J.R. Physiological time series analysis using approximate entropy and sample entropy. *Am. J. Physiol. Heart Circ. Physiol.* **2000**, *278*, 2039–2049.

12. Lake, D.E.; Richman, J.S.; Griffin, M.P.; Moorman, J.R. Sample entropy analysis of neonatal heart rate variability. *Am. J. Physiol. Regul. Integr. Comp. Physiol.* **2002**, *283*, 789–797.

13. Ahamed, V.T.; Karthick, N.G.; Joseph, P.K. Effect of mobile phone radiation on heart rate variability. *Comput. Biol. Med.* **2008**, *38*, 709–712.

14. Al-Angari, H.M.; Sahakian, A.V. Use of sample entropy approach to study heart rate variability in obstructive sleep apnea syndrome. *IEEE Trans. Biomed. Eng.* **2007**, *54*, 1900–1904.

15. Song, Y.; Crowcroft, J.; Zhang, J. Automatic epileptic seizure detection in EEGs based on optimized sample entropy and extreme learning machine. *J. Neurosci. Methods* **2012**, *210*, 132–146.

16. Alcaraz, R.; Rieta, J.J. Sample entropy of the main atrial wave predicts spontaneous termination of paroxysmal atrial fibrillation. *Med. Eng. Phys.* **2009**, *31*, 917–922.

17. Ramdani, S.; Seigle, B.; Lagarde, J.; Bouchara, F.; Bernard, P.L. On the use of sample entropy to analyze human postural sway data. *Med. Eng. Phys.* **2009**, *31*, 1023–1031.

18. Manis, G.; Nikolopoulos, S. Speeding up the computation of approximate entropy. In *11th Mediterranean Conference on Medical and Biomedical Engineering and Computing 2007*; Springer: Berlin/Heidelberg, Germany, 2007.

19. Manis, G. Fast computation of approximate entropy. *Comput. Methods Programs Biomed.* **2008**, *91*, 48–54.

20. Yu-Hsiang, P.; Wang, Y.H.; Liang, S.F.; Lee, K.T. Fast computation of sample entropy and approximate entropy in biomedicine. *Comput. Methods Programs Biomed.* **2011**, *104*, 382–396.

21. Jiang, Y.; Mao, D.; Xu, Y. A fast algorithm for computing Sample Entropy. *Adv. Adapt. Data Anal.* **2011**, *3*, 167–186.

22. Pincus, S.M.; Goldberger, A.L. Physiological time-series analysis: What does regularity quantify. *Am. J. Physiol. Heart Circ. Physiol.* **1994**, *266*, 1643–1656.

23. Aktaruzzaman, M.; Sassi, R. Parametric estimation of sample entropy in heart rate variability analysis. *Biomed. Signal Process. Control* **2014**, *14*, 141–147.

24. Alcaraz, R.; Abásolo, D.; Hornero, R.; Rieta, J.J. Optimal parameters study for sample entropy-based atrial fibrillation organization analysis. *Comput. Methods Programs Biomed.* **2010**, *99*, 124–132.

25. Goldberger, A.L.; Amaral, L.A.N.; Glass, L.; Hausdorff, J.M.; Ivanov, P.C.; Mark, R.G.; Mietus, J.E.; Moody, G.B.; Peng, C.K.; Stanley, H.E. PhysioBank, PhysioToolkit, and PhysioNet: Components of a New Research Resource for Complex Physiologic Signals. *Circulation* **2000**, *101*, e215–e220.

26. Fusheng, Y.; Bo, H.; Qingyu, T. Approximate Entropy and Its Application to Biosignal Analysis. In *Nonlinear Biomedical Signal Processing: Dynamic Analysis and Modeling, Volume 2*; Akay, M., Ed.; Wiley-IEEE Press: New York, NY, USA, 2000; pp. 72–91.

27. Grassberger, P. An Optimized Box-Assisted Algorithm for Fractal Dimensions. *Phys. Lett. A* **1990**, *148*, 63–68.

28. Stuart, B.; Mark, K. Multidimensional Trees, Range Searching, and a Correlation Dimension Algorithm of Reduced Complexity. *Phys. Lett. A* **1989**, *140*, 327–330.

*Article*

# A Novel Multivariate Sample Entropy Algorithm for Modeling Time Series Synchronization

David Looney, Tricia Adjei and Danilo P. Mandic *

Department of Electrical and Electronic Engineering, Imperial College London, London SW7 2AZ, UK;
daveloon@gmail.com (D.L.); t.adjei15@imperial.ac.uk (T.A.)
* Correspondence: d.mandic@imperial.ac.uk

Received: 29 November 2017; Accepted: 19 January 2018; Published: 24 January 2018

**Abstract:** Approximate and sample entropy (AE and SE) provide robust measures of the deterministic or stochastic content of a time series (regularity), as well as the degree of structural richness (complexity), through operations at multiple data scales. Despite the success of the univariate algorithms, multivariate sample entropy (mSE) algorithms are still in their infancy and have considerable shortcomings. Not only are existing mSE algorithms unable to analyse within- and cross-channel dynamics, they can counter-intuitively interpret increased correlation between variates as decreased regularity. To this end, we first revisit the embedding of multivariate delay vectors (DVs), critical to ensuring physically meaningful and accurate analysis. We next propose a novel mSE algorithm and demonstrate its improved performance over existing work, for synthetic data and for classifying wake and sleep states from real-world physiological data. It is furthermore revealed that, unlike other tools, such as the correlation of phase synchrony, synchronized regularity dynamics are uniquely identified via mSE analysis. In addition, a model for the operation of this novel algorithm in the presence of white Gaussian noise is presented, which, in contrast to the existing algorithms, reveals for the first time that increasing correlation between different variates reduces entropy.

**Keywords:** multivariate sample entropy; time series synchronization; structural complexity

---

## 1. Introduction

Multivariate data analysis tools are essential for characterizing the interaction between the variates of complex systems; applications are wide-ranging and include those in biology [1], climatology [2] and finance [3]. Standard methods for estimating interdependencies between multiple data channels are almost invariably linear; typical examples are cross-correlation and coherence (correlation in the frequency domain). More advanced methods, such as Granger causality [4], offer insight into the temporal ordering of interactions and are widely used in, for example, neuroscience applications, for which directionality information is of value. Partial directed coherence extends the concept of Granger causality to the frequency domain [5].

While progress has been made on nonlinear extensions of the above second-order algorithms, information-theoretic measures, such as mutual information and transfer entropy [6], are intrinsically suited to cater for higher-order interactions. Traditional entropy methods are limited by their requirements for large numbers of samples and sensitivity to noise [7]. To this end, the approximate entropy (AE) algorithm [7] was developed to provide a statistically valid measure of entropy for real-world univariate data that may be noisy and short. It represents the probability that similar patterns (delay vectors—DVs) in a time series will remain similar once the pattern lengths are increased (extended DVs), thereby providing a natural measure of the *regularity* of a time series. An extension, the sample entropy (SE) algorithm [8], improves the bias issues experienced by AE by omitting self-matches from the similarity calculation stage. In general, a stochastic time series is characterized by high AE/SE, while a regular signal exhibits low AE/SE. Additionally, multiscale extensions of the algorithm (multiscale sample entropy—MSE) [9,10] have expanded its capabilities beyond regularity estimation to evaluating the structural richness of a signal-generating system, its *complexity*. Both the SE and MSE algorithms have been widely used in physiological applications [11–13].

Recently, two multivariate extensions of the SE and MSE algorithms have been developed to evaluate the regularity and complexity for any number of data channels in a rigorous and unified way [14,15]. The key difference between the two multivariate algorithms is the manner in which multivariate DVs are extended. In the first extension, termed the naïve method [14], extended DV subspaces are generated for each variate, and self-matches are computed for each subspace separately. A shortcoming of the naïve method is its failure to account for inter-channel couplings, such as correlation between the variates. This was reflected in the "full method" [15]. Critically, a multivariate MSE (mMSE) method that caters fully for cross-channel dynamics enables the modeling of any complexity that exists between the variates, offering greater insight for physical systems that are typically multivariate and correlated. In [15], the benefits of a multivariate approach over univariate algorithms were demonstrated for applications spanning the categorization of wind regimes and the analysis of human gait recordings.

Despite the clear improvements of the full multivariate method over existing work, there are several concerns regarding its operation. Firstly, the way in which the DVs are extended is such that distances are directly calculated between elements of the different variates, potentially obscuring the physical meaning of the analysis for heterogeneous data. Secondly, such inconsistencies in the alignment of extended DVs can lead to inaccuracies. Finally, empirical results obtained using the full method for white Gaussian noise (WGN) and $1/f$ noise imply that regularity decreases and complexity increases with increasing correlation. However, viewing increased correlation between the variates as a decrease in cross-channel regularity is not consistent with physical intuition, motivating this work. It must also be noted that it has been suggested that the size of the time delay employed in the SE algorithm can have an effect on the computed entropy [16]; however, for the purposes of consistency across all analyses reported in this study, a unity time delay is employed.

To address these concerns, we propose a novel multivariate sample entropy (mSE) and its multiscale extension for complexity analysis (mMSE). At the core of the algorithm is our novel treatment of multivariate DVs, which ensures (i) element-by-element distances are not computed directly between different variates, and (ii) alignments between multivariate DVs remain consistent and independent of the number of data channels. Simulation results for WGN illustrate how the proposed algorithm interprets increased correlation between the variates as an increase in regularity—a missing result that is in agreement with physical intuition. To support this, we have derived a model for the performance of the algorithm in the presence of bivariate WGN; its numerical outcomes are in agreement with simulations. It is furthermore shown via a random alignment operation applied to the variates that this makes it possible to comprehensively distinguish between within- and cross-channel dynamics, thus providing additional insight into inter- versus intra-modal properties.

The algorithm is validated on synthetic and real data for biological applications. Through simulations, we reveal a previously unstudied feature of multivariate SE algorithms—their ability to detect synchronized regularity dynamics. Unlike other measures of synchronization, which assume temporal locking of phase information (phase synchrony [17,18]) or the existence of a functional mapping between the variates (generalized synchrony [19]), synchronized regularity can exist between time series that are generated by independent processes.

## 2. Sample Entropy

Motivated by the shortcomings of standard entropy measures for short and noisy time series, the approximate entropy technique was introduced in [7]. It characterizes the likelihood that similar patterns within a time series, the signal DVs, will remain similar when the pattern lengths are increased. A robust extension, which neglects self-matches, called the SE, has been developed [8] and is described below:

1. For lag $\tau$ and embedding dimension $m$, generate DVs:

$$X_m(i) = [x_i, x_{i+1}, \ldots, x_{i+\tau(m-1)}] \tag{1}$$

where $i = 1, 2, \ldots, (N - \tau(m-1))$.

2. For a given DV, $X_m(i)$, and a threshold, $r$, count the number of instances, $\Phi_m(i, r)$, for which $d\{X_m(i), X_m(j)\} \le r, i \ne j$, where $d\{\cdot\}$ denotes the maximum norm.

3. Define the frequency of occurrence as

$$\Phi_m(r) = \frac{1}{N - \tau(m-1) + 1} \sum_{i=1}^{N-\tau(m-1)+1} \Phi_m(i, r) \tag{2}$$

4. Extend the embedding dimension ($m \to m+1$) of the DVs in step (1), and repeat steps (2) and (3) to obtain $\Phi_{m+1}(r)$.

5. The SE is defined as the negative logarithm of the values for different embedding dimensions, that is,

$$SE(m, r, \tau) = -ln\left[\frac{\Phi_{m+1}(r)}{\Phi_m(r)}\right] \tag{3}$$

In general, the less predictable or the more irregular a time series, the higher its SE. A block diagram of the algorithm is shown in Figure 1.

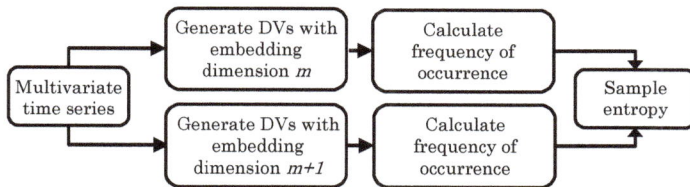

**Figure 1.** Block diagram of the multivariate sample entropy (mSE) algorithm.

### 2.1. Multiscale Sample Entropy

The MSE calculates the SE over multiple time scales in the data [9,10]. A key insight is given by the fact that SE for WGN (which has no structure) decreases for increasing scale factors, while the SE for $1/f$ noise, which has a self-similar infinitely repeating behaviour (contains structure), remains constant with scale. In this way, the multiscale extension reveals long-range signal correlations—dynamics closely linked to the signal complexity. If the SE of a time series remains high over multiple scale factors, it is said to exhibit high complexity.

## 3. Multivariate Sample Entropy

### 3.1. Existing Algorithms

Two multivariate extensions of the SE algorithm have been proposed in [14,15]. In both cases, the *un-extended* multivariate DVs are generated on the basis of the approach outlined by Cao et al. [20]. Given a $P$-variate time series, $x_{k,i}$ where $k = 1, \ldots, P$ is the channel index and $i = 1, \ldots, N$ is the sample index, the multivariate DVs are given by

$$X_M(i) = [x_{1,i}, \ldots, x_{1,i+(m_1-1)\tau_1},$$
$$x_{2,i}, \ldots, x_{2,i+(m_2-1)\tau_2}, \tag{4}$$
$$\ldots,$$
$$x_{P,i}, \ldots, x_{P,i+(m_P-1)\tau_P}]$$

where $M = [m_1, m_2, \ldots, m_P]$ is the multivariate embedding dimension vector and $\o = [\tau_1, \tau_2, \ldots, \tau_P]$ is the multivariate time-lag vector. Figure 2 shows a bivariate time series and its multivariate DVs with embedding dimension $M = [1, 1]$.

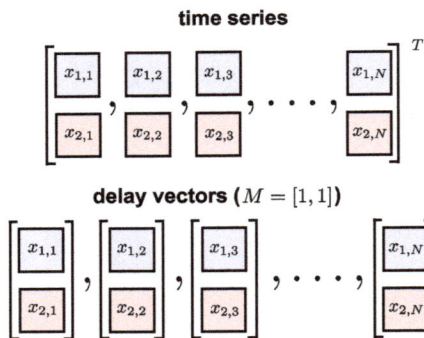

**Figure 2.** A bivariate time series (upper panel) and its delay vectors with embedding dimension $M = [1, 1]$ (lower panel).

To extend the multivariate DVs, the naïve approach in [14] generates $P$ DV subspaces, in which the length of a specific variate is extended but other variates are left unchanged. The extended DV subspace for a variate $k$ is given by

$$X_M^{e_k}(i) = [x_{1,i}, \ldots, x_{1,i+(m_1-1)\tau_1}, \ldots,$$
$$x_{k,i}, \ldots, x_{k,i+(m_k-1)\tau_k}, x_{k,i+(m_k)\tau_k}, \tag{5}$$
$$\ldots,$$
$$x_{P,i}, \ldots, x_{P,i+(m_P-1)\tau_P}]$$

for $i = 1, \ldots, N$ and $e_k = e_1, \ldots, e_P$. For the bivariate case, this approach to DV extension is illustrated in Figure 3. Distances are then calculated between DV pairs in each of the $P$ subspaces to obtain $P$ estimates for the frequency of occurrence, and the final value is obtained by averaging the subspace estimates. The next steps of the algorithm are the same as for the univariate case. A significant shortcoming of the naïve method is that, by measuring distances between the DVs within separate subspaces, it does not cater fully for inter-channel couplings. This was illustrated in [15], where the naïve method could not distinguish between different degrees of correlated variates.

To address these issues, the full method was proposed in [15], in which pair-wise distances are calculated between all DVs across all the subspaces, as described in Equation (5) and illustrated for the bivariate case in Figure 3. This enables the enhanced modeling of inter-channel couplings, as exemplified by its ability to distinguish between different degrees of correlation in the variates. The two approaches are compared in Figure 4 for bivariate WGN with a length of 5000 samples; the mSE parameters were $M = [1, 1]$, $\tau = [1, 1]$ and $r = 0.15$, each of the variates was standardized to unit variance and zero mean, and the correlation between the variates was $\rho = 0.95$. We observe that while the naïve algorithm [14] cannot distinguish between correlated and uncorrelated WGN, the full algorithm [15] exhibits different SE values for each scale factor. As the full algorithm described in [15] represents the most recent multivariate extension to date; in the sequel we refer to it as the *existing method*.

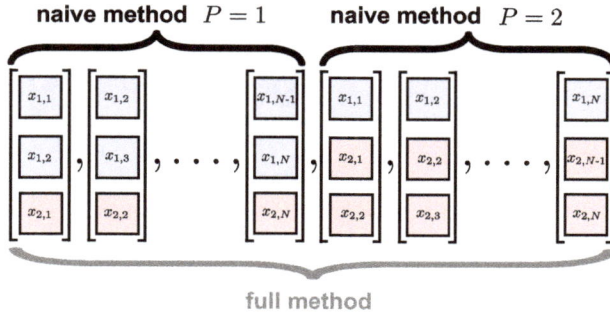

**Figure 3.** The *extended* multivariate delay vectors (DVs) of the multivariate DVs shown in Figure 2 using existing techniques. The *naïve method* [14] calculates the pair-wise distances within the two subspaces ($P = 1$ and $P = 2$) separately, while the *full method* [15] calculates the pair-wise distances across all the delay vectors.

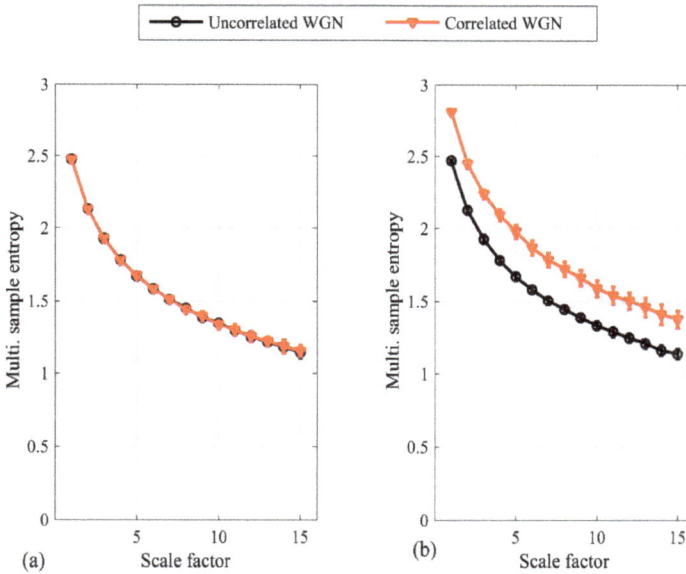

**Figure 4.** Multivariate sample entropy (mSE) estimate for bivariate white Gaussian noise (WGN), correlated and uncorrelated, using the naïve method [14] (**a**); and the full or existing method [15] (**b**). The error bars denote ± standard deviation. We observe that only the full/existing method can distinguish between correlated and uncorrelated variates.

Despite providing clear improvements over the previous mMSE algorithms, there are still issues regarding the existing method that need to be addressed:

1. Inability to cater adequately for heterogeneous data sources.
2. Inconsistencies in the alignment of DVs.

3.    Counter-intuitive representation of multivariate dynamics.

Concerning item 1, in the standard method, the DV distances are directly calculated between elements of different variates. This is illustrated in Figure 3 for the bivariate case. If the variates are heterogeneous, that is, if they reflect different components of a biological system (e.g., cardiac and neural activity), then operations involving element-by-element comparisons will hinder the physical meaning of the analysis.

Concerning item 2 above, Figure 3 shows that one of the distances calculated for the extended DVs of a bivariate signal is between $[x_{1,1}, x_{1,2}, x_{2,1}]^T$ and $[x_{1,2}, x_{1,3}, x_{2,2}]^T$. In this instance, as is also the case in the operation of the univariate algorithm, the delays between corresponding elements are the same ($+1$ in this instance). However, when the distances are calculated between $[x_{1,1}, x_{1,2}, x_{2,1}]^T$ and $[x_{1,1}, x_{2,1}, x_{2,2}]^T$, this results in different delays between the DV elements: $0$ ($\{x_{1,1}, x_{1,1}\}$), $-1$ ($\{x_{1,2}, x_{2,1}\}$) and $+1$ ($\{x_{2,1}, x_{2,2}\}$). This inconsistency in the alignment of the DV elements may cause contradictory results and affect the accuracy of the analysis.

Concerning item 3 above, Figure 4 illustrates that the multivariate SE increases across all scale factors for increasing correlation between the variates. Considering the algorithm output at the first scale factor only, the result indicates that regularity decreases with increasing correlation—a result that is not consistent with physical intuition. Instead, it is expected that as variates become more independent, it will be characterized by a decrease in regularity. Likewise, considering the algorithm output across all scale factors, the result indicates that complexity increases with increasing correlation. *No precise definition of complexity exists; however an intuitive expectation is that the number of required mathematical descriptors for a system should increase with complexity.* For this reason, it can be assumed that as the variates become more correlated/dependent, the system will be characterized by a decrease in complexity, the "complexity loss theory".

### 3.2. Synchronized Regularity

The ability of multivariate SE algorithms to distinguish between coupled regularity dynamics has so far been confined to correlated variates [15] (see also Figure 4). In such studies, the couplings between the variates are static—they remain unchanged across time. Additionally, such relationships are adequately modeled by correlation estimation, not requiring the unique capabilities of a multivariate SE algorithm.

We instead propose a multivariate benchmark test in which the *variates are generated from independent random processes but their regularity is synchronized.* In other words, the within-channel regularity changes at the same time within each variate. We consider the bivariate signal

$$x_{1,i} = \lambda_{1,i}^{\frac{1}{2}} v_{1,i} + (1 - \lambda_{1,i})^{\frac{1}{2}} u_{1,i} \tag{6}$$

$$x_{2,i} = \lambda_{2,i}^{\frac{1}{2}} v_{2,i} + (1 - \lambda_{2,i})^{\frac{1}{2}} u_{2,i} \tag{7}$$

where $i = 1, \ldots, N$, $N$ is the number of samples, $\lambda_{1,i}$ and $\lambda_{2,i} \in [0,1]$ are mixing parameters, $v_{1,i}$ and $v_{2,i}$ are independent realizations of WGN (zero mean and unit variance), and $u_{1,i}$ and $u_{2,i}$ are independent realizations of $1/f$ noise (zero mean and unit variance). We note that no correlation, phase or generalized synchronization exists between the variates, as they are all independent. Instead, the parameter $\lambda_i$ enforces synchronized regularity; when $\lambda_{1,i} = \lambda_{2,i} = 0$, the SE in both variates is low as each contains $1/f$ noise, and when $\lambda_{1,i} = \lambda_{2,i} = 1$, the SE in both variates is high as each contains WGN. A dynamic mixing parameter can be generated by

$$\lambda_{k,i} = \left\{ \begin{array}{ll} 1 & \text{if } z_i(\beta) > 0 \\ 0 & \text{if } z_i(\beta) < 0 \end{array} \right. \tag{8}$$

where $k \in \{1, 2\}$, $i = 1, \ldots, N$ and $z_i(\beta)$ is a random noise time series with a spectral distribution that decays at rate $\beta$. When $\beta = 0$, the time series is WGN; when $\beta = -1$, it is $1/f$ noise; and when $\beta = -2$, it is Brownian noise. Figure 5 shows a bivariate time series generated using the dynamic mixing model described by Equations (6)–(8) with $\beta = -2$; we observe that the within-channel regularities are synchronized.

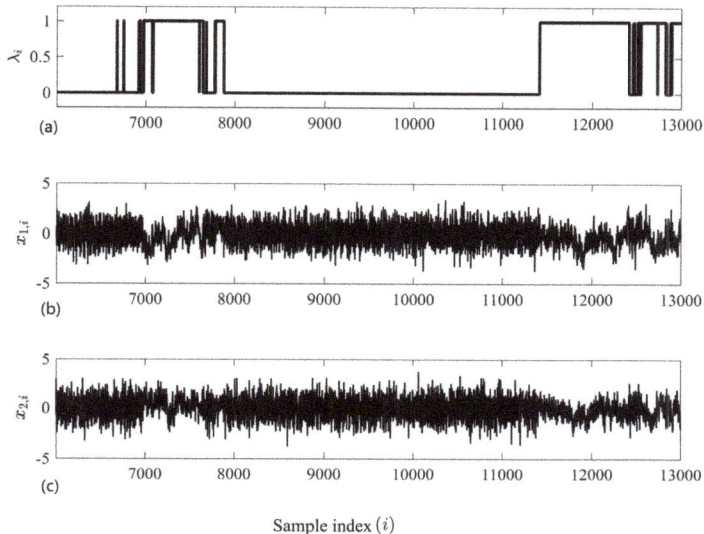

**Figure 5.** A bivariate time series exhibiting synchronized regularity. (**a**): The mixing parameter, which in this instance is the same for each variate. (**b**): The first variate. (**c**): The second variate. We observe that when $\lambda_i = 1$ (Equations (6) and (7)), both variates exhibit $1/f$-type structures (i.e., for $i = 7000, \ldots, 7500$), and when $\lambda_i = 0$ (Equations (6) and (7)), both variates exhibit WGN-type structures (i.e., for $i = 8000, \ldots, 11,000$).

The existing mMSE algorithm was applied to distinguish between the scenarios in which (i) the variates are independent, that is, $\lambda_{1,i} \neq \lambda_{2,i}$; and (ii) the regularity of the variates is synchronized, that is, $\lambda_{1,i} = \lambda_{2,i}$. In both scenarios, $\beta = -1.6$ (see Equation (8)) was used to generate the mixing parameters. The signal length was $N = 15,000$ samples; $r = 0.4$ and $M = [1, 1]$. Figure 6 shows the mMSE for each scenario; a two-tailed two-sample $t$-test was applied to determine the degree of separation at each scale factor (38 degrees of freedom). We observe that a statistically significant separation was determined at scale factor 4 ($p < 0.05$) and at all subsequent scale factors. It is also noted that the *synchronized regularity caused a decrease in the mSE*; this result is in agreement with physical intuition but contradicts results obtained with the same algorithm for correlated variates (see Figure 4b), where coupled regularity dynamics was found to increase the multivariate SE. As real-world processes can be expected to exhibit different forms of coupled regularity dynamics simultaneously, this compromises the accuracy of the algorithm in applications.

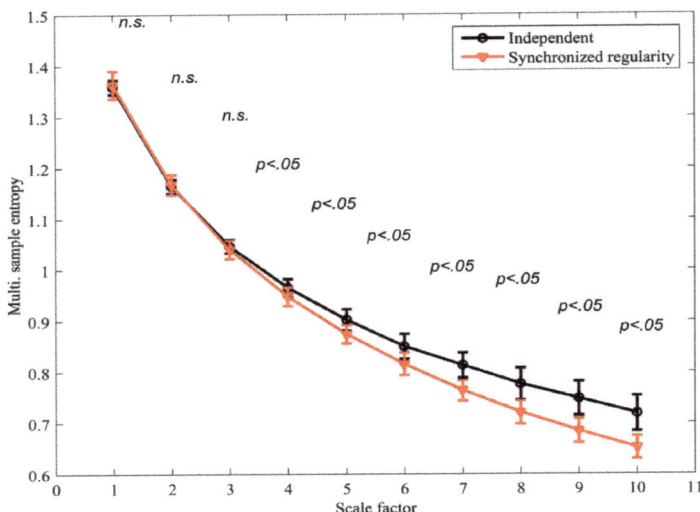

**Figure 6.** Multivariate multiscale sample entropy (mMSE) extension using the existing algorithm for signals with (i) independent mixing parameters, and (ii) synchronized mixing parameters; see model described by Equations (6)–(8). The error bars denote ± standard deviation. A two-tailed two-sample *t*-test was applied to determine the degree of separation between the two scenarios at each scale factor; instances in which the distributions were found to overlap are denoted by "not-significant" (n.s.).

### 3.3. The Proposed Algorithm

At the core of the proposed algorithm, which addresses the above issues, is the manner in which multivariate DVs are extended to calculate the SE. To avoid the direct calculation of distances between the elements of different variates and DV misalignment, and to induce a desired mode-of-operation for correlated data, generating an extended set of multivariate DVs from those in Equation (4) is proposed in the following manner:

$$
\begin{aligned}
X_M^e(i) = \big[ & x_{1,i}, \ldots, x_{1,i+(m_1-1)\tau_1}, x_{1,i+(m_1)\tau_1} \cdots, \\
& x_{k,i}, \ldots, x_{k,i+(m_k-1)\tau_k}, x_{k,i+(m_k)\tau_k} \cdots, \\
& x_{P,i}, \ldots, x_{P,i+(m_P-1)\tau_P}, x_{P,i+(m_P)\tau_P} \big]
\end{aligned}
\tag{9}
$$

for $i = 1, \ldots, N$. For the bivariate case, the proposed DV extension is illustrated in Figure 7. We observe that distances are only calculated between elements of corresponding variates, making the approach perfectly suited for heterogeneous data.

Although the distances between DVs were not directly calculated between the elements of different variates, the estimation of multivariate SE on the basis of the proposed approach can still detect coupled regularity dynamics as well as existing methods, and in some instances, its performance enables better separation (see the following section). To support the claim of a desired mode-of-operation for correlated data, in the Appendix A, we present a model for the operation of the proposed multivariate algorithm in the presence of bivariate WGN; the model reveals how an increasing correlation between the variates reduces the multivariate SE. This is demonstrated in Figure 8 for the multiscale operation of the algorithm for bivariate WGN. The algorithm was also applied to time series with synchronized regularity between the variates generated using the same mixing model as before (see Equation (8) with $\beta = -1.6$); the results are shown in Figure 9. We note that, unlike the existing algorithm, a significant separation between the two time series also exists at the first scale factor (cf. Figure 4).

It is important to note an increase in the degree of coupling between the variates, either via correlation or synchronized regularity; in both cases this causes a decrease in the SE. It is natural to expect that real-world systems will exhibit different forms of coupling simultaneously and that an algorithm that behaves in a consistent way for each form of coupling is better equipped to model changes in multivariate regularity and complexity.

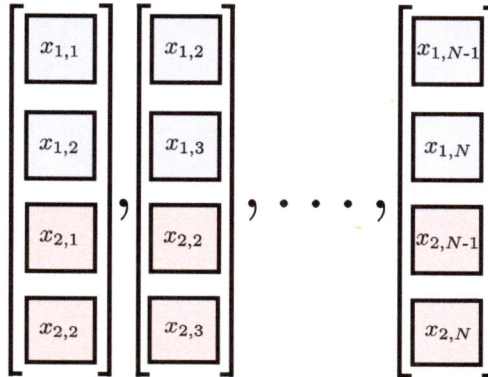

**Figure 7.** The *extended* multivariate delay vectors (DVs) of the multivariate DVs shown in Figure 2 using the proposed DV extension method (see Equation (9)).

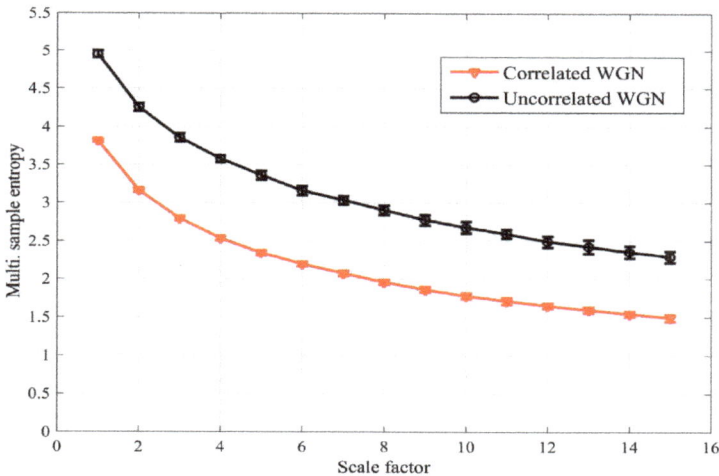

**Figure 8.** Multivariate sample entropy (mSE) estimate for bivariate white Gaussian noise (WGN), correlated and uncorrelated, using the proposed algorithm. The error bars denote ± standard deviation. We observe that increasing correlation reduces the SE (cf. Figure 4).

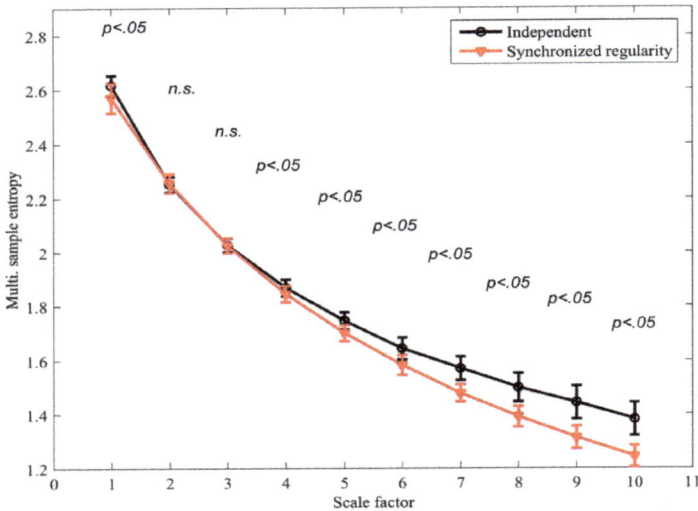

**Figure 9.** The multivariate multiscale sample entropy (mMSE) extension using the proposed algorithm for signals with (i) independent mixing parameters, and (ii) synchronized mixing parameters (see model described by Equations (6)–(8)). The error bars denote ± standard deviation. A two-tailed two-sample *t*-test was also applied to determine the degree of separation between the two scenarios at each scale factor; instances in which the distributions were found to overlap are denoted by "not-significant" (n.s.). Compare with the existing approaches in Figure 6.

### 3.4. Multivariate Surrogates

For a given measure or index, it is common to employ signal surrogates to provide a baseline or reference value. Surrogates are a set of time series that share certain characteristics of the original signal but lack the property whose effect on the measure we wish to study [1]. For instance, in univariate sample analysis, it is common to generate surrogates that retain the spectrum shape of the original signal but that have randomized phase values, in this way creating similar signals with high irregularity and no structure (low complexity). In the same way, for greater insight, the SE of the original signal can be compared with the values obtained for the surrogates.

For the multivariate case, it is desirable to remove any interdependencies between the variates in order to distinguish between within- and cross-channel regularity and complexity. We propose to utilize the multivariate surrogate approach used in previous synchronization measures, for example, in the study of asymmetry [21] and mutual information [22], whereby one of the variates is temporally shifted with respect to the other (random alignment). Figure 10 shows the mSE, obtained with parameters $M = [1, 1]$, $\tau = [1, 1]$ and $r = 0.15$, for correlated bivariate WGN ($\rho = 0.6$; 5000 samples), as well as the average values obtained for 30 surrogates created by randomly shifting the second variate. We observe that the mSE of the original signal is significantly lower than that obtained for the surrogates, indicating the presence of coupled dynamics between the variates.

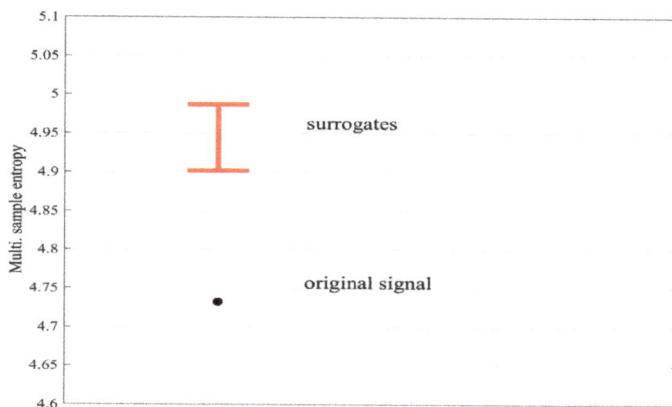

**Figure 10.** The multivariate sample entropy (mSE) for the original signal with correlated variates (black circle) and its multivariate surrogates (red error bars, which denote ± standard deviation). In the case of the surrogates, the interdependency (correlation) between the variates is removed and thus its average mSE value is higher, indicating the presence of coupled dynamics in the original signal.

## 4. Simulations

The performance of the proposed algorithm was illustrated over the following case studies: (i) detecting joint regularity dynamics between recordings of cardiac function, and (ii) classifying multichannel recordings of brain function during wake and sleep states. In all cases, unless stated otherwise, the time series were scaled to zero mean and unit variance prior to mSE analysis.

### 4.1. Detecting Synchronized Cardiac Behaviour

We firstly considered a real-world synchronized regularity scenario. A healthy male subject (aged 32) performed a series of breathing exercises while his cardiac function was monitored via the electrocardiogram (ECG); electrodes were placed on the chest and ECG data were recorded using the gtec g.USBamp, a certified and U.S. Food and Drug Administration (FDA)-listed medical device, with a sampling frequency of 1200 Hz. The time difference between R peaks, sharp dominant peaks in the ECG waveform, was calculated and the R–R interval time series was generated using cubic spline interpolation at regular time intervals of 0.25 s.

For a given 30 s period, the instruction was either to (i) breathe normally (unconstrained), or (ii) breathe at a fixed rate of 15 breathing cycles (inhale/exhale) per minute aided by a metronome (constrained). The instruction was alternated from period to period. The periods of constrained breathing had the effect of inducing different regularity into the R–R interval through the phenomenon of respiratory sinus arrhythmia (RSA), the modulation of the cardiac output by the respiration effort [23]. Two 300 s trials were recorded; in the first, the subject started breathing in a constrained fashion, and in the second, the subject started breathing in an unconstrained fashion. Figure 11 shows a segment of the trial in which the subject started breathing in a constrained fashion. Prior to analysis, a high-pass filter with a cutoff frequency of 0.1 Hz was applied to the R–R interval.

Figure 12 compares the results of the proposed algorithm ($M = [1, 1]$, $\tau = [1, 1]$, and $r = 0.4$) with those obtained by cross-correlation, with the aim to determine the degree of synchronization between the two trials at different time-lags. Spectral analysis of the proposed mSE results revealed that synchronized regularity occurred at 60 s intervals (approximately $1/60 = 0.0167$ Hz)—this clearly indicates joint regularity between the recording trials whenever the subject performed the same breathing instruction every $2 \times 30$ s. Furthermore, the approach revealed the correct time-lag between the trials, with the minimum mSE value occurring at lag −32 s. We observe that the cross-correlation approach was unable to reveal the synchronization in regularity; spectral analysis of the results showed no peak at 0.017 Hz.

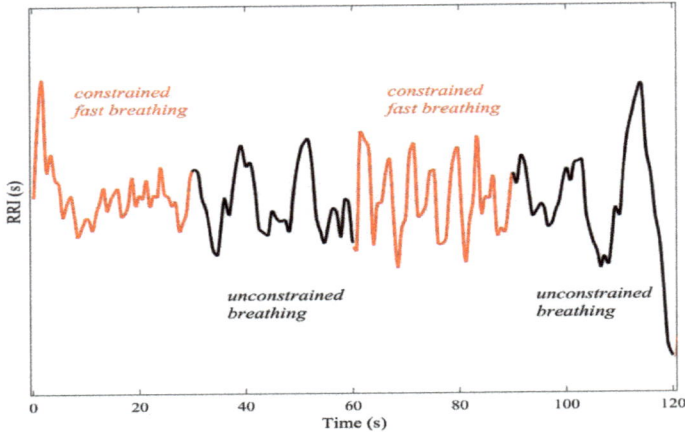

**Figure 11.** The R–R interval as the subject performed different breathing instructions (constrained/ unconstrained) every 30 s. As a result of the phenomenon of respiratory sinus arrhythmia, the breathing instructions induced periods of different regularity into the R–R interval.

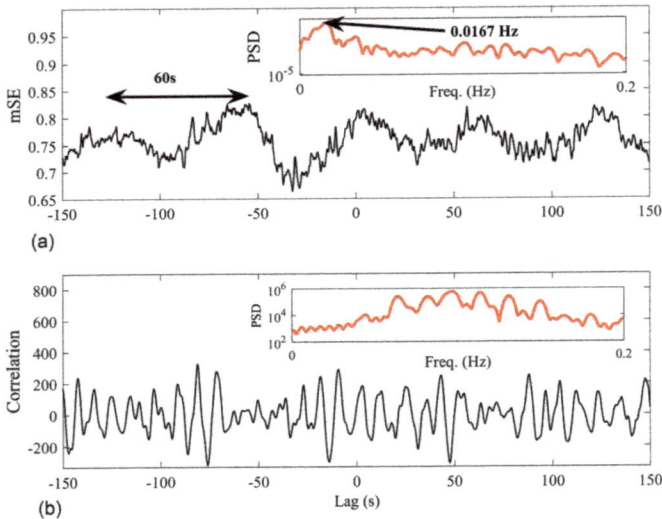

**Figure 12.** Multivariate sample entropy (mSE; (**a**)) and cross-correlation (**b**) results when applied to detect synchronized activity between two trials of R–R interval recordings. In the upper corner of each panel in red, the spectral analysis of each approach reveals the dominant synchronization period; only the mSE approach reveals synchronized regularity dynamics at 0.017 Hz (1/60).

### 4.2. Classifying Sleep States

The proposed algorithm was next applied to recordings of brain function during wake and sleep states. The data, electroencephalogram (EEG) recordings extracted from whole-night polysomnography recordings, were obtained from (http://www.tcts.fpms.ac.be/~devuyst/Databases/DatabaseSpindles/). Each 5 s segment of the EEG was independently annotated by an expert according to the Rechtschaffen and Kales criteria (wake; Rapid Eye Movement (REM); sleep stages S1, S2, S3 and S4). The task was to distinguish between the wake state and the state of slow-wave sleep (stages S3 and S4). Eight 30 min excerpts were available; however only four of these were considered. For two excerpts (2 and 7), there were insufficient wake periods (no sustained wake periods exceeding 1 min); for another excerpt (3), the sampling frequency was too low for meaningful analysis (50 Hz); and for another (8) there were no periods of slow-wave sleep. The excerpts analyzed were 1, 4, 5 and 6; excerpt 1 had a sampling frequency of 100 Hz, and the two channels of the EEG were obtained using the bipolar electrode configurations C3-A1 and FP1-A1; excerpts 4, 5 and 6 had a sampling frequency of 200 Hz, and the two channels of the EEG were obtained using the bipolar electrode configurations CZ-A1 and FP1-A1.

The multivariate SE was estimated ($M = [1,1]$, $\tau = [1,1]$, and $r = 0.4$) between corresponding EEG segments of length 2.5 s with a 50% overlap, with segments containing voltage values exceeding $\pm 100$ µV rejected to discard artefacts. The difference between the multivariate SE values for the states of wake and slow-wave sleep were determined using the Bhattacharyya distance, given by

$$D = \frac{1}{4} \ln \left( \frac{1}{4} \left( \frac{\sigma_w^2}{\sigma_s^2} + \frac{\sigma_s^2}{\sigma_w^2} + 2 \right) \right) + \frac{1}{4} \left( \frac{(\mu_s - \mu_w)^2}{\sigma_s^2 + \sigma_w^2} \right) \tag{10}$$

where $\mu_w$ and $\sigma_w$ denote the mean and standard deviation of the feature values for the wake state, and $\mu_s$ and $\sigma_s$ are the same values for the slow-wave sleep state. The degrees of separation using the existing and the proposed algorithm are shown in Table 1; we observe that for all excerpts, the degrees of separation between the states of wake and slow-wave sleep were greater using the proposed algorithm (the value of $D$ is larger).

**Table 1.** The first two columns denote the degree of separation, calculated using the Bhattacharyya distance ($D$; see Equation (10)), in multivariate sample entropy (mSE) using the existing and proposed methods for bivariate electroencephalogram (EEG) recordings, between the states of wake (W) and slow-wave sleep (SWS), as annotated by an expert. The last columns denote the percentage of the entire recording annotated as wake (% W) and slow-wave sleep (% SWS).

|  | Existing | Proposed | % W | % SWS |
|---|---|---|---|---|
| Excerpt 1 | $D = 1.28$ | $D = 1.66$ | 14.5 | 19.9 |
| Excerpt 4 | $D = 1.99$ | $D = 3.54$ | 27.1 | 6.5 |
| Excerpt 5 | $D = 1.10$ | $D = 2.06$ | 10.8 | 28.6 |
| Excerpt 6 | $D = 1.48$ | $D = 1.61$ | 3.3 | 32.3 |

## 5. Conclusions

We have introduced an mSE algorithm that, unlike existing approaches, caters for heterogeneous data and yields improved insight into coupled regularity dynamics. This is exemplified, in part, by its consistent treatment of correlation and synchronized regularity, both of which exhibit low mSE values; this is in agreement with intuition concerning regularity and the complexity loss theory. Simulation results for the proposed algorithm reveal greater separation between neural data during different sleep stages. Unlike standard tools, which are invariably linear, it is also shown how the approach is sensitive to higher-order synchronization. Multivariate surrogate-generation techniques have been shown to enhance the significance of the results. The concept of synchronized regularity has been illuminated and a benchmark test proposed, which opens a new avenue of research in a number of applications.

**Acknowledgments:** The authors wish to thank Professor Mary Morrell of the Sleep Unit, Royal Brompton Hospital, Imperial College London, United Kingdom, for all her assistance and guidance concerning the application of our work on sleep. Our collaboration has been supported by a UK Biomedical Research Unit Pump Priming grant.

**Author Contributions:** David Looney performed the data analysis reported in this paper, under the supervision of Danilo P. Mandic. Tricia Adjei revised this paper. All authors read and approved the final manuscript.

**Conflicts of Interest:** The authors declare no conflict of interest.

## Abbreviations

The following abbreviations are used in this manuscript:

AE      Approximate entropy
DV      Delay vector
ECG     Electrocardiogram
EEG     Electroencephalogram
mMSE    Multivariate multiscale sample entropy
MSE     Multiscale sample entropy
mSE     Multivariate sample entropy
RSA     Respiratory sinus arrhythmia
SE      Sample entropy
WGN     White Gaussian noise

## Appendix A. Multivariate SE for WGN

One of the key SE results is that for WGN. We here revisit the model for the operation of the univariate algorithm, presented in [10], and then derive a model for the case of bivariate WGN.

In the case of the univariate algorithm when $m = 1$, the denominator in Equation (3) indicates the probability that the distance between two data points is less than or equal to $r$, that is, $|x_i - x_j| \leq r$, which we denote by

$$P_{\{m=1,r\}} = P(|x_i - x_j| \leq r) \tag{A1}$$

The physical meaning of the numerator in Equation (3) is the probability that the pair-wise distance for DVs with embedding dimension $m = 2$, the extended DVs, is less than or equal to $r$. As the elements in the DVs are independent random variables, the probability can be rewritten as follows (for simplicity, we here assume $\tau = 1$):

$$\begin{aligned} P_{\{m=2,r\}} &= P(|x_i - x_j| \leq r) \times P(|x_{i+1} - x_{j+1}| \leq r) \\ &= P_{\{m=1,r\}} \, P_{\{m=1,r\}} \end{aligned} \tag{A2}$$

Generalizing this result for any embedding dimension, for univariate WGN, the SE is given by

$$\begin{aligned} \mathrm{SE}(m,r) &= \log_e \frac{P_{\{m,r\}}}{P_{\{m+1,r\}}} \\ &= \log_e \frac{P_{\{m=1,r\}} \, P_{\{m=1,r\}} \cdots P_{\{m=1,r\}}}{P_{\{m=1,r\}} \, P_{\{m=1,r\}} \cdots P_{\{m=1,r\}} P_{\{m=1,r\}}} \\ &= -\log_e P_{\{m=1,r\}} \end{aligned} \tag{A3}$$

Thus, the SE for WGN is defined by $P_{\{m=1,r\}}$, which, for zero mean and unit variance, is given by

$$P_{\{m=1,r\}} = P(|x_i - x_j| \le r) =$$

$$\int_{-\infty}^{\infty} \left\{ \int_{x_i-r}^{x_i+r} p(x_j)dx_j \right\} p(x_i)dx_i =$$

$$\frac{1}{2\pi} \int_{-\infty}^{\infty} \left\{ \int_{x_i-r}^{x_i+r} e^{-x_j^2/2}dx_j \right\} e^{-x_i^2/2}dx_i = \tag{A4}$$

$$\frac{1}{2\pi} \int_{-\infty}^{\infty} \left\{ \mathrm{erf}\left(\frac{x_i+r}{\sqrt{2}}\right) - \mathrm{erf}\left(\frac{x_i-r}{\sqrt{2}}\right) \right\} e^{-x_i^2/2}dx_i$$

The term within curly brackets $\{\cdot\}$ denotes the probability that a point at $x_i$ is within distance $r$ of another randomly selected point. This is represented by the area under the normal curve in Figure A1. Numerically evaluating the model for $r = 0.15$ gives 2.47, which matches the simulation result for the same parameters (see first scale factor in Figure 4b).

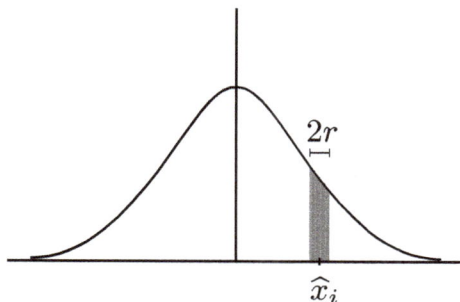

**Figure A1.** The area under the Gaussian curve, centred at $\hat{x}_i$, denotes the probability that the distance between this point and any other randomly selected point is less than or equal to $r$.

For the case of bivariate WGN, the proposed multivariate algorithm calculates the SE as

$$\mathrm{mSE}(M, r) = \log_e \frac{P_{\{M,r\}}}{P_{\{M^e,r\}}} \tag{A5}$$

where $P_{\{M,r\}}$ is the probability that the pair-wise distance for DVs with embedding dimension $M$ is less than or equal to $r$, and $P_{\{M^e,r\}}$ is the probability for the extended case as described by Equation (9). Similarly to the univariate case, independence between the multivariate DVs can be assumed, and the above can be rewritten as

$$\mathrm{mSE}(M, r) = -\log_e P_{\{M=[1,1],r\}} \tag{A6}$$

where the probability is given by

$$P_{\{M=[1,1],r\}} = P(|x_{1,i} - x_{1,k}| \le r, |x_{1,j} - x_{1,l}| \le r) =$$

$$\int_{-\infty}^{\infty} \int_{-\infty}^{\infty} \left\{ \int_{x_{1,k}=x_{1,i}-r}^{x_{1,i}+r} \int_{x_{1,l}=x_{1,j}-r}^{x_{1,j}+r} p(x_{1,k}, x_{2,l}) \delta x_{1,k} \delta x_{2,l} \right\} \tag{A7}$$

$$\times \, p(x_{1,i}, x_{2,j}) \delta x_{1,i} \delta x_{2,j}$$

for which, assuming variates with zero mean and unit variance, the joint probability is given by

$$p(x_{1,i}, x_{2,j}) =$$

$$\frac{1}{2\pi\sqrt{1-\rho^2}} e^{\left(-\frac{1}{2(1-\rho^2)}\left[x_{1,i}^2 + x_{2,j}^2 - 2\rho x_{1,i} x_{2,j}\right]\right)} \tag{A8}$$

and $\rho$ is the correlation coefficient. Similarly to the univariate case, the term within curly brackets $\{\cdot\}$ denotes the probability that a point at $x_{1,i}, x_{2,j}$ is within distance $r$ of another randomly selected bivariate point; this is represented by the volume under the bivariate surface in Figure A2.

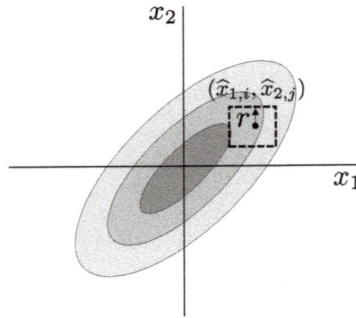

**Figure A2.** The volume under the surface of a normal bivariate distribution, centred at $(\widehat{x}_{1,i}, \widehat{x}_{2,j})$, denotes the probability that the distance between this point and any other randomly selected point is less than or equal to $r$.

Unlike the univariate case, however, a closed-form solution for the term enclosed in brackets does not exist for $\rho \neq 0$. However, it is possible to approximate the cumulative distribution function:

$$F(\widehat{x}_{1,i}, \widehat{x}_{2,j}; \rho) = P(x_{1,k} > \widehat{x}_{1,i}, x_{2,l} > \widehat{x}_{2,j}) \tag{A9}$$

as follows [24]:

$$\begin{aligned} F(\widehat{x}_{1,i}, \widehat{x}_{2,j}) &\approx L(\widehat{x}_{1,i}, \widehat{x}_{2,j}; \rho) = \\ F(-\widehat{x}_{1,i}) &[F\{\zeta(\widehat{x}_{1,i}, \widehat{x}_{2,j}; \rho)\} - \tfrac{1}{2}\rho^2(1-\rho^2)^{-1} \\ \zeta(\widehat{x}_{1,i}, \widehat{x}_{2,j}; \rho) &p\{\zeta(\widehat{x}_{1,i}, \widehat{x}_{2,j}; \rho)\}[1 + \mu(\widehat{x}_{1,i}) - \mu^2(\widehat{x}_{1,i})]] \end{aligned} \tag{A10}$$

where

$$\zeta(\widehat{x}_{1,i}, \widehat{x}_{2,j}; \rho) = \frac{\rho\mu(\widehat{x}_{1,i}) - \widehat{x}_{2,j}}{\sqrt{(1-\rho^2)}} \tag{A11}$$

and

$$\mu(\widehat{x}_{1,i}) = \frac{\widehat{x}_{1,i}}{F(-\widehat{x}_{1,i})} \tag{A12}$$

Using this approximation, for which the accuracy decreases for increasing $\rho$, we can derive an approximation to the probability in Equation (A8) as

$$P_{\{M=[1,1],r\}} \approx R_a - (R_b + R_c) \tag{A13}$$

where

$$R_a =$$
$$L(\widehat{x}_{1,i} - r, \widehat{x}_{2,j} - r, \rho) - L(\widehat{x}_{1,i} + r, \widehat{x}_{2,j} + r, \rho) \tag{A14}$$
$$R_b =$$
$$L(\widehat{x}_{1,i} - r, \widehat{x}_{2,j} - r, \rho) - L(\widehat{x}_{1,i} + r, \widehat{x}_{2,j} - r, \rho) \tag{A15}$$
$$R_c =$$
$$L(\widehat{x}_{1,i} - r, \widehat{x}_{2,j} - r, \rho) - L(\widehat{x}_{1,i} - r, \widehat{x}_{2,j} + r, \rho) \tag{A16}$$

The above expressions are illustrated in Figure A3. Combining the probability approximation in Equation (A13) with Equation (A6), we obtain a model that approximates the behaviour of the proposed algorithm for bivariate WGN for any correlation between the variates. A comparison of the numerical evaluation of the model and simulation results (averaged over 100 realizations of bivariate WGN; $M = [1, 1]$ and $r = 0.15$; variate lengths of 3000 samples) for different correlation coefficient values ($\rho$) is shown in Figure A4. We observe that the model results match the simulation results, particularly for small values of $\rho$. However, the model is less accurate for larger values of $\rho$, as the accuracy of the approximation of the cumulative distribution function from [24], on which the model is based, decreases.

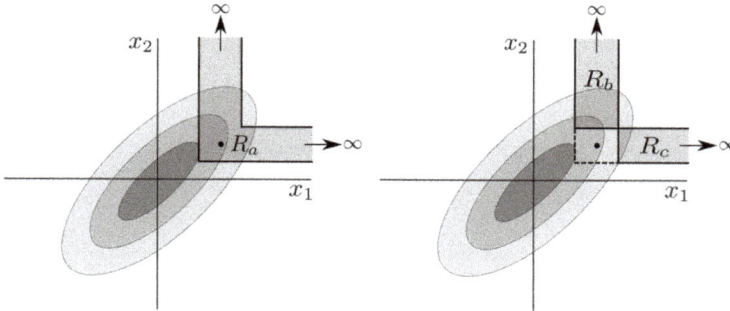

**Figure A3.** Approximating the bivariate cumulative distribution function at the locations shown can be used to approximate $P_{\{M=[1,1],r\}}$ (see Equation (A13)).

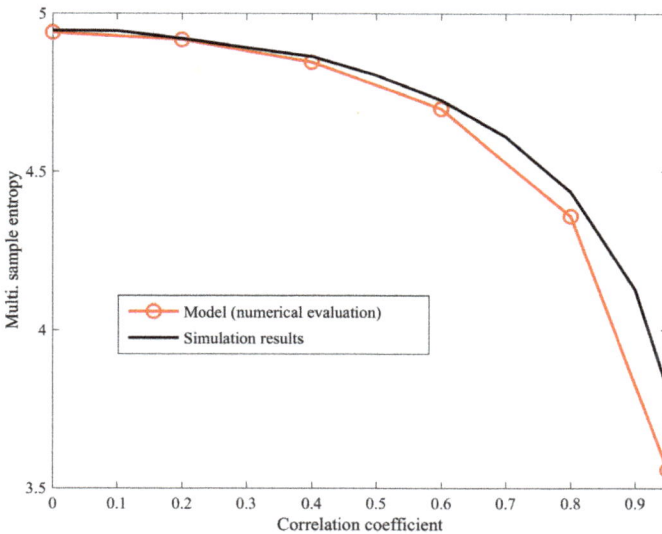

**Figure A4.** Numerical evaluation of the derived model (Equation (A13)) and simulation results for bivariate white Gaussian noise (WGN) with varying degrees of correlation ($\rho$). As expected, the accuracy of the model decreases for larger correlation values.

## References

1. Pereda, E.; Quiroga, R.Q.; Bhattacharya, J. Nonlinear multivariate analysis of neurophysiological signals. *Prog. Neurobiol.* **2005**, *77*, 1–37.

2.  Burke, D.J.; O'Malley, M.J. A study of principal component analysis applied to spatially distributed wind power. *IEEE Trans. Power Syst.* **2011**, *26*, 2084–2092.
3.  Filis, G.; Degiannakis, S.; Floros, C. Dynamic correlation between stock market and oil prices: The case of oil-importing and oil-exporting countries. *Int. Rev. Financ. Anal.* **2011**, *20*, 152–164.
4.  Granger, C.W.J. Investigating Causal Relations by Econometric Models and Cross-spectral Methods. *Econometrica* **1969**, *37*, 424–438.
5.  Sameshima, K.; Baccalá, L.A. Using partial directed coherence to describe neuronal ensemble interactions. *J. Neurosci. Methods* **1999**, *94*, 93–103.
6.  Schreiber, T. Measuring Information Transfer. *Phys. Rev. Lett.* **2000**, *85*, 461–464.
7.  Pincus, S.M. Approximate entropy as a measure of system complexity. *Proc. Natl. Acad. Sci. USA* **1991**, *88*, 2297–2301.
8.  Richman, J.S.; Moorman, J.R. Physiological time-series analysis using approximate entropy and sample entropy. *Am. J. Physiol. Heart Circ. Physiol.* **2000**, *278*, H2039–H2049.
9.  Costa, M.; Goldberger, A.L.; Peng, C.K. Multiscale entropy analysis of complex physiologic time series. *Phys. Rev. Lett.* **2002**, *89*, 068102.
10. Costa, M.; Goldberger, A.L.; Peng, C.K. Multiscale entropy analysis of biological signals. *Phys. Rev. E* **2005**, *71*, 021906.
11. Bruce, E.N.; Bruce, M.C.; Vennelaganti, S. Sample entropy tracks changes in electroencephalogram power spectrum with sleep state and aging. *J. Clin. Neurophysiol.* **2009**, *26*, 257–266.
12. Costa, M.; Ghiran, I.; Peng, C.K.; Nicholson-Weller, A.; Goldberger, A.L. Complex dynamics of human red blood cell flickering: Alterations with in vivo aging. *Phys. Rev. E* **2008**, *78*, 020901.
13. Hornero, R.; Abásolo, D.; Escudero, J.; Gómez, C. Nonlinear analysis of electroencephalogram and magnetoencephalogram recordings in patients with Alzheimer's disease. *Philos. Trans. R. Soc. Lond. A Math. Phys. Eng. Sci.* **2009**, *367*, 317–336.
14. Ahmed, M.U.; Li, L.; Cao, J.; Mandic, D.P. Multivariate multiscale entropy for brain consciousness analysis. In Proceedings of the IEEE Engineering in Medicine and Biology Society (EMBC), Boston, MA, USA, 30 August–3 September 2011; pp. 810–813.
15. Ahmed, M.U.; Mandic, D.P. Multivariate multiscale entropy: A tool for complexity analysis of multichannel data. *Phys. Rev. E* **2011**, *84*, 061918.
16. Kaffashi, F.; Foglyano, R.; Wilson, C.G.; Loparo, K.A. The effect of time delay on Approximate & Sample Entropy calculations. *Phys. D Nonlinear Phenom.* **2008**, *237*, 3069–3074.
17. Rosenblum, M.G.; Pikovsky, A.S.; Kurths, J. Phase Synchronization of Chaotic Oscillators. *Phys. Rev. Lett.* **1996**, *76*, 1804–1807.
18. Rosenblum, M.G.; Pikovsky, A.S. Detecting direction of coupling in interacting oscillators. *Phys. Rev. E* **2001**, *64*, 045202.
19. Rulkov, N.F.; Sushchik, M.M.; Tsimring, L.S.; Abarbanel, H.D.I. Generalized synchronization of chaos in directionally coupled chaotic systems. *Phys. Rev. E* **1995**, *51*, 980–994.
20. Cao, L. Dynamics from multivariate time series. *Phys. D Nonlinear Phenom.* **1998**, *121*, 75–88.
21. Bhattacharya, J.; Pereda, E.; Petsche, H. Effective detection of coupling in short and noisy bivariate data. *IEEE Trans. Syst. Man Cyberne. Part B Cybern.* **2003**, *33*, 85–95.
22. Netoff, T.I.; Schiff, S.J. Decreased neuronal synchronization during experimental seizures. *J. Neurosci.* **2002**, *22*, 7297–7307.
23. Hayano, J.; Yasuma, F.; Okada, A.; Akiyoshi, M.; Mukai, S.; Fujinami, T. Respiratory Sinus Arrhythmia. *Circulation* **1996**, *94*, 842–847.
24. Cox, D.R.; Wermuth, N. A simple approximation for bivariate and trivariate normal integrals. *Int. Stat. Rev.* **1991**, *59*, 262–269.

*entropy*

MDPI

*Article*

# Hierarchical Cosine Similarity Entropy for Feature Extraction of Ship-Radiated Noise

Zhe Chen [1], Yaan Li [1,*], Hongtao Liang [2] and Jing Yu [1]

[1] School of Marine Science and technology, Northwestern Polytechnical University, Xi'an 710072, China; chenzhe@mail.nwpu.edu.cn (Z.C.); yujing@nwpu.edu.cn (J.Y.)
[2] School of Physics and Information Technology, Shaanxi Normal University, Xi'an 710119, China; lianghongtao.789@163.com
* Correspondence: liyaan@nwpu.edu.cn; Tel.: +86-29-8849-5817

Received: 30 April 2018; Accepted: 31 May 2018; Published: 1 June 2018

**Abstract:** The classification performance of passive sonar can be improved by extracting the features of ship-radiated noise. Traditional feature extraction methods neglect the nonlinear features in ship-radiated noise, such as entropy. The multiscale sample entropy (MSE) algorithm has been widely used for quantifying the entropy of a signal, but there are still some limitations. To remedy this, the hierarchical cosine similarity entropy (HCSE) is proposed in this paper. Firstly, the hierarchical decomposition is utilized to decompose a time series into some subsequences. Then, the sample entropy (SE) is modified by utilizing Shannon entropy rather than conditional entropy and employing angular distance instead of Chebyshev distance. Finally, the complexity of each subsequence is quantified by the modified SE. Simulation results show that the HCSE method overcomes some limitations in MSE. For example, undefined entropy is not likely to occur in HCSE, and it is more suitable for short time series. Compared with MSE, the experimental results illustrate that the classification accuracy of real ship-radiated noise is significantly improved from 75% to 95.63% by using HCSE. Consequently, the proposed HCSE can be applied in practical applications.

**Keywords:** hierarchical cosine similarity entropy; multiscale entropy; sample entropy; feature extraction; complexity

## 1. Introduction

Ship-radiated noise is the main signal source of passive sonar for underwater target detection and recognition. Extracting useful features from ship-radiated noise can effectively improve the performance of passive sonar. Thus, the research of feature extraction techniques has received considerable attention [1–3].

The propulsion parts of a ship, such as engines, turbines and propellers, are the principal sources of ship-radiated noise [4]. The emitted noise is propagated in the ocean channel and eventually received by hydrophones. Because of the complicated production mechanism and the influence of stochastic ocean medium, the received ship-radiated noise is usually nonstationary, non-Gaussian and nonlinear. It is challenging to extract useful features from such a signal. The physical features, such as the blade rate, propeller shaft frequency and the number of blades, have been studied in past decades by utilizing frequency domain based techniques, such as the power spectrum density (PSD), short-time Fourier transform (STFT) and wavelet transform [5–8]. These traditional feature extraction methods have achieved great effectiveness in practical engineering applications, but there are still some limitations. For example, the PSD is not able to reflect local properties of a signal, while the wavelet transform is limited by the selection of wavelet basis function, improper choice of wavelet basis function may lead to distortion [3,7]. Moreover, physical features are not sufficient to classify ships that have similar propulsion components [1].

Although ship-radiated noise is nonlinear, this characteristic is neglected by traditional linear processing techniques. Therefore, there is a growing demand to extract the nonlinear features of ship-radiated noise [9]. Entropy is a powerful nonlinear analysis tool that can analyze complex mechanical systems [10–12]. To date, various statistical entropy algorithms for quantifying the complexity of signals have been developed, such as permutation entropy [13,14], approximate entropy (ApEn) [15] and sample entropy (SE) [16]. One of the most commonly used structural complexity estimators, SE, which is obtained by calculating the conditional probability of occurrences of similar patterns, has attracted considerable attention in numerous fields such as biological signal analysis [17], fault diagnosis [18] and acoustic signal processing [9]. However, despite its great success, there exist some drawbacks: (I) High computational cost. Calculating SE needs to estimate the probability of occurrence of similar patterns found in the reconstructed m and (m + 1) dimensional phase-space, any two vectors in the phase-space are similar if their Chebyshev distance is lower than a predefined tolerance $r_{SE}$. The similarity checking can result in the computation cost quadratically increasing as the data-length increases [19], which may limit the SE's performance in some real-time applications; (II) Sensitive to erratic noise [20]. In SE, the similarity checking is based on the Chebyshev distance, which is amplitude dependent, leading to that high peaks existing in the time series can directly affect the entropy estimates. Unluckily, the unwanted erratic noise is likely to present in real data. For example, when a hydrophone is recording the underwater sound, it is unavoidable to be influenced by waves, which will produce undesirable outliers; (III) Undefined entropy value. The SE is effective when data-length is more than $10^m$ [21], where $m$ denotes the embedding dimension. However, as the data-length decreases, SE may produce undefined entropy value because of few similar patterns existing in the reconstructed m and (m + 1) dimensional phase-space [22]; (IV) Single-scale based [16,19–21]. Since complex signals often show structures on multiple temporal scales, the SE method is not able to estimate structural complexity of such a time series comprehensively and precisely.

In order to overcome the above mentioned shortcomings, many efforts have been made. Recently, there have been some solutions to compute SE quickly [19,23,24]. Furthermore, cosine similarity entropy (CSE) [20] was proposed to deal with issue (II) and (III), whereby the angular distance is employed instead of the Chebyshev distance, and the conditional entropy is replaced with the Shannon entropy. Regarding the fourth issue, the multiscale entropy (MSE) was introduced by Costa et al. [25–27] to measure complexity over a range of temporal scales. The method shows that the entropy of white noise decreases as the scale factor increases, which agrees with the fact that the white noise is not structurally complex. Therefore, the MSE algorithm offers a better interpretation of the complexity of a signal. Even though, the MSE only takes the lower frequency components of a time series into consideration, while the higher frequency components are ignored [28]. To remedy this, a new multiscale decomposition technique, the hierarchical decomposition was developed by Jiang et al. [28]. By further combining the hierarchical decomposition with SE, the scheme (i.e., hierarchical sample entropy (HSE)) is capable of analyzing a signal more adequately than MSE. Unfortunately, since SE is not perfect, its limitations are bound to be retained in HSE.

In this paper, the hierarchical cosine similarity entropy (HCSE) is proposed for feature extraction of ship-radiated noise. The presented algorithm takes advantages of both hierarchical decomposition and CSE. By analyzing synthetic signals, a set of parameters for computing the HCSE is recommended. The simulation results indicate that HCSE overcomes some limitations in SE and MSE. By employing HCSE to extract features of experimental data (see Section 4.2 for detailed description), the classification performance is significantly improved. The remainder of this paper is organized as follows: Section 2 provides a description of the proposed HCSE; parameter selection for HCSE is studied in Section 3; the HCSE is employed to analyze synthetic signals and experimental data in Section 4; the paper is concluded in Section 5.

## 2. Hierarchical Cosine Similarity Entropy

### 2.1. Cosine Similarity Entropy

For a time series $\{x_i\}_{i=1}^N$, the CSE is computed as follows:

(1)   Given embedding dimension $m$ and time delay $\tau$, the embedding vectors are constructed as:

$$x_i^{(m)} = \left[x_i, x_{i+\tau}, \cdots, x_{i+(m-1)\tau}\right] \tag{1}$$

(2)   Calculate angular distance for all pairwise vectors. The angular distance is derived from the cosine similarity and cosine distance. The cosine similarity of two diverse vectors $x_i^{(m)}$ and $x_j^{(m)}$ is defined as:

$$CosSim_{i,j}^{(m)} = \frac{x_i^{(m)} \cdot x_j^{(m)}}{\left|x_i^{(m)}\right| \cdot \left|x_j^{(m)}\right|}. \tag{2}$$

Notice that the value of cosine similarity is ranging from $-1$ to $1$, the cosine distance is then defined as $CosDis_{i,j}^{(m)} = 1 - CosSim_{i,j}^{(m)}$ to provide a positive distance metric. However, both the cosine similarity and the cosine distance violate the triangle inequality property, neither of them is a proper distance metric [20]. The angular distance, which is defined as $AngDis_{i,j}^{(m)} = arccos(CosSim_{i,j}^{(m)})/\pi$, obeys the axioms of a valid distance metric and thus be an appropriate selection for measuring distance of vectors. It is deserved to mention that the distance metric is one of the differences between SE and CSE. In SE, the Chebyshev distance is used for quantifying distance of vectors, which can be written as $CheDis_{i,j}^{(m)} = max(\left|x_i^{(m)}(k) - x_j^{(m)}(k)\right|)$, $k = 1, 2, \cdots, m$. It can be concluded that the Chebyshev distance is amplitude based and sensitive to outliers, while the angular distance is more stable.

(3)   Given a tolerance $r_{CSE}$, any two diverse vectors $x_i^{(m)}$ and $x_j^{(m)}$ are regarded as similar patterns if $AngDis_{i,j}^{(m)} \leq r_{CSE}$. Count the number of similar patterns of each vector $P_i^{(m)}$, the local and global probability of occurrences of similar patterns can be calculated as:

$$A_i^{(m)} = \frac{1}{N-m} P_i^{(m)}, \tag{3}$$

$$A^{(m)} = \frac{1}{N-m+1} \sum_{i=1}^{N-m+1} A_i^{(m)}. \tag{4}$$

(4)   The CSE is finally defined in the form of Shannon entropy:

$$CSE = -\left[A^{(m)} log_2\left(A^{(m)}\right) + \left(1 - A^{(m)}\right) log_2\left(1 - A^{(m)}\right)\right]. \tag{5}$$

By employing Shannon entropy to estimate CSE, the entropy values of CSE range from 0 to 1. It means that a time series is structurally complex when CSE approaches 1 and structurally simple when CSE approaches 0. Unlike SE, undefined entropy value is unlikely to occur in CSE unless $A^{(m)} = 0$, which means that none vectors are similar.

### 2.2. Hierarchical Decomposition

The MSE algorithm only takes the lower frequency components of a time series into consideration, while the higher frequency components are ignored. To remedy this, the hierarchical decomposition was introduced by Jiang et al. [28]. For a time series $\{x_i\}_{i=1}^N$, where $N = 2^n$, it can be decomposed by following procedures:

(1)   Averaging operator $Q_0$ and difference operator $Q_1$ are defined by:

$$Q_0(x) := \left(\frac{x_{2i-1} + x_{2i}}{2}\right), i = 1, 2, \cdots, 2^{n-1}, \tag{6}$$

$$Q_1(x) := \left(\frac{x_{2i-1} - x_{2i}}{2}\right), i = 1, 2, \cdots, 2^{n-1}, \tag{7}$$

where the operator $Q_0$ and $Q_1$ are the low and high pass filters of the Harr wavelet [28], respectively. For simplicity, $Q_0$ and $Q_1$ can be written in matrix form:

$$Q_j = \begin{pmatrix} \frac{1}{2} & \frac{(-1)^j}{2} & 0 & 0 & \cdots & 0 & 0 \\ 0 & 0 & \frac{1}{2} & \frac{(-1)^j}{2} & \cdots & 0 & 0 \\ 0 & 0 & 0 & 0 & \cdots & \frac{1}{2} & \frac{(-1)^j}{2} \end{pmatrix}_{2^{n-1} \times 2^n} , j = 1, 2. \tag{8}$$

(2)   Let $e$ be a nonnegative integer and $L_k$ equals to 0 or 1, where $k = 1, 2, \cdots, n$. For a given $e$, there is a unique vector $[L_1, L_2, \cdots, L_n]$ that fulfills Equation (9). Then, the hierarchical decomposition of a time series can be defined by Equation (10):

$$e = \sum_{k=1}^{n} L_k 2^{n-k}, \tag{9}$$

$$x_{n,e} = Q_{L_n} \circ Q_{L_{n-1}} \circ \cdots Q_{L_1}(x), \tag{10}$$

where $x_{n,e}$ denotes the hierarchical component (i.e., the subsequence) of the original time series. To illustrate the decomposition process more clearly, the hierarchical components can be arranged in a tree diagram (see Figure 1). In Figure 1, the original time series is represented as $x_{0,0}$ at the root node. After an average and difference operation, the root node $x_{0,0}$ has a left child node $x_{1,0}$ and a right child node $x_{1,1}$, which correspond to the lower and higher frequency components of $x_{0,0}$, respectively. Analogously, each node $x_{n,e}$ has the left child node $x_{n+1,2e}$ and the right child node $x_{n+1,2e+1}$. In fact, nodes $x_{0,0}$, $x_{1,0}$ and $x_{2,0}$ are equal to the coarse-graining process (which is the multiscale decomposition method used in MSE) at scale 1, 2 and 4, respectively. In other words, the hierarchical decomposition not only preserves the advantages of coarse-graining, but also additionally focuses on the higher frequency components in diverse scales. Hence, it is able to provide more information of the time series than coarse-graining.

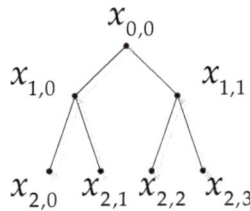

**Figure 1.** Hierarchical decomposition of $x$ (scale = 3).

### 2.3. Hierarchical Cosine Similarity Entropy

Within the HCSE algorithm, only the hierarchical decomposition is required to proceed prior to entropy estimation. Then, each subsequence at the node is served as an input of CSE to measure its complexity.

### 3. Parameters Selection for HCSE

To compute the HCSE algorithm, parameters such as tolerance $r_{CSE}$, embedding dimension $m$, data length $N$ and scale factor $s$ must be properly selected. The selection of these parameters is studied in the next subsections using uncorrelated random noise and long-term correlated noise: the White

Gaussian noise (WGN) and the $1/f$ noise. Because time lag$\tau$ is analogous to down sampling to some extent, it is typically set as $\tau = 1$ for structural preservation [29].

### 3.1. Selection of Tolerance $r_{CSE}$

Notice that the range of angular distance is from 0 to 1, the boundary values of $r_{CSE}$ should also be 0 to 1. We varied $r_{CSE}$ from 0.01 to 0.99 with a step length of 0.02 to observe how CSE values change against $r_{CSE}$ (see Figure 2). The results in Figure 2 were obtained by 30 independent trials, in which the embedding dimension, time lag and data-length were chosen as the recommended $m = 2$, $\tau = 1$ and $N = 10,000$, respectively [20,29]. It can be seen that the mean CSE values of both WGN and $1/f$ noise are firstly increased with an increase in $r_{CSE}$ from 0.01 to 0.49, and then decrease as the $r_{CSE}$ increases from 0.51 to 0.99. By comparing the mean CSE values of WGN with $1/f$ noise, it is found that they are more discriminative between $r_{CSE} = 0.07 \sim 0.21$. Therefore, the tolerance should be selected within the range, in this paper, $r_{CSE} = 0.07$ is chosen for subsequent analysis.

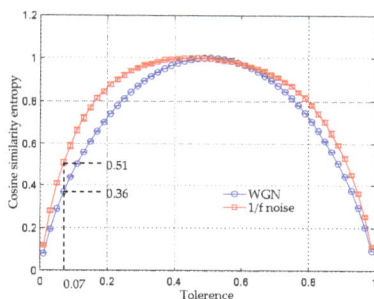

**Figure 2.** Selection of tolerance for CSE.

### 3.2. Selection of Embedding Dimension $m$

In this subsection, the relation of CSE values and a varied $m$ is studied by conducting 30 independent trials. $r_{CSE} = 0.07$, $\tau = 1$ and $N = 10,000$ were selected to calculate the CSE. For comparison, the SE of WGN and $1/f$ noise were also computed with diverse $m$. Parameters for computing SE were chosen as $m = 2$, $\tau = 1$ and $r_{SE} = 0.15 \cdot \rho$ [16–19], where $\rho$ denotes the standard deviation (SD) of the analyzed time series. The average entropy values with their SD error bar over a varying embedding dimension are shown in Figure 3. Figure 3a provides the results of the SE in which the mean SE values remain constant for different $m$ and the SD of SE values increases as the $m$ increases. It can be seen that only a small range of $m = [1,2,3]$ and $m = [1,2,3,4]$ were plotted for the WGN and the $1/f$ noise, respectively. This is because that the SE algorithm produces undefined entropy beyond the above mentioned range.

In Figure 3b, the mean CSE values of both synthetic signals decrease as $m$ increases. They approach to 0 when $m \geq 6$. In addition, the SD of entropies remains constantly small. The results in Figure 3b can be explained by the conclusion in [30], that is, as the embedding dimension $m$ increases, the trajectory of phase-space tends to be more and more deterministic, meaning that lower and lower complexity. The situation $m = 1$ is not given in Figure 3b, this is because that the angular distance is valid for vectors with at least two elements. Therefore, $m = 2$ is the minimum embedding dimension for calculating the CSE. By comparing Figure 3a with Figure 3b, it can be seen that the CSE algorithm can provide more stable entropy estimation in a broader range of $m$ than SE. Since the CSE value approaches to 0 for a large $m$, a smaller embedding dimension, such as $m = 2$ and $m = 3$, is recommended to compute the algorithm. We selected $m = 2$ for subsequent calculation.

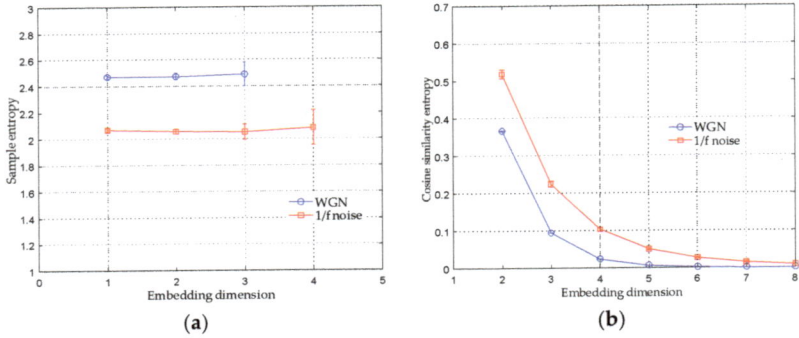

**Figure 3.** The mean entropy values with their SD error bar over a varying embedding dimension. (**a**) the results of SE; (**b**) the results of CSE.

### 3.3. Selection of Data-Length N

We examined the relationship between CSE values and data-length in this subsection. For comparison, the SE was also applied to compute the same synthetic signals. Parameters for computing the SE and the CSE were selected as $m = 2$, $\tau = 1$, $r_{SE} = 0.15 \cdot \rho$ and $r_{CSE} = 0.07$. The average entropy values with their SD error bar over a varying data-length $N$ are plotted in Figure 4. The data-length $N$ was varied from 10 to 2000 with a step length of 10. In Figure 4a, because of producing undefined entropy, the SE of the WGN and the $1/f$ noise are invalid when $N \leq 200$; when $200 \leq N \leq 700$, both synthetic signals acquire unstable entropy estimates with a large SD; their entropies become stable when $N \geq 700$. The results in Figure 4a correspond well to that in [21], where it is shown that the SE algorithm requires sufficient samples.

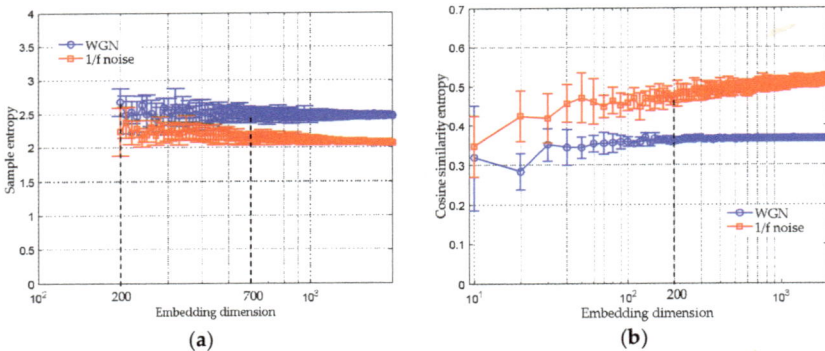

**Figure 4.** The mean entropy values with their SD error bar over a varying data-length. (**a**) the results of SE; (**b**) the results of CSE.

In Figure 4b, the CSE algorithm is valid even for 10 samples; when $100 \leq N \leq 200$, the CSE values of both synthetic signals have a large SD; their entropy estimates become stable when $N \geq 200$. Comparing Figure 4a with Figure 4b, it can be found that the WGN and the $1/f$ noise are more distinguishable by employing the CSE rather than the SE. In contrast to SE, the CSE is more stable in processing short-time series. Regarding the issue of selecting proper data-length $N$ for computing the CSE, it should be chosen according to actual needs as long as it is larger than 200.

### 3.4. Selection of Scale Factor s

It is also necessary to determine the scale factor $s$ appropriately. There is a doubling reduction in the data-length when the scale factor increases by 1. Therefore, the selected $s$ has to ensure that each subsequence has a data-length larger than 200 (as discussed in Section 3.3). Without loss of generality, $s = 5$ was chosen for multiscale analysis in the subsequent study.

## 4. Feature Extraction of Synthetic Signals and Real Ship-Radiated Noise

We applied the proposed HCSE to analyze synthetic signals and real ship-radiated noise, the MSE was also utilized to compute the same signals for comparison. Parameters for computing the HCSE and the MSE were set as $m = 2$, $\tau = 1$, $r_{SE} = 0.15 \cdot \rho$, $s = 5$ and $r_{CSE} = 0.07$. Considering that the hierarchical decomposition demands a data-length of $N = 2^n$, the data-length $N$ was selected as 8192.

### 4.1. HCSE Analysis for Synthetic Signals

It is necessary to firstly apply the proposed HCSE method to analyze signals with known characteristics and complexity levels. In this subsection, the uncorrelated WGN and the long-term correlated $1/f$ noise were analyzed. The results are obtained from 30 independent realizations.

Figure 5 offers the mean HCSE values of WGN. It is shown that the mean HCSE value of every node is approximately equal to 0.365, implying that each subsequence is as complex as the original WGN.

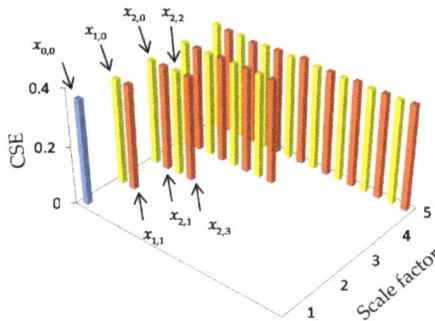

**Figure 5.** HCSE analysis of the WGN. (the yellow bar and the red bar, respectively, represent the lower and the higher frequency components, as is the same for the subsequent results).

Figure 6 depicts the HCSE analysis results of $1/f$ noise. In Figure 6, the $1/f$ noise is denoted by $f_{0,0}$, while its lower frequency components and higher frequency components at scale 2 are represented by $f_{1,0}$ and $f_{1,1}$, respectively. In Figure 6a, as the scale factor $s$ increases, the mean HCSE values of node $f_{s-1,0}$ remain constant at 0.517, while the other hierarchical components have an equal HCSE value of 0.365, which is equal to the WGN. The subtrees with root node $f_{1,0}$ and $f_{1,1}$ are also plotted in Figure 6b,c, respectively. Comparing Figure 5 with Figure 6c, it can be observed that the subtree of $f_{1,1}$ looks pretty much like that of the WGN. Similar results can also be found by comparing Figure 6a,b. Hence, Figures 5 and 6 verifies the assumption in [28] that $f_{1,0}$ is still $1/f$ noise, while $f_{1,1}$ is approximately equal to the WGN.

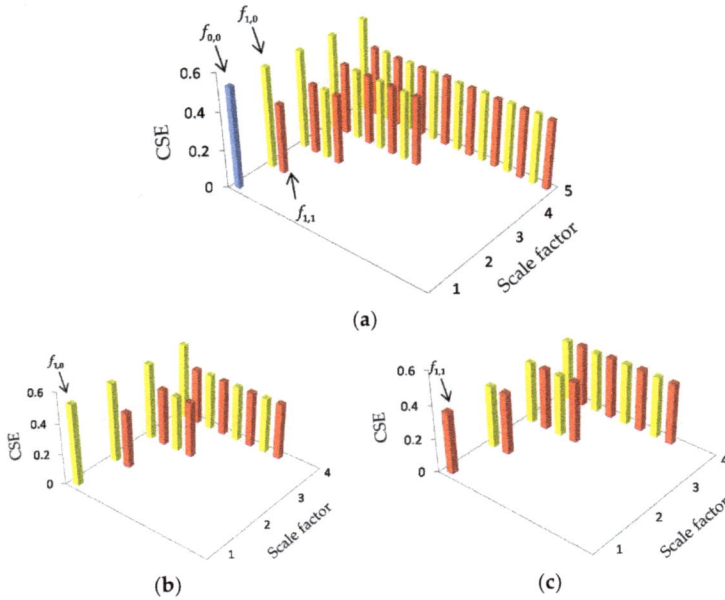

**Figure 6.** HCSE analysis of the 1/f noise. (**a**) The results of original 1/f noise; (**b**) the results of $f_{1,0}$; (**c**) the results of $f_{1,1}$.

The average SE values with their SD error bar over a varying scale factor are shown in Figure 7. The WGN achieves the highest mean entropy value at scale 1. The entropy curve decreases as the scale factor increases. Its SE value finally falls to 1.68 when $s = 5$. With respect to the 1/f noise, the mean entropies remain constant between 1.89 and 1.98 for the whole range of scale factors.

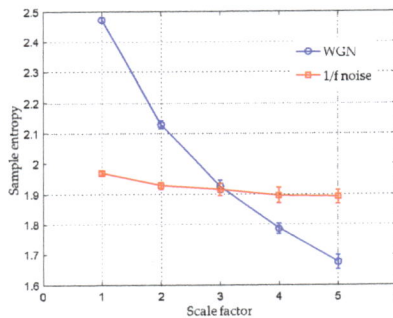

**Figure 7.** MSE analysis for synthetic signals.

Comparing the MSE analysis result with that of the HCSE, it is shown that HCSE can provide information of the higher frequency components of a signal, which is not provided by MSE. Hence, HCSE is capable of extracting features of a signal more comprehensively and precisely. Furthermore, the HCSE analysis result of the WGN corresponds well with the fact that the WGN is not structurally complex and it also agree well with previous claim in [28] that different hierarchical components of WGN are still WGN.

### 4.2. Feature Extraction of Real Ship-Radiated Noise

We utilized the proposed HCSE to extract features of four types of ship radiated noise, which were recorded in the South China Sea. The depth of the experimental area is about 4000 m, and the seabed is approximately flat. The data acquisition was carried out under the level 1 sea state to avoid serious influence of ocean ambient noise. The sensitivity and frequency response of the omnidirectional hydrophone are 170 dB re 1 v/μpa and 0.1 Hz–80 kHz, respectively. The hydrophone, which was carried by a research ship, was deployed at a depth of 30 m. In order to eliminate the self noise of the research ship, its engines were shut down and its speed reduced to approximately zero. Then, four different target ships, which were 2.5 km away from the research ship, moved towards the hydrophone at an average speed of 10 knots. When one target ship was moving, the other ships were rested. When its distance to the hydrophone is less than 1 km, the target ship would slow down and stop. The data was recorded at a sampling rate of 16 kHz. It should be pointed out that four target ships have different size, tonnage and propulsion equipment, so that they can be classified into four categories. In the subsequent study, the radiated noise of four different target ships are represented as type A, B, C and D, respectively. Each type contains 819,200 sample points, which are cut equally into 100 pieces for analysis. Figure 8 depicts the normalized waveforms of four types of ship-radiated noise. In order to show more details, detail view is also provided in each picture. Spectrograms of four types of ship-radiated noise are given in Figure 9, which represents the energy distribution against time and over frequencies, the amount of acoustic power is represented as the intensity at each time frequency point. The spectrogram is a method to recognize diverse vessels, because different types of ships may have different acoustic energy distribution against frequencies. It is found that type A and B can be distinguished well, since type B have obvious higher energy in high frequency area (1.5–2 kHz). However, the acoustic energy signatures of type C and D are too similar to discriminate them. Hence, it is necessary to extract other features of the ship, such as entropy.

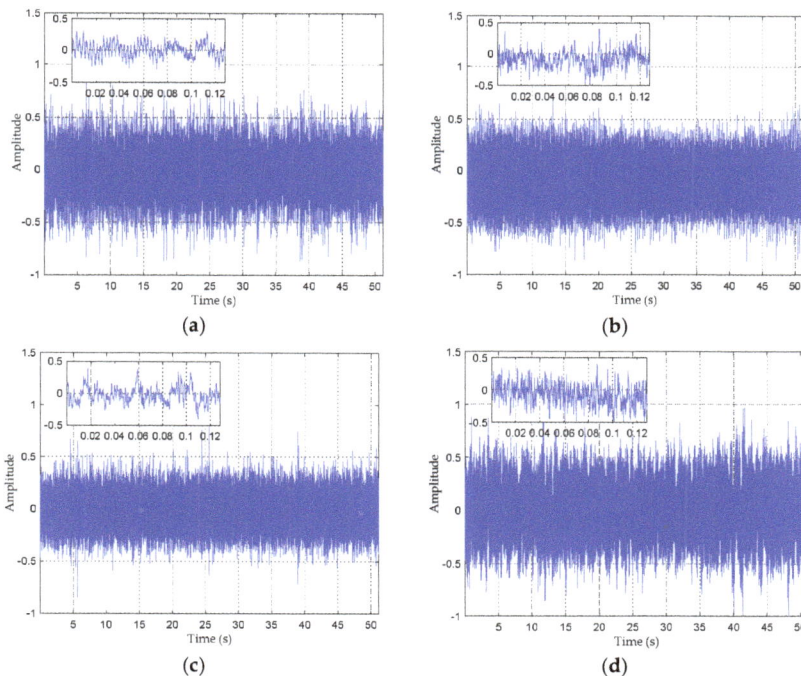

**Figure 8.** Recordings of four types of ship-radiated noise. (**a**) type A; (**b**) type B; (**c**) type C; (**d**) type D.

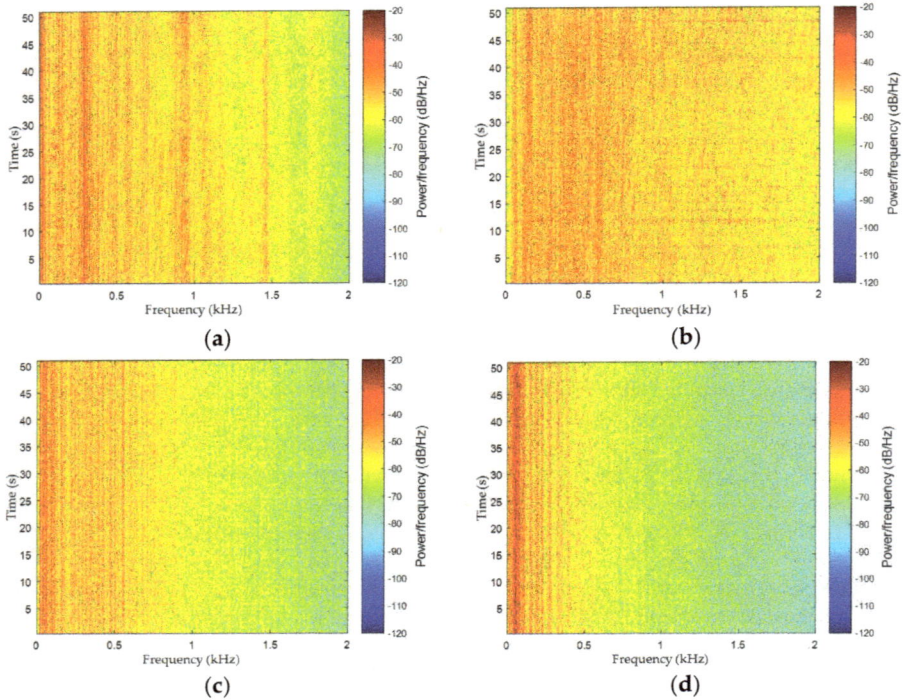

**Figure 9.** Spectrograms of four types of ship-radiated noise. (**a**) type A; (**b**) type B; (**c**) type C; (**d**) type D.

The feature extraction results of MSE are shown in Figure 10 and Table 1. The average SE values with their SD error bar over a varying scale factor are plotted in Figure 10. It can be seen that, for all four types, there is an increasing mean SE values as the scale factor increases. The MSE features seem to be effective for classifying type A, B and C, because their entropies over diverse scales are visually and statistically discernable. However, type C and D have a similar entropy distribution over different scales, and the SE estimation for type D is unstable, they may not be distinguished well by using MSE.

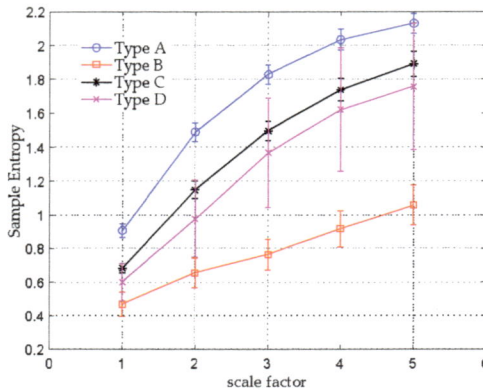

**Figure 10.** Feature extraction results of the MSE.

**Table 1.** Feature extraction results of the MSE.

| Type | Mean SE Values | | | | |
|------|---------|---------|---------|---------|---------|
| | Scale 1 | Scale 2 | Scale 3 | Scale 4 | Scale 5 |
| A | 0.907 | 1.486 | 1.827 | 2.03 | 2.130 |
| B | 0.467 | 0.654 | 0.764 | 0.917 | 1.058 |
| C | 0.683 | 1.150 | 1.496 | 1.738 | 1.893 |
| D | 0.599 | 0.977 | 1.366 | 1.620 | 1.757 |

Figure 11, Tables 2 and 3 provide the HCSE feature extraction results. Unlike MSE, the HCSE results in decreasing entropies at nodes $x_{s-1,0}$ with an increasing scale factor. As mentioned before, the hierarchical components at nodes $x_{s-1,0}$ are equal to the coarse-grained subsequences at scale $2^{s-1}$. Thus, similar with the MSE analysis results, type C and D can not be recognized well by only observing the CSE values at nodes $x_{s-1,0}$. Fortunately, except for the lower frequency components, the HCSE is able to provide information of the higher frequency components of a signal. Comparing Figure 11 with Figure 5, it is seen that the subtrees with root node $x_{1,1}$ of type A, B, and C (i.e., the bold represented parts in Tables 2 and 3) look pretty much like that of the WGN, where the CSE values of every node is approximately equal to 0.365. In the corresponding area, type D achieves obvious higher entropies at nodes $x_{1,1}$, $x_{3,6}$, $x_{4,8}$, $x_{4,10}$, $x_{4,12}$ and $x_{4,14}$, where the CSE values are larger than 0.39. The result means that type D has a more complex structure than the other 3 types. Type C and D are distinguishable by comparing entropies in that area.

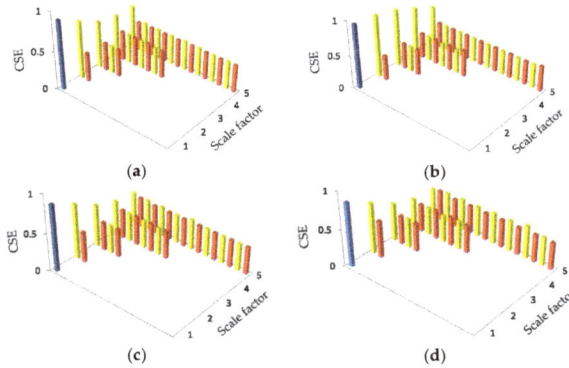

**Figure 11.** HCSE analysis for real ship-radiated noise.

**Table 2.** Mean HCSE values for type A and B.

| Type | Scales | | | | | Type | Scales | | | | |
|------|-------|-------|-------|-------|-------|------|-------|-------|-------|-------|-------|
| | 1 | 2 | 3 | 4 | 5 | | 1 | 2 | 3 | 4 | 5 |
| | 0.896 | 0.743 | 0.604 | 0.548 | 0.534 | | 0.956 | 0.920 | 0.836 | 0.702 | 0.566 |
| | | 0.380 | 0.366 | 0.380 | 0.364 | | | 0.372 | 0.383 | 0.370 | 0.363 |
| | | | 0.365 | 0.376 | 0.386 | | | | 0.385 | 0.366 | 0.365 |
| | | | 0.366 | 0.376 | 0.370 | | | | 0.386 | 0.371 | 0.364 |
| | | | | 0.378 | 0.381 | | | | | 0.370 | 0.365 |
| | | | | 0.379 | 0.370 | | | | | 0.376 | 0.364 |
| | | | | 0.366 | 0.376 | | | | | 0.370 | 0.364 |
| A | | | | 0.366 | 0.365 | B | | | | 0.402 | 0.366 |
| | | | | | 0.387 | | | | | | 0.367 |
| | | | | | 0.372 | | | | | | 0.365 |
| | | | | | 0.380 | | | | | | 0.364 |
| | | | | | 0.370 | | | | | | 0.367 |
| | | | | | 0.364 | | | | | | 0.363 |
| | | | | | 0.364 | | | | | | 0.363 |
| | | | | | 0.370 | | | | | | 0.367 |
| | | | | | 0.364 | | | | | | 0.386 |

Table 3. Mean HCSE values for type C and D.

| Type | Scales | | | | | Type | Scales | | | | |
|---|---|---|---|---|---|---|---|---|---|---|---|
| | **1** | **2** | **3** | **4** | **5** | | **1** | **2** | **3** | **4** | **5** |
| | 0.870 | 0.714 | 0.553 | 0.447 | 0.411 | | 0.856 | 0.680 | 0.504 | 0.401 | 0.378 |
| | | **0.402** | 0.370 | 0.371 | 0.384 | | | **0.489** | 0.385 | 0.368 | 0.400 |
| | | | **0.370** | 0.370 | 0.391 | | | | **0.389** | 0.367 | 0.411 |
| | | | **0.366** | 0.370 | 0.365 | | | | **0.377** | 0.385 | 0.371 |
| | | | | 0.370 | 0.392 | | | | | 0.367 | 0.417 |
| | | | | 0.374 | 0.367 | | | | | 0.387 | 0.373 |
| | | | | **0.365** | 0.368 | | | | | 0.394 | 0.407 |
| C | | | | **0.365** | 0.362 | D | | | | 0.379 | 0.370 |
| | | | | | 0.398 | | | | | | 0.417 |
| | | | | | 0.368 | | | | | | 0.374 |
| | | | | | 0.370 | | | | | | 0.410 |
| | | | | | 0.362 | | | | | | 0.371 |
| | | | | | 0.363 | | | | | | 0.442 |
| | | | | | 0.363 | | | | | | 0.373 |
| | | | | | 0.363 | | | | | | 0.392 |
| | | | | | 0.365 | | | | | | 0.367 |

## 4.3. Feature Classification

To test the validity of the proposed algorithm, the extracted features were input into the widely used probability neural network (PNN) [31] for training and testing. 20 pieces of ship-radiated noise were set as training samples and the other 80 pieces were used for testing. As shown in Tables 4 and 5, the classification results agree well with the feature extraction results in Section 4.2. For example, the MSE can perfectly classify type A, B and C, but no type D is correctly recognized, and the classification accuracy turns out to be 75%. Compared with MSE, HCSE achieves a little lower accuracy in classifying type A, but the classification performance of type D is remarkably improved. The classification accuracy finally reaches 95.63%, which is 20.63% higher than MSE.

Table 4. PNN classfication results of MSE.

| Type | Recognized as | | | | Accuracy |
|---|---|---|---|---|---|
| | **A** | **B** | **C** | **D** | |
| A | 80 | 0 | 0 | 0 | 100% |
| B | 0 | 80 | 0 | 0 | 100% |
| C | 0 | 0 | 80 | 0 | 100% |
| D | 0 | 15 | 65 | 0 | 0% |
| In total | 80 | 95 | 145 | 0 | 75% |

Table 5. PNN classfication results of HCSE.

| Type | Recognized as | | | | Accuracy |
|---|---|---|---|---|---|
| | **A** | **B** | **C** | **D** | |
| A | 69 | 0 | 11 | 0 | 86.25% |
| B | 0 | 80 | 0 | 0 | 100% |
| C | 0 | 0 | 80 | 0 | 100% |
| D | 1 | 0 | 2 | 77 | 96.25% |
| In total | 70 | 80 | 93 | 77 | 95.63% |

## 5. Conclusions

The classification performance of passive sonar can be improved by extracting the features of ship-radiated noise. Traditional feature extraction methods neglected the nonlinear features in ship-radiated noise, such as entropy. For the purpose of extracting useful nonlinear features of ship radiated noise, the HCSE method is proposed in this paper. The presented algorithm takes strength of both hierarchical decomposition and CSE. The advantages of the proposed method are as follows from the simulation and experimental results:

(1) The undefined entropy is unlikely to occur in HCSE by utilizing Shannon entropy rather than conditional entropy and employing angular distance instead of Chebyshev distance. As a consequence, the HCSE method is valid when data-length $N = 10$, while the MSE method is invalid when $N \leq 200$.

(2) The HCSE is suitable for short time series. It can provide stable entropy estimation when $N \geq 200$, while the MSE demands $N \geq 700$.

(3) The HCSE analysis result of the WGN is in consistent with the fact that WGN is not structurally complex, and it also agrees well with claim that hierarchical components of WGN are still WGN.

(4) The HCSE method can extract the features of a signal more comprehensively and precisely, because it takes both lower and higher frequency components into consideration. Compared with MSE, the classification accuracy of real ship-radiated noise is significantly improved from 75% to 95.63% by using HCSE.

**Author Contributions:** Z.C. designed the project; Y.L. provided the data; Z.C. wrote the manuscript, Y.L., H.L. and J.Y. help to revise the manuscript. All co-authors reviewed and approved the final manuscript.

**Funding:** This research was funded by National Natural Science Foundation of China (No. 51178157, No. 51409214, No. 11574250 and No. 51709228).

**Conflicts of Interest:** The authors declare no conflict of interest.

## References

1. Wang, Q.; Zeng, X.Y.; Wang, L.; Wang, H.T.; Cai, H.H. Passive moving target classification via spectra multiplication method. *IEEE Signal Process. Lett.* **2017**, *24*, 451–455. [CrossRef]

2. Das, A.; Kumar, A.; Bahl, R. Marine vessel classification based on passive sonar data: The cepstrum-based approach. *IET Radar Sonar Navig.* **2013**, *7*, 87–93. [CrossRef]

3. Wang, S.G.; Zeng, X.Y. Robust underwater noise targets classification using auditory inspired time-frequency analysis. *Appl. Acoust.* **2014**, *78*, 68–76. [CrossRef]

4. Hodges, R.P. *Underwater Acoustics: Analysis, Design and Performance of Sonar*, 1st ed.; Wiley: West Sussex, UK, 2010; pp. 183–184.

5. Waite, A.D. *Sonar for Practicing Engineers*, 3rd ed.; Wiley: West Sussex, UK, 2002; pp. 126–127.

6. Li, Y.X.; Li, Y.A.; Chen, Z.; Chen, X. Feature extraction of ship-radiated noise based on permutation entropy of the intrinsic mode function with the highest energy. *Entropy* **2016**, *18*, 393. [CrossRef]

7. Li, Y.X.; Li, Y.A.; Chen, X.; Yu, J. A novel feature extraction method for ship-radiated noise based on variational mode decomposition and multi-scale permutation entropy. *Entropy* **2017**, *19*, 342.

8. Li, Y.X.; Li, Y.A.; Chen, X.; Yu, J. Denoising and feature extraction algorithms using NPE combined with VMD and their applications in ship-radiated noise. *Entropy* **2017**, *9*, 256. [CrossRef]

9. Shashidhar, S.; Li, Y.A.; Guo, X.J.; Chen, X.; Zhang, Q.F.; Yang, K.D.; Yang, Y.X. A complexity-based approach for the detection of weak signals in ocean ambient noise. *Entropy* **2016**, *18*, 101.

10. Kim, H.J.; Bae, M.; Jin, D. On a robust MaxEnt process regression model with sample-selection. *Entropy* **2018**, *20*, 262. [CrossRef]

11. Villecco, F.; Pellegrino, A. Entropic measure of epistemic uncertainties in multibody system models by axiomatic design. *Entropy* **2017**, *19*, 291. [CrossRef]

12. Villecco, F.; Pellegrino, A. Evaluation of uncertainties in the design process of complex mechanical systems. *Entropy* **2017**, *19*, 475. [CrossRef]

13. Bandt, C.; Pompe, B. Permutation entropy: A natural complexity measure for time series. *Phys. Rev. Lett.* **2002**, *88*, 174102. [CrossRef] [PubMed]

14. Gao, Y.D.; Villecco, F.; Li, M.; Song, W.Q. Multi-scale permutation entropy based on improved LMD and HMM for rolling bearing diagnosis. *Entropy* **2017**, *19*, 176. [CrossRef]

15. Pincus, S.M. Approximate entropy as a measure of system complexity. *Proc. Natl. Acad. Sci. USA* **1991**, *88*, 2297–2301. [CrossRef] [PubMed]
16. Richman, J.S.; Moorman, J.R. Physiological time-series analysis using approximate entropy and sample entropy. *Am. J. Physiol.* **2000**, *278*, H2039–H2049. [CrossRef] [PubMed]
17. Yentes, J.M.; Hunt, N.; Schmid, K.K.; Kaipust, J.P.; McGrath, D.; Stergiou, N. The appropriate use of approximate entropy and sample entropy with short data sets. *Ann. Biomed. Eng.* **2013**, *41*, 349–365. [CrossRef] [PubMed]
18. Han, M.; Pan, J. A fault diagnosis method combined with LMD, sample entropy and energy ratio for roller bearings. *Measurement* **2015**, *76*, 7–19. [CrossRef]
19. Manis, G.; Aktaruzzaman, M.; Sassi, R. Low computational cost for sample entropy. *Entropy* **2018**, *20*, 61. [CrossRef]
20. Chanwimalueang, T.; Mandic, D. Cosine similarity entropy: Self-correlated-based complexity analysis of dynamical systems. *Entropy* **2017**, *19*, 652. [CrossRef]
21. Alcaraz, R.; Abasolo, D.; Hornero, R.; Rieta, J. Study of sample entropy ideal computational parameters in the estimation of atrial fibrillation organization from the ECG. *Comput. Cardiol.* **2010**, *37*, 1027–1030.
22. Wu, S.D.; Wu, C.W.; Lin, S.G.; Wang, C.C.; Lee, K.Y. Time series analysis using composite multiscale entropy. *Entropy* **2013**, *15*, 1069–1084. [CrossRef]
23. Yu-Hsiang, P.; Wang, Y.H.; Liang, S.F.; Lee, K.T. Fast computation of sample entropy and approximate entropy in biomedicine. *Comput. Methods Programs Biomed.* **2011**, *104*, 382–396.
24. Jiang, Y.; Mao, D.; Xu, Y. A fast algorithm for computing Sample entropy. *Adv. Adapt. Data Anal.* **2011**, *3*, 167–186. [CrossRef]
25. Costa, M.; Goldberger, A.L.; Peng, C.K. Multiscale entropy analysis of complex physiologic time series. *Phys. Rev. Lett.* **2002**, *89*, 068102. [CrossRef] [PubMed]
26. Costa, M.; Goldberger, A.L.; Peng, C.K. Multiscale entropy analysis of biological signals. *Phys. Rev. E* **2005**, *71*, 021906. [CrossRef] [PubMed]
27. Costa, M.; Goldberger, A.L.; Peng, C.K. Multiscale entropy to distinguish physiologic and synthetic RR time series. *Comput. Cardiol.* **2002**, *29*, 137–140. [PubMed]
28. Jiang, Y.; Peng, C.K.; Xu, Y.S. Hierarchical entropy analysis for biological signals. *J. Comput. Appl. Math.* **2011**, *236*, 728–742. [CrossRef]
29. Kaffashi, F.; Foglyano, R.; Wilson, C.G.; Loparo, K.A.; Kenneth, A.L. The effect of time delay on approximate entropy & sample entropy calculations. *Phys. D* **2008**, *237*, 3069–3074.
30. Taken, F. Detecting strange attractors in turbulence. In *Dynamical Systems and Turbulence*; Rand, D., Young, L.S., Eds.; Springer: Berlin/Heidelberg, Germany, 1981; pp. 366–381.
31. Specht, D.F. Probability neural networks and the polynomial Adaline as complementary techniques for classification. *IEEE Trans. Neural Netw.* **1990**, *1*, 111–121. [CrossRef] [PubMed]

*Article*

# Coarse-Graining Approaches in Univariate Multiscale Sample and Dispersion Entropy

## Hamed Azami * and Javier Escudero

School of Engineering, Institute for Digital Communications, The University of Edinburgh,
Edinburgh EH9 3FB, UK; javier.escudero@ed.ac.uk
* Correspondence: hamed.azami@ed.ac.uk; Tel.: +44-748-147-8684

Received: 1 December 2017; Accepted: 16 February 2018; Published: 22 February 2018

**Abstract:** The evaluation of complexity in univariate signals has attracted considerable attention in recent years. This is often done using the framework of Multiscale Entropy, which entails two basic steps: coarse-graining to consider multiple temporal scales, and evaluation of irregularity for each of those scales with entropy estimators. Recent developments in the field have proposed modifications to this approach to facilitate the analysis of short-time series. However, the role of the downsampling in the classical coarse-graining process and its relationships with alternative filtering techniques has not been systematically explored yet. Here, we assess the impact of coarse-graining in multiscale entropy estimations based on both Sample Entropy and Dispersion Entropy. We compare the classical moving average approach with low-pass Butterworth filtering, both with and without downsampling, and empirical mode decomposition in Intrinsic Multiscale Entropy, in selected synthetic data and two real physiological datasets. The results show that when the sampling frequency is low or high, downsampling respectively decreases or increases the entropy values. Our results suggest that, when dealing with long signals and relatively low levels of noise, the refine composite method makes little difference in the quality of the entropy estimation at the expense of considerable additional computational cost. It is also found that downsampling within the coarse-graining procedure may not be required to quantify the complexity of signals, especially for short ones. Overall, we expect these results to contribute to the ongoing discussion about the development of stable, fast and robust-to-noise multiscale entropy techniques suited for either short or long recordings.

**Keywords:** complexity; multiscale dispersion and sample entropy; refined composite technique; intrinsic mode dispersion and sample entropy; moving average; Butterworth filter; empirical mode decomposition; downsampling

---

## 1. Introduction

A system is complex when it entails a number of components intricately entwined altogether (e.g., the subway network of the New York City) [1]. Following Costa's framework [2,3], the complexity in univariate signals denotes "meaningful structural richness", which may be in contrast with regularity measures defined from entropy metrics such as sample entropy (SampEn), permutation entropy, (PerEn), and dispersion entropy (DispEn) [3–6]. In fact, these entropy techniques assess repetitive patterns and return maximum values for completely random processes [3,5,7]. However, a completely ordered signal with a small entropy value or a completely disordered signal with maximum entropy value is the least complex [3,5,8]. For instance, white noise is more irregular than $1/f$ noise (pink noise), although the latter is more complex because $1/f$ noise contains long-range correlations and its $1/f$ decay produces a fractal structure in time [3,5,8].

From the perspective of physiology, some diseased individuals' recordings, when compared with those for healthy subjects, are associated with the emergence of more regular behavior, thus leading to lower entropy values [3,9]. In contrast, certain pathologies, such as cardiac arrhythmias, are associated with highly erratic fluctuations with statistical characteristics resembling uncorrelated noise. The entropy values of these noisy signals are higher than those of healthy individuals, even though the healthy individuals' time series show more physiologically complex adaptive behavior [3,10].

In brief, the concept of complexity for univariate physiological signals builds on the following three hypotheses [3,5]:

- The complexity of a biological or physiological time series indicates its ability to adapt and function in an ever-changing environment.
- A biological time series requires operating across multiple temporal and spatial scales and so its complexity is similarly multiscaled and hierarchical.
- A wide class of disease states, in addition to ageing, which decrease the adaptive capacity of the individual, appear to degrade the information carried by output variables.

Therefore, the multiscale-based methods focus on quantifying the information expressed by the physiological dynamics over multiple temporal scales.

To provide a unified framework for the evaluation of impact of diseases in physiological signals, multiscale SampEn (MSE) [3] was proposed to quantify the complexity of signals over multiple temporal scales. The MSE algorithm includes two main steps: (1) coarse-graining technique—i.e., combination of moving average (MA) filter and downsampling (DS) process—; and (2) calculation of SampEn of the coarse-grained signals at each scale factor $\tau$ [3]. A low-pass Butterworth (BW) filter was used as an alternative to MA to limit aliasing and avoid ripples [11]. To differentiate it from the original MSE, we call this method $MSE_{BW}$ herein.

Since their introduction, MSE and $MSE_{BW}$ have been widely used to characterize physiological and non-physiological signals [12]. However, they have several main shortcomings [12–14]. First, the coarse-graining process causes the length of a signal to be shortened by the scale factor $\tau$ as a consequence of the downsampling in the process. Therefore, when the scale factor increases, the number of samples in the coarse-grained sequence decreases considerably [14]. This may yield an unstable estimation of entropy. Second, SampEn is either undefined or unreliable for short coarse-grained time series [13,14].

To alleviate the first problem of MSE, intrinsic mode SampEn (InMSE) [15] and refined composite MSE (RCMSE) [14] were developed [15]. The coarse-graining technique is substituted by an approach based on empirical mode decomposition (EMD) in InMSE. The length of coarse-grained series obtained by InMSE is equal to that of the original signal, leading to more stable entropy values. Nevertheless, EMD-based approaches have certain limitations such as sensitivity to noise and sampling [16]. At the scale factor $\tau$, RCMSE considers $\tau$ different coarse-grained signals, corresponding to different starting points of the coarse-graining process [14]. Therefore, RCMSE yields more stable results in comparison with MSE. Nevertheless, both InMSE and RCMSE may lead to undefined values for short signals as a consequence of using SampEn in the second step of their algorithms [13]. Additionally, the SampEn-based approaches may not be fast enough for some real-time applications.

To deal with these deficiencies, multiscale DispEn (MDE) based on our introduced DispEn was developed [13]. Refined composite MDE (RCMDE) was then proposed to improve the stability of the MDE-based values [13]. It was found that MDE and RCMDE have the following advantages over MSE and RCMSE: (1) they are noticeably faster as a consequence of using DispEn with computational cost of $O(N)$—where $N$ is the signal length—, compared with the $O(N^2)$ for SampEn; (2) they result in more stable profiles for synthetic and real signals; (3) MDE and RCMDE discriminate different kinds of physiological time series better than MSE and RCMSE; and (4) they do not yield undefined values [13].

The aim of this research is to contribute to the understanding of different alternatives to coarse-graining in complexity approaches. To this end, we first revise the frequency responses for the three main filtering processes (i.e., MA, BW, and EMD) used in such methods. The role of downsampling in the classical coarse-graining process, which has not been systematically explored yet, is then investigated in the article. We assess the impact of coarse-graining in multiscale entropy estimations based on both SampEn and DispEn. To compare these methods, several synthetic data and two real physiological datasets are employed. For the sake of clarity, a flowchart of the alternatives to the coarse-graining method in addition to the datasets used in this article is shown in Figure 1.

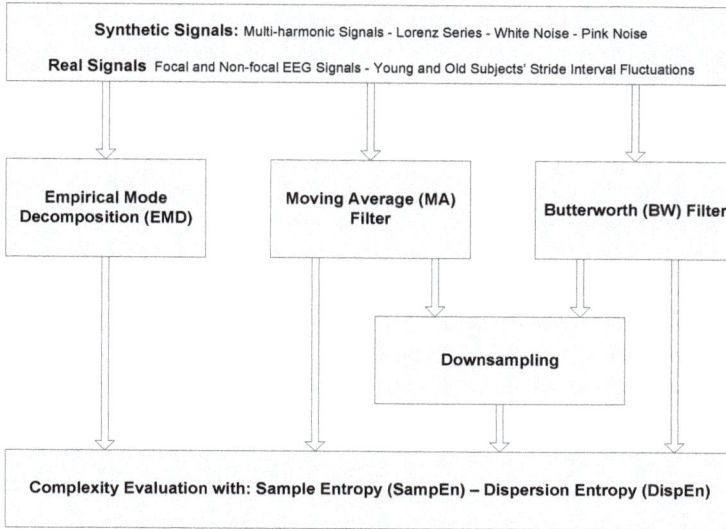

**Figure 1.** Flowchart of the alternatives to the coarse-graining method and the datasets used in this study.

## 2. Multiscale Entropy-Based Approaches

The MSE- and MDE-based methods include two main steps: (1) coarse-graining process; and (2) calculation of SampEn and DispEn at each scale $\tau$. For simplicity, we detail only the DispEn-based complexity algorithms. Likewise, the SampEn-based algorithms are defined.

### 2.1. MDE Based on Moving Average (MA) and Butterworth (BW) Filters with and without Downsampling (DS)

#### 2.1.1. Coarse-Graining Approaches

A coarse-graining technique with DS denotes a decimation by scale factor $\tau$. Decimation is defined as two steps [17,18]: (1) reducing high-frequency time series components with a digital low-pass filter; and (2) DS the filtered time series by $\tau$; that is, keep only one every $\tau$ sample points.

Assume that we have a univariate signal of length $L$: $\mathbf{u} = \{u_1, u_2, \ldots, u_i, \ldots, u_L\}$. In the coarse-graining process, the original signal $\mathbf{u}$ is first filtered by an MA—a low-pass finite-impulse response (FIR) filter—as follows:

$$v_\ell^{(\tau)} = \frac{1}{\tau} \sum_{k=0}^{\tau-1} u_{\ell+k}, \quad 1 \le \ell \le L - \tau + 1. \tag{1}$$

The frequency response of the MA filter is as follows [19]:

$$\left| H\left(e^{j2\pi f}\right) \right| = \frac{1}{\tau} \frac{sin(\pi f \tau)}{sin(\pi f)}, \tag{2}$$

where $f$ denotes the normalized frequency ranging from 0 to 0.5 cycles per sample (normalized Nyquist frequency). The frequency response of the MA filter has several shortcomings: (1) a slow roll-off of the main lobe; (2) large transition band; (3) and important side lobes in the stop-band. To alleviate these problems, a low-pass BW filter was proposed [11]. This filter provides a maximally flat (no ripples) response [19]. The squared magnitude of the frequency of BW filter is defined as follows:

$$\left| H\left(e^{j2\pi f}\right) \right|^2 = \frac{1}{1 + (f/f_c)^{2n}},$$ (3)

where $f_c$ and $n$ denote the normalized cut-off frequency and filter order, respectively [11,19]. Herein, $n = 6$ and $f_c = \frac{0.5}{\tau}$ [11]. The original signal $\mathbf{u}$ is filtered by BW filter. In fact, the low-pass filters eliminate the fast temporal scales (higher frequency components) to take into account progressively slower time scales (lower frequency components).

Next, the time series filtered by either MA or BW is downsampled by the scale factor $\tau$. Assume the downsampled signal is $\mathbf{x}^{(\tau)} = \{x_j^{(\tau)}\}$ $(1 \leq j \leq \left\lfloor \frac{L}{\tau} \right\rfloor = N)$.

In this study, we consider the coarse-graining process with and without DS. MSE and MDE with MA filter and without DS are respectively named $\text{MSE}_{\text{MA}}$ and $\text{MDE}_{\text{MA}}$. $\text{MSE}_{\text{MA}}$ and $\text{MDE}_{\text{MA}}$ with DS are termed MSE and MDE herein.

### 2.1.2. Calculation of DispEn or SampEn at Every Scale Factor

The DispEn or SampEn value is calculated for each coarse-grained signal $\mathbf{x}^{(\tau)} = \{x_j^{(\tau)}\}$. It is worth noting that MDE is more than the combination of the coarse-graining [3] with DispEn: the mapping based on the normal cumulative distribution function (NCDF) used in the calculation of DispEn [6] for the first temporal scale is maintained across all scales. That is, in MDE and RCMDE, $\mu$ and $\sigma$ of NCDF are respectively set at the average and standard deviation (SD) of the original signal and they remain constant for all scale factors. This approach is similar to keeping the threshold $r$ constant fixed (usually 0.15 of the SD of the original signal) in the MSE-based algorithms [3]. In a number of studies (e.g., [3,20]), it was found that keeping $r$ constant is preferable to recalculating the threshold $r$ at each scale factor separately.

### 2.2. Refined Composite Multiscale Dispersion Entropy (RCMDE)

At scale factor $\tau$, RCMDE considers $\tau$ different coarse-grained signals, corresponding to different starting points of the coarse-graining process. Then, for each of these shifted series, the relative frequency of each dispersion pattern is calculated. Finally, the RCMDE value is defined as the Shannon entropy value of the averages of the rates of appearance of dispersion patterns of those shifted sequences [13]. The MA filter used in RCMDE and RCMSE may be substituted by the BW filter, respectively called $\text{RCMDE}_{\text{BW}}$ and $\text{RCMSE}_{\text{BW}}$ here.

### 2.3. Intrinsic Mode Dispersion Entropy (InMDE)

Due to the advantages of DispEn over SampEn for short signals, intrinsic mode DispEn (InMDE) based on the algorithm of InMSE is proposed herein. The algorithm of InMDE includes the following two key steps:

1.  Calculation of the sum of the intrinsic mode functions (IMFs) obtained by EMD: In this step, the original signal $\mathbf{u}$ is decomposed to $IMF_\alpha$ $(1 \leq \alpha \leq \tau_{\max} - 1)$ and a residual signal $IMF_{\tau_{\max}} = \mathbf{u} - \sum_{\alpha=1}^{\tau_{\max}-1} IMF_\alpha$. It is worth noting that the first IMF, $IMF_1$, shows the highest frequency component in a signal, while the last IMF, $IMF_{\tau_{\max}}$, reflects the trend of the time series. Next, the cumulative sums of IMFs (CSI) for each scale factor $\tau$ are defined as follows [15]:

$$\mathbf{CSI}^{(\tau)}(\mathbf{x}) = \sum_{\lambda=\tau}^{\tau_{\max}} IMF_\lambda,$$ (4)

where $IMF_\lambda$ denotes the $\lambda^{th}$ IMF obtained by EMD. Thus, $\mathbf{CSI}^{(1)}$ is equal to the original signal $\mathbf{u}$.

2. Calculation of DispEn of $\mathbf{CSI}^{(\tau)}(\mathbf{x})$ at each scale factor: The DispEn value is calculated at each scale factor. Like MDE and RCMDE, $\mu$ and $\sigma$ of NCDF are respectively set at the average and SD of the original signal and they remain constant for all scale factors in InMDE.

It is worth noting that InMSE and InMDE do not downsample the filtered signals. That is, the number of samples for each $\mathbf{CSI}^{(\tau)}(\mathbf{x})$ is equal to that for the original signal, leading to more reliable results for higher scale factors. The complexity metrics for univariate signals and their characteristics are summarized in Table 1. The Matlab codes used in this study are described in Appendix.

**Table 1.** Characteristics of the complexity metrics for univariate signals.

| Methods | Filtering | Downsampling | Applicability of Refined Composite |
|---|---|---|---|
| MSE [2] and MDE [13] | Moving average | yes | yes |
| $\text{MSE}_{MA}$ and $\text{MDE}_{MA}$ | Moving average | no | no |
| $\text{MSE}_{BW}$ [11] and $\text{MDE}_{BW}$ | Butterworth | yes | yes |
| $\text{MSE}_{BW}$ [11] and $\text{MDE}_{BW}$ without downsampling | Butterworth | no | no |
| InMSE [15] and InMDE | Cumulative sums of IMFs | no | no |

### 2.4. Parameters of the Multiscale Entropy Approaches

For all the SampEn-based methods, we set $d = 1$, $m = 2$, and $r = 0.15$ of the SD of the original signal [3]. For all the DispEn-based approaches, we set $d = 1$ and $c = 6$. For more information about $c$ and $d$, please refer to [6,13].

For the DispEn-based complexity measures without DS, as the length of coarse-grained signals is equal to that of the original signal, it is advisable to follow $c^m < L$. For the SampEn-based complexity approaches without DS, it is recommended to have at least $10^m$ (or preferably $20^m$) sample points for the embedding dimension $m$ [21,22].

For the DispEn-based multiscale approaches with DS, since the decimation process causes the length of a signal decreases to $\left\lfloor \frac{L}{\tau_{max}} \right\rfloor$, $c^m < \left\lfloor \frac{L}{\tau_{max}} \right\rfloor$ is recommended. Similarly, for the SampEn-based complexity techniques with DS, $10^m < \left\lfloor \frac{L}{\tau_{max}} \right\rfloor$ [3] is recommended.

On the other hand, in RCDME, we consider $\tau$ coarse-grained time series with length $\left\lfloor \frac{L}{\tau_{max}} \right\rfloor$. Therefore, the total sample points calculated in RCDME is $\tau \times \left\lfloor \frac{L}{\tau_{max}} \right\rfloor \approx L$. Thus, RCDME follows $c^m < L$, leading to more reliable results, especially for short signals. Likewise, it is advisable to have at least $10^m$ (or preferably $20^m$) sample points for RCMSE with embedding dimension $m$.

### 3. Evaluation Signals

In this section, the synthetic and real signals used in this study to evaluate the behaviour of the multiscale entropy approaches are described.

### 3.1. Synthetic Signals

White noise is more irregular than pink noise ($1/f$ noise), although the latter is more complex because pink noise contains long-range correlations and its $1/f$ decay produces a fractal structure in time [3,5,8]. Therefore, white and pink noise are two important signals to evaluate the multiscale entropy techniques [3,5,8,23–25].

In order to investigate the change in the behavior of a nonlinear system, the Lorenz attractor is used. Further details can be found in [26,27]. To evaluate the effect of filtering and downsampling processes on different frequency components of time series, multi-harmonic signals are employed [16]. Finally, to inspect the effect of noise on multiscale approaches, white noise was added to the Lorenz and multi-harmonic time series.

*3.2. Real Biomedical Datasets*

Multiscale entropy techniques are broadly used to characterize physiological recordings [2,3,12,25]. To this end, electroencephalograms (EEGs) [28] and stride internal fluctuations [29] are used to distinguish different kinds of dynamics of time series.

### 3.2.1. Dataset of Focal and Non-Focal Brain Activity

The ability of complexity measures to discriminate focal from non-focal signals is evaluated by the use of an EEG dataset (publicly-available at [30]) [28]. The dataset includes five patients and, for each patient, there are 750 focal and 750 non-focal bivariate time series. The length of each signal was 20 s with sampling frequency of 512 Hz (10,240 samples). For more information, please, refer to [28]. All subjects gave written informed consent that their signals from long-term EEG might be used for research purposes [28]. Before computing the entropies, the EEG signals were digitally band-pass filtered between 0.5 Hz and 150 Hz using a fourth-order Butterworth filter.

### 3.2.2. Dataset of Stride Internal Fluctuations

To compare multiscale entropy methods, stride interval recordings are used [29,31]. The time series were recorded from five young, healthy men (23–29 years old) and five healthy old adults (71–77 years old). All the individuals walked continuously on level ground around an obstacle-free path for 15 min. The stride interval was measured by the use of ultra-thin, force sensitive resistors placed inside the shoe. For more information, please refer to [29].

## 4. Results and Discussion

*4.1. Synthetic Signals*

### 4.1.1. Frequency Responses of Cumulative Sums of IMFs (CSI), and Moving Average (MA) and Butterworth (BW) Filters

To investigate the frequency responses of MA, BW, and CSI, we used 200 realizations of white noise with length 512 sample points following [32,33]. The average Fourier spectra obtained by MA, BW, and CSI at different scale factors (i.e., 2, 4, 6, 8, and 10) are depicted in Figure 2. The results show that BW, MA, and CSI can be considered as low-pass filters with different cut-off frequencies. The results for MA and BW filters are in agreement with their theoretical frequency responses shown in Equations (2) and (3), respectively. The results for CSI are also in agreement with the fact that $IMF_1$ corresponds to a half-band high-pass filter and $IMF_\lambda$ ($\lambda \geq 2$) can be considered as a filter bank of overlapping bandpass filters [33].

The magnitude of the frequency response for BW, compared with MA, is flatter in the passband, side lobes in its stopband are not present, and the roll-off is faster. Therefore, the filter's frequency response leads to a more accurate elimination of the components with frequency above cut-off frequencies. This fact reduces aliasing while the filtered signals are downsampled. The behavior of the frequency response for CSI is similar to that for BW. However, the cut-off frequencies obtained by CSI are considerably smaller than those for BW. This fact results in very low entropy values at high scale factors.

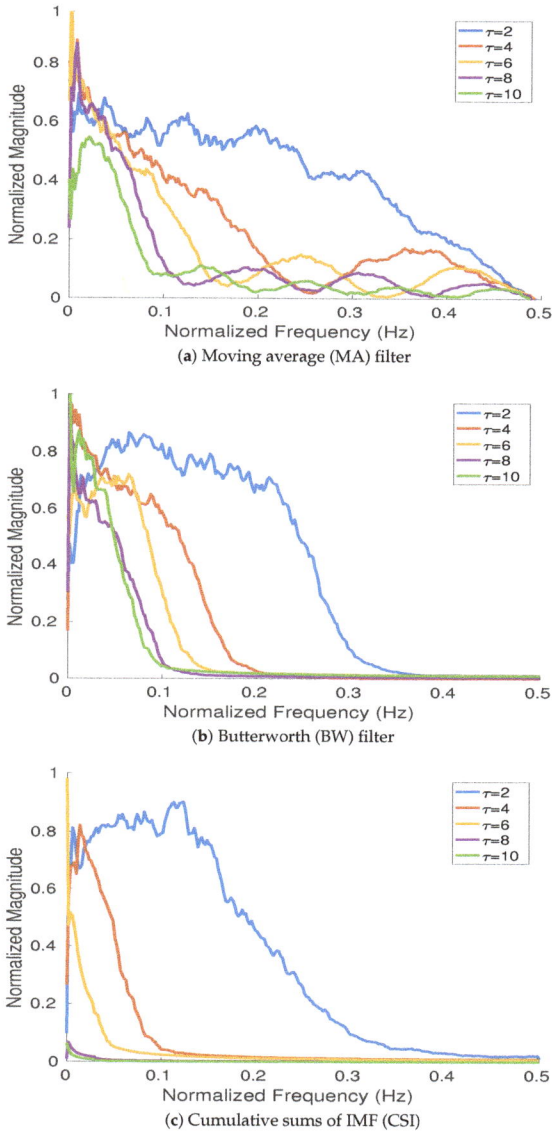

**Figure 2.** Magnitude of the frequency response for (**a**) MA, (**b**) BW, and (**c**) CSI at different scale factors ($\tau$ =2, 4, 6, 8, and 10) computed from 200 realizations of white noise with length 512 sample points.

### 4.1.2. Effect of Different Low-Pass Filters on Multi-Harmonic and Lorenz Series

To understand the effect of MA, BW, and CSI on multi-harmonic signals, we use $b_i = \cos(2\pi 10i) + \cos(2\pi 20i) + \cos(2\pi 50i)$ with sampling frequency 200 Hz and length 20 s. The first second of the signal **b** is depicted in Figure 3. To show the frequency components of **b** and their amplitude values, we used the combination of Hilbert transform and recently introduced variational mode decomposition (VMD). VMD is a generalization of the classic Wiener filter into adaptive, multiple bands [16]. After decomposing the original signals into its IMFs using VMD, we employ the Hilbert transform to find the instantaneous frequency of each IMF [16,34].

The frequency components of **b** and their corresponding amplitudes are depicted in Figure 3a. The Hilbert transform of **b** filtered by 4-sample MA (Figure 3b) illustrates that the harmonic $\cos(2\pi 50i)$ is completely eliminated, in agreement with the fact that MA is a low-pass filter with cut-off frequency $\frac{f_s}{2\tau}$ and completely eliminates the frequency component $f_z$ at $\frac{f_s}{\tau}$ (here at $50 = \frac{200}{4}$) based on Equation (2) [11].

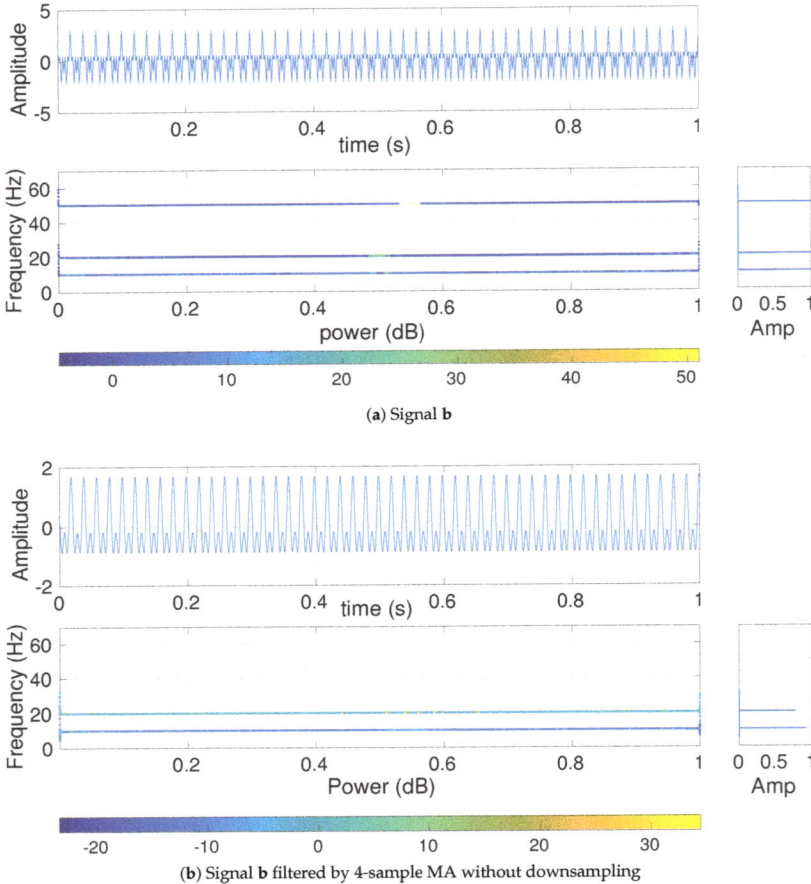

(a) Signal **b**

(b) Signal **b** filtered by 4-sample MA without downsampling

**Figure 3.** Hilbert transform of the decomposed VMD-based IMFs obtained from (a) $b_i = \cos(2\pi 10i) + \cos(2\pi 20i) + \cos(2\pi 50i)$ and (b) **b** filtered by 20-sample MA (scale 20).

The MDE values for **b**, depicted in Figure 4a, show that the largest changes in entropy values occur at temporal scale 4 and 10 (based on $50 = \frac{200}{4}$ and $20 = \frac{200}{10}$—please see the red double arrows in Figure 4). In fact, the largest changes in entropy values are related to the main frequency components of a multi-harmonic time series. To investigate the effect of noise on MDE values, we created $g_i = b_i + \eta$, where $\eta$ denotes a uniform random variable between 0 to 1. The MDE values for **g**, plotted in Figure 4b, illustrate a decrease at temporal scales from 1 to 19 and then the entropy values become approximately constant. This is in agreement with the fact that the smaller scale factors correspond to higher frequency components, whereas smaller scales correspond to lower frequencies [35]. Comparing Figures 4a and b shows that after filtering the effect of white noise by MA, the profiles for **b** and **g** are very close (temporal scales 19 to 25). This suggests that white noise affects lower temporal scales. It is worth noting that the behavior of MDE$_{BW}$ and that of MDE$_{MA}$ is similar.

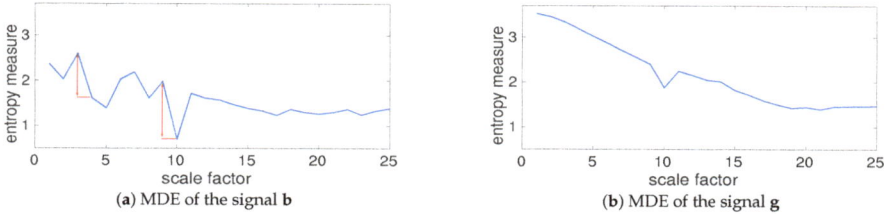

(a) MDE of the signal **b**

(b) MDE of the signal **g**

**Figure 4.** MDE values for (a) $b_i = \cos(2\pi 10i) + \cos(2\pi 20i) + \cos(2\pi 50i)$ and (b) $g_i = b_i + \eta$. The largest changes in entropy values (the red double arrows) occur at temporal scale 4 and 10 (respectively correspond to $50 = \frac{200}{4}$ and $20 = \frac{200}{10}$).

However, the effect of CSI at scale 2 on **b** is shown in Figure 5. The results, compared with those for MA (see Figure 4b), illustrate similar behavior of CSI at scale 2 and MA at scale 4 in terms of the elimination of the highest frequency component of **b**. This is in agreement with the fact that, at a specific scale factor, the cut-off frequency for CSI is considerably lower than that for MA or BW (see Figure 2).

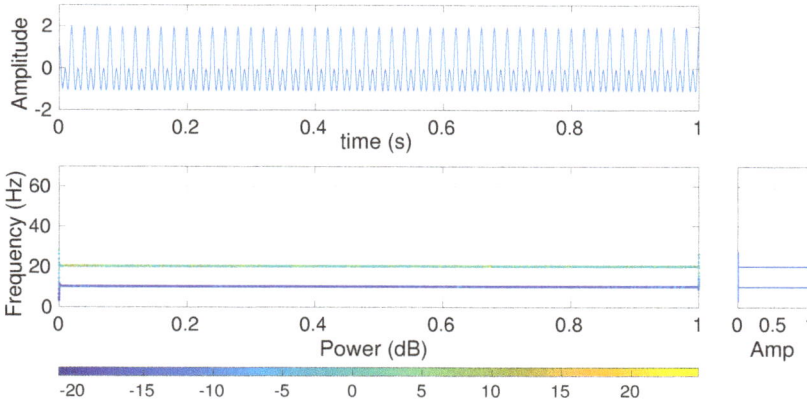

**Figure 5.** Hilbert transform of the decomposed VMD-based IMFs obtained from the signal **b** for CSI at scale 2.

We also generated the Lorenz signal **o** with length 10,000 sample points and sampling frequency ($f_s$) 300 Hz. To have a nonlinear behavior, $\lambda = 10$, $\beta = \frac{8}{3}$, and $\rho = 99.96$ were set [26,27]. The signal **o** and **o** filtered by MA at scale 10 are shown in Figure 6. The MDE-based values for **o** are depicted in Figure 7a. The Nyquist frequency of the signal is ($\frac{300}{2} = 150$) Hz and is close to its highest frequency component (around 150 Hz). Note that choosing a lower sampling frequency may result in aliasing. As the main frequency components of this time series are around 20–30 Hz, the MA filter is not able to completely eliminate the main frequency components of this signal at scale 10. It leads to the amplitude values of the filtered signal at scale 10 (without downsampling) being very close to those of the original time series **o**.

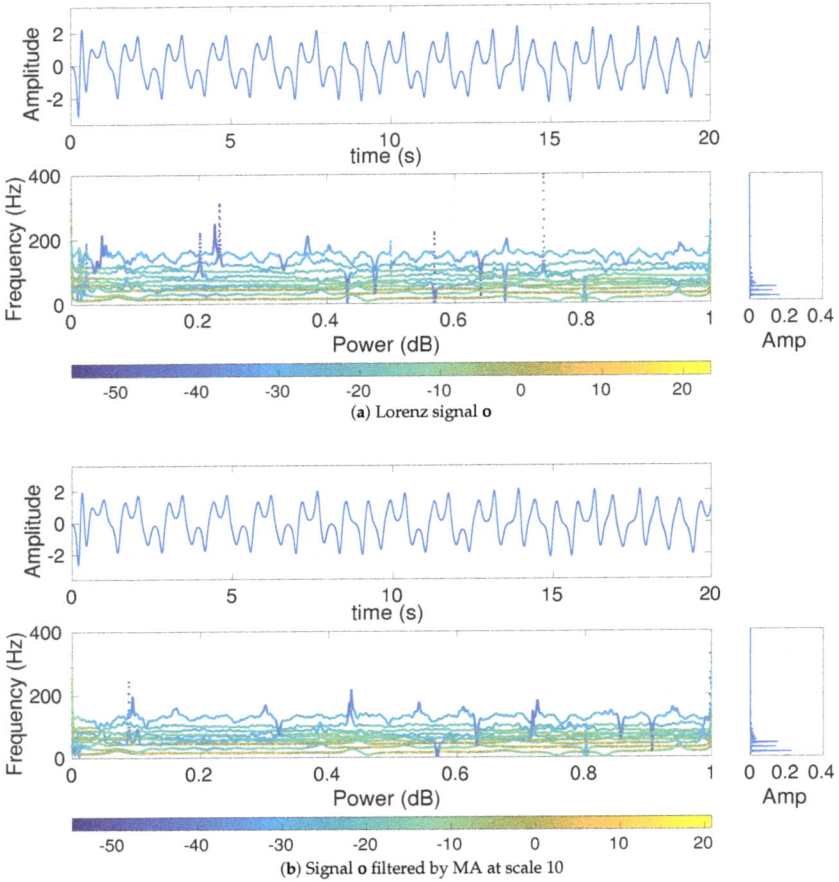

**Figure 6.** Hilbert transform of the decomposed VMD-based IMFs obtained from (**a**) the Lorenz signal **o** and (**b**) **o** filtered by MA at scale 10.

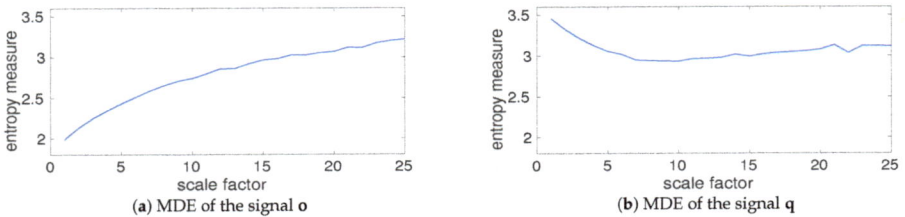

**Figure 7.** MDE results for (**a**) the Lorenz signal **o** and (**b**) $q_i = o_i + \eta$.

To inspect the effect of additive noise on MDE values, we created $q_i = o_i + \eta$, where $\eta$ is a random variable between 0 to 1. The MDE values for **q**, plotted in Figure 7b, illustrate a decrease at low temporal scale and then an increase at high time scale factors. It is also found that the MDE values of **o** and **q** are approximately equal at scales between 18 to 25. This is also consistent with the fact that lower scale factors correspond to higher frequency components, whereas larger scales correspond to lower frequencies [35].

### 4.1.3. Effect of Downsampling and Sampling Frequency on Multiscale Entropy Methods

To investigate the effect of downsampling (without low-pass filtering) on multiscale entropy approaches, we created the signal $s_i = \cos(2\pi i)$ with length 300 sample points and sampling frequency 10 Hz, and (b) $w_i = \cos(2\pi i)$ with length 300 sample points and sampling frequency 100 Hz. The signals and their downsampled series by a factor of 12 are depicted in Figure 8.

When the sampling frequency of a time series is close to its main frequency components (see **s**—Figure 8a), the downsampled signal may have a lower frequency component in comparison with the original signal. It shows the effect of aliasing in the time series. Accordingly, the downsampled signals are more regular (have smaller entropy values). It is confirmed by the fact that the DispEn of **s** and its corresponding downsampled series are 2.0267 and 1.6058, respectively.

On the other hand, when the sampling frequency is high (see **w**—Figure 8b), the amplitude values of downsampled signal are approximately equal to those of the original signal. However, as the number of sample points decreases by 12, the rate of change along sample points is 12 times larger than that for the original signal. Thus, the original signal is more regular than its corresponding downsampled series. It is confirmed by the fact that the DispEn of **w** and its corresponding downsampled series are respectively 1.9618 and 2.5539.

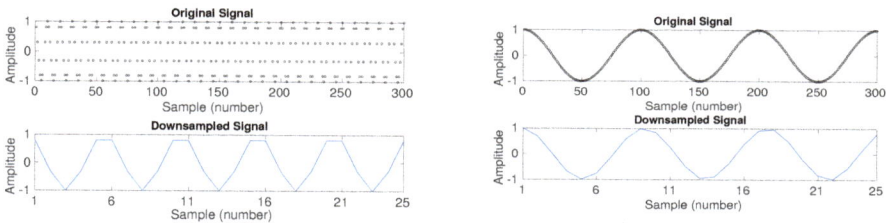

(a) Signal $s_i = \cos(2\pi i)$ with sampling frequency 10 Hz  (b) Signal $w_i = \cos(2\pi i)$ with sampling frequency 100 Hz

**Figure 8.** Downsampling the signal (**a**) $s_i = \cos(2\pi i)$ with length 300 sample points and sampling frequency 10 Hz, and (**b**) $w_i = \cos(2\pi i)$ with length 300 sample points and sampling frequency 100 Hz. The factor of downsampling is 12.

### 4.1.4. Multiscale Entropy Methods vs. Noise

All of the complexity methods are used to distinguish the dynamics of white from pink noise. The mean and SD of results for the signals with length 8000 (long series) and 400 (short series) sample points are respectively depicted in Figures 9 and 10. The results obtained by the complexity techniques with DS show that the entropy values decrease monotonically with scale factor $\tau$ for white noise. However, for pink noise, the entropy values become approximately constant over larger-scale factors. These are in agreement with the fact that, unlike white noise, $1/f$ noise has structure across temporal scale factors [3,5]. The profiles for $MDE_{MA}$ and $MSE_{MA}$ without DS, $MDE_{BW}$ and $MSE_{BW}$ without DS, InMSE, and InMDE decrease along the temporal scales as there is not a DS process to increase the rate of changes to increase entropy values. It should be mentioned that, as the crossing point of profiles for white and pink noise is at scale 23, $\tau_{max}$ for the MA-based coarse graining is equal to 50. Furthermore, $\tau_{max}$ for InMSE and InMDE is 10, as the entropy values at high scales are close to 0.

Entropy values obtained by MSE, RCMSE, $MSE_{BW}$, and $RCMSE_{BW}$ are undefined at high scale factors. Comparing Figures 9 and 10 demonstrates that the longer the signals, the more robust the multiscale entropy estimations. The results also show that InMDE, compared with InMSE, better discriminates white from pink noise.

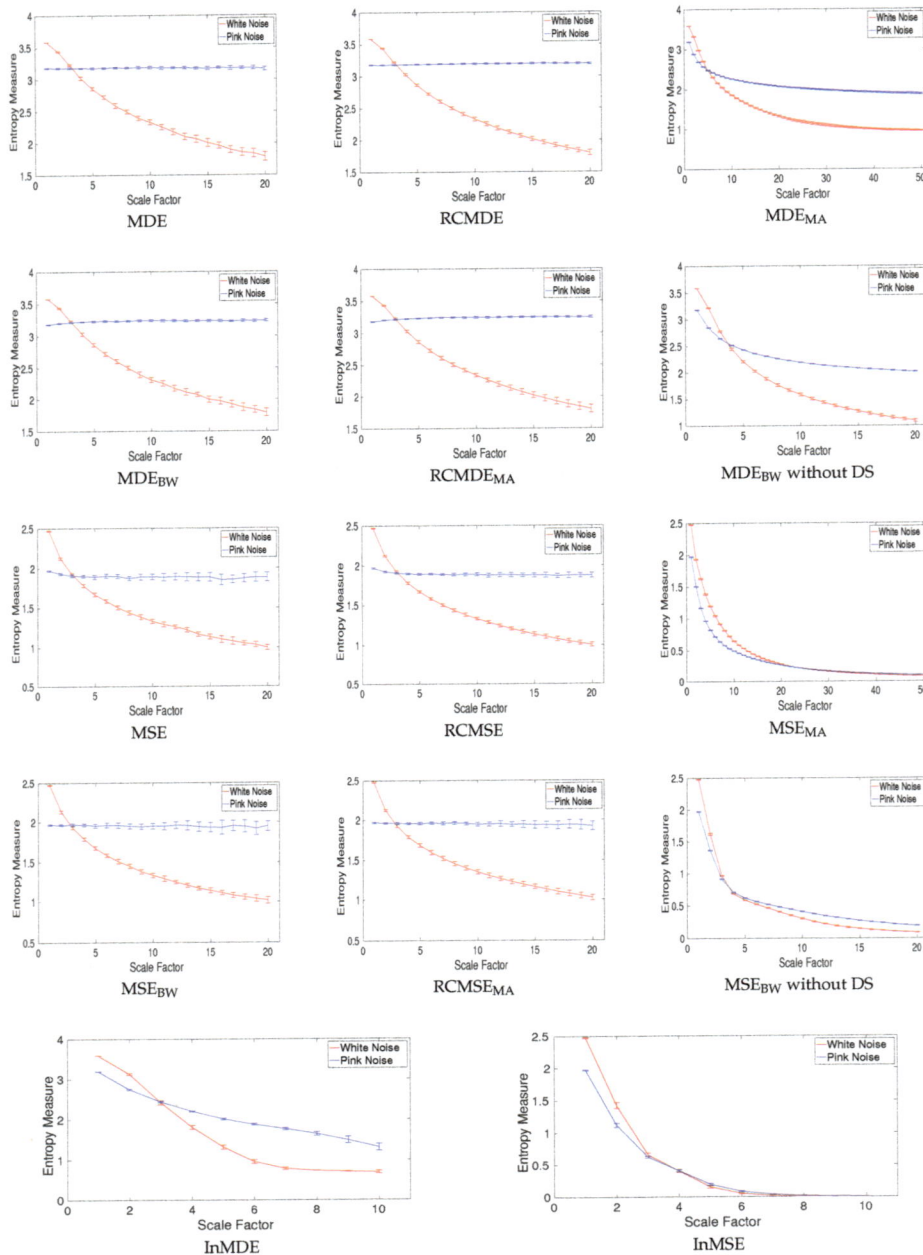

**Figure 9.** Mean value and SD of results obtained by the complexity measures computed from 40 different realizations of pink and white noise with length 8000 samples. Red and blue demonstrate white and pink noise, respectively.

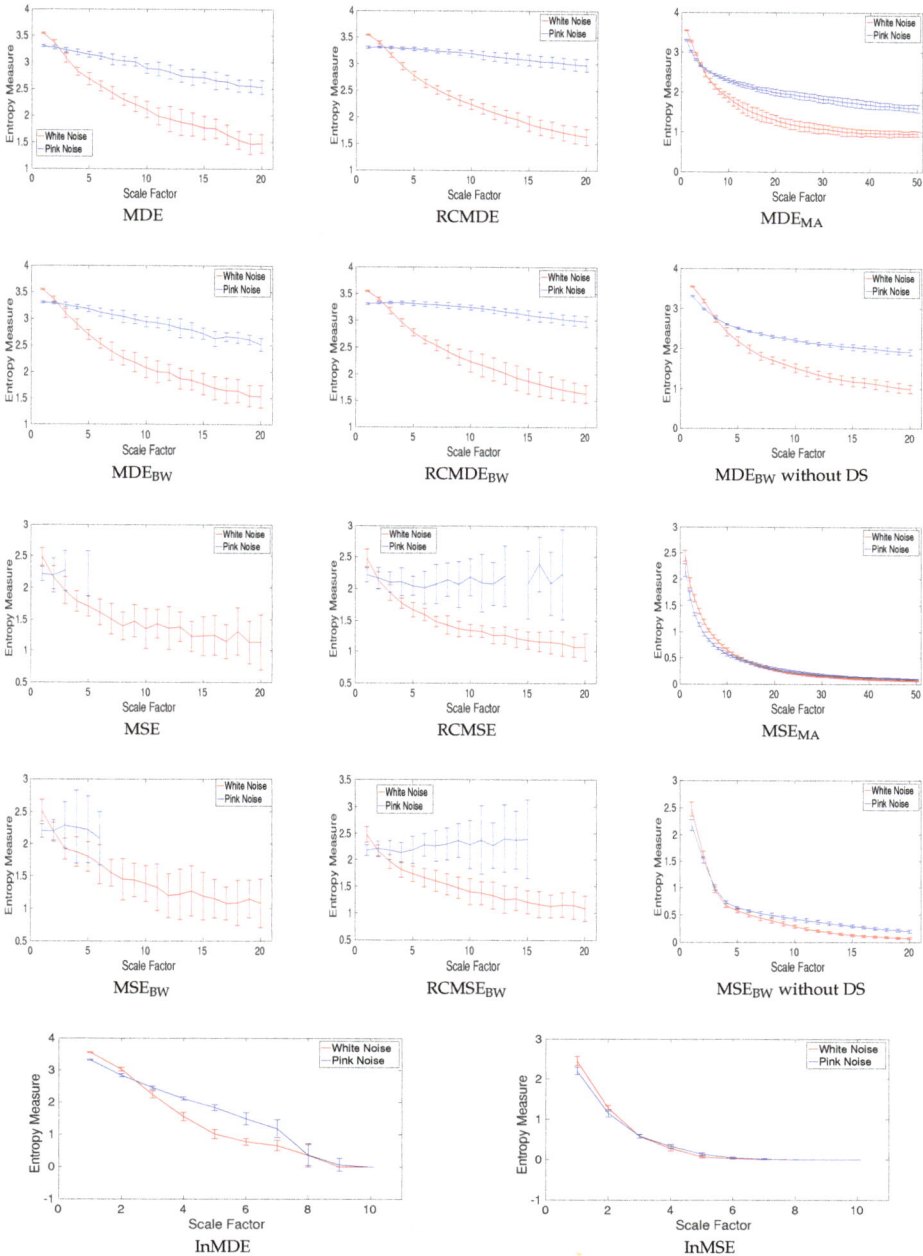

**Figure 10.** Mean value and SD of results obtained by the complexity measures computed from 40 different realizations of pink and white noise with length 400 samples. Entropy values obtained by MSE, RCMSE, MSE$_{BW}$, and RCMSE$_{BW}$ are undefined at several high scale factors. Red and blue demonstrate white and pink noise, respectively.

To compare the results obtained by the complexity algorithms, we used the coefficient of variation (CV) defined as the SD divided by the mean. We use such a metric as the SDs of signals may increase or decrease proportionally to the mean. The CV values at scale 10, as a trade-off between low and high scale factors, for noise signals with length 8000 and 400 sample points are respectively illustrated in Tables 2 and 3. Of note is that we consider scale 25 and 5 for the $MSE_{MA}$ and $MDE_{MA}$, and InMSE and InMDE profiles, respectively. The refined composite technique decreases the CVs for all the MSE- and MDE-based algorithms, showing its advantage to improve the stability of results for short and long noise. The smallest CVs for long pink and white noise are our developed $MDE_{BW}$ without DS and $RCMDE_{BW}$ methods, respectively. The smallest CVs for short pink and white noise are achieved by $RCMDE_{BW}$ and RCMDE, respectively. Overall, the smallest CVs are obtained by the DispEn-based complexity measures.

**Table 2.** CV values obtained by the complexity measures at scale factor 10 for forty realizations of pink and white noise with length 8000 sample points. Note that the scales 25 and 5 are considered for $MSE_{MA}$ and $MDE_{MA}$, and InMSE and InMDE, respectively.

| Noise | MDE | RCMDE | $MDE_{MA}$ (Scale 25) | $MDE_{BW}$ | $RCMDE_{BW}$ | $MDE_{BW}$ without DS | InMDE (Scale 5) |
|---|---|---|---|---|---|---|---|
| Pink | 0.0058 | 0.0038 | 0.0069 | 0.0044 | 0.0038 | 0.0031 | 0.0091 |
| White | 0.0174 | 0.0124 | 0.0246 | 0.0166 | 0.0115 | 0.0182 | 0.0394 |

| Noise | MSE | RCMSE | $MSE_{MA}$ (Scale 25) | $MSE_{BW}$ | $RCMSE_{BW}$ | $MSE_{BW}$ without DS | InMSE (Scale 5) |
|---|---|---|---|---|---|---|---|
| Pink | 0.0186 | 0.0105 | 0.0131 | 0.0176 | 0.0124 | 0.0130 | 0.0982 |
| White | 0.0201 | 0.0133 | 0.0135 | 0.0219 | 0.0203 | 0.0308 | 0.1330 |

**Table 3.** CV values obtained by the complexity measures at scale factor 10 for forty realizations of pink and white noise with length 400 sample points. Note that the scales 25 and 5 are considered for $MSE_{MA}$ and $MDE_{MA}$, and InMSE and InMDE, respectively.

| Noise | MDE | RCMDE | $MDE_{MA}$ (Scale 25) | $MDE_{BW}$ | $RCMDE_{BW}$ | $MDE_{BW}$ without DS | InMDE (Scale 5) |
|---|---|---|---|---|---|---|---|
| Pink | 0.0317 | 0.0194 | 0.0473 | 0.0320 | 0.0141 | 0.0204 | 0.0522 |
| White | 0.0726 | 0.0415 | 0.1116 | 0.0929 | 0.0876 | 0.0726 | 0.1435 |

| Noise | MSE | RCMSE | $MSE_{MA}$ (Scale 25) | $MSE_{BW}$ | $RCMSE_{BW}$ | $MSE_{BW}$ without DS | InMSE (Scale 5) |
|---|---|---|---|---|---|---|---|
| Pink | undefined | 0.1327 | 0.0434 | undefined | 0.2008 | 0.0822 | 0.2351 |
| White | 0.2385 | 0.0738 | 0.0605 | 0.2024 | 0.1736 | 0.1060 | 0.3779 |

### 4.1.5. Effect of Refined Composite on Nonlinear Systems without Noise

To understand the effect of the refined composite technique on nonlinear signals without noise, we created 40 realizations of two Lorenz signals with lengths of 450 and 4500 sample points and sampling frequency ($f_s$) 150 Hz. To have a nonlinear behavior, the values of $\lambda = 10$, $\beta = \frac{8}{3}$, and $\rho = 28$ were used in the Lorenz system [26,27]. The results obtained by MSE, MDE, RCMSE, and RCMDE are depicted in Figure 11 and are in agreement with [25,27]. Of note is that the entropy values for $RCMSE_{BW}$ and $RCMDE_{BW}$ are similar to those for RCMSE and RCMDE, respectively. Thus, these results are not shown herein.

To investigate the effect of the refined composite technique on the stability of results, the CVs for the multiscale approaches at scale 5 are calculated. The smallest CVs, illustrated in Table 4 are obtained by MDE and RCMDE approaches. The results also suggest that the refined composite does not improve the stability of profiles for the signal with length 4500 samples (long signals). For the Lorenz series with length 450 sample points, RCMSE and RCMDE lead to smaller CV values in comparison with MSE and MDE, in that order, showing the importance of the refined composite method to characterize small time series.

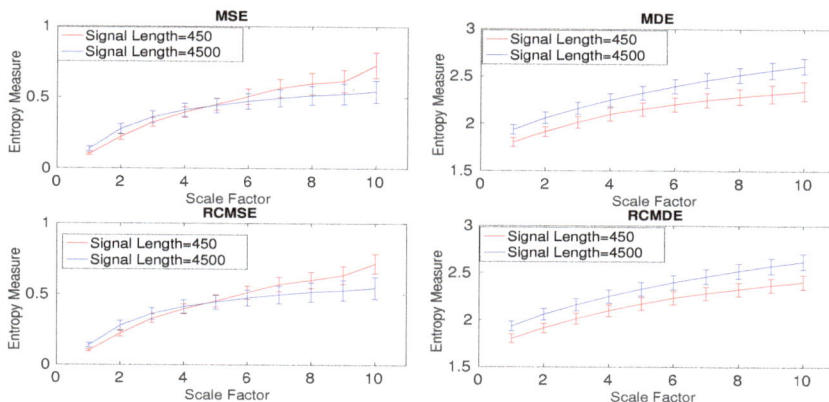

**Figure 11.** Mean and SD of the results obtained by the MSE, MDE, RCMSE, and RCMDE for the Lorenz series with lengths 450 and 4500 sample points.

**Table 4.** CVs of MSE, RCMSE, MDE, and RCMDE values for the 40 different realizations of the Lorenz signals with length 450 and 4500 samples at scale five.

| Signal Length | MSE | MDE | RCMSE | RCMDE |
|---|---|---|---|---|
| 450 sample points | 0.1000 | 0.0898 | 0.0700 | 0.0309 |
| 4500 sample points | 0.1156 | 0.0310 | 0.1134 | 0.0312 |

## 4.2. Real Signals

### 4.2.1. Dataset of Focal and Non-Focal Brain Activity

For the focal and non-focal EEG dataset, the results obtained by MSE, MDE, RCMSE, RCMDE, $MSE_{BW}$, $MDE_{BW}$, InMSE, and InMDE, depicted in Figure 12, show that the non-focal signals are more complex than the focal ones. This fact is in agreement with previous studies [28,36].

The results for $RCMSE_{BW}$ and $RCMDE_{BW}$ were respectively similar to those for $MSE_{BW}$ and $MDE_{BW}$. Thus, they are not shown herein. Note that, for MDE and RCMDE, $\tau_{max}$ and $m$, respectively, were 30 and 3. It should also be mentioned that the average entropy values over two channels for these bivariate EEG signals are reported for the univariate complexity techniques.

To compare the results, the CV values obtained by the univariate multiscale approaches, except InMSE and InMDE, are calculated at scale factor 15. These are shown in Table 5. The CV values for MDE, RCMDE, MSE, and RCMSE illustrate that the refined composite approach does not enhance the stability of the MDE and MSE profiles. Overall, the smallest CV values are achieved by DispEn-based complexity methods.

**Table 5.** CVs of MSE, RCMSE, $MSE_{BW}$, MDE, RCMDE, and $MDE_{BW}$ values for the focal and non-focal EEGs at scale 15.

| Signals | MSE | RCMSE | $MSE_{BW}$ | MDE | RCMDE | $MDE_{BW}$ |
|---|---|---|---|---|---|---|
| Focal EEGs | 0.0229 | 0.0229 | 0.0224 | 0.0083 | 0.0089 | 0.0083 |
| Non-focal EEGs | 0.0178 | 0.0191 | 0.0172 | 0.0111 | 0.0121 | 0.0109 |

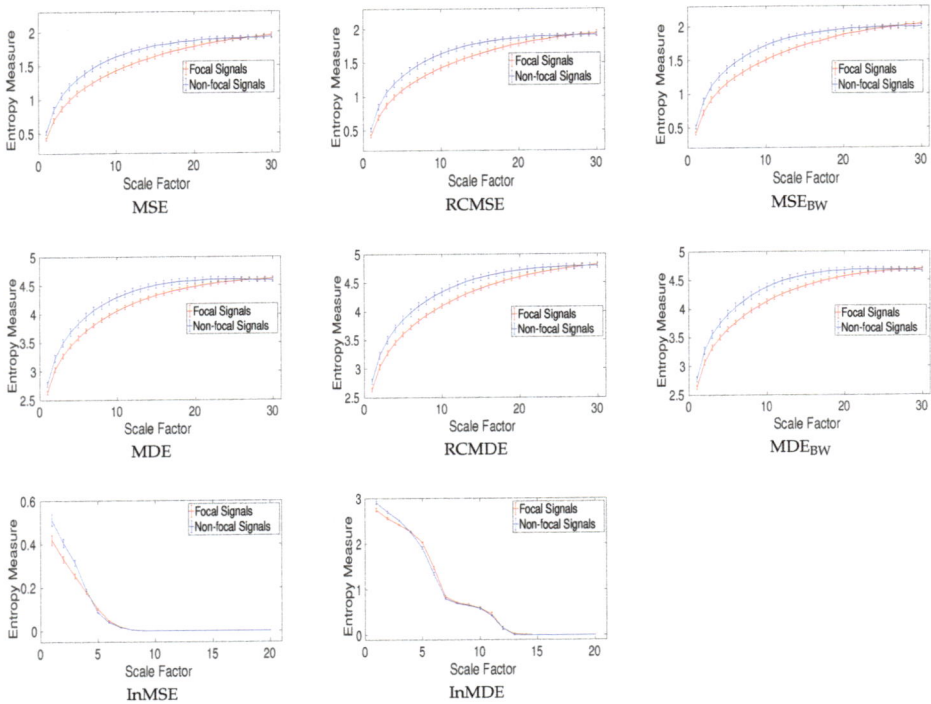

**Figure 12.** Mean value and SD of results obtained by the MSE, MDE, RCMSE, RCMDE, MSE$_{BW}$, MDE$_{BW}$, InMSE, and InMDE computed from the focal and non-focal EEGs.

### 4.2.2. Dataset of Stride Internal Fluctuations

In Figure 13, the mean and SD of the RCMDE$_{BW}$, RCMDE, MDE$_{MA}$, MDE$_{BW}$ without DS, InMDE, RCMSE$_{BW}$, RCMSE, MSE$_{MA}$, MSE$_{BW}$ without DS, and InMSE values computed from young and old subjects' stride internal fluctuations are illustrated. As the number of samples for these time series are between 400 to 800 sample points, we do not use MSE, MDE, MSE$_{BW}$, and MDE$_{BW}$.

For each scale factor, the average of entropy values for elderly subjects is smaller than that for young ones, in agreement with those obtained by the other entropy-based methods [37] and the fact that recordings from healthy young subjects correspond to more complex states because of their ability to adapt to adverse conditions, whereas aged individuals' signals present complexity loss [3,5,38]. The results also suggest that, when dealing with short signals, the complexity measures without downsampling (i.e., MSE$_{MA}$, MDE$_{MA}$, and MSE$_{BW}$ and MDE$_{BW}$ without DS) are appropriate to distinguish different kinds of dynamics of real signals.

The CV values at those scales whose profiles do not have an overlap are illustrated in Table 6. It is found that MDE$_{BW}$ without DS leads to the smallest CV values.

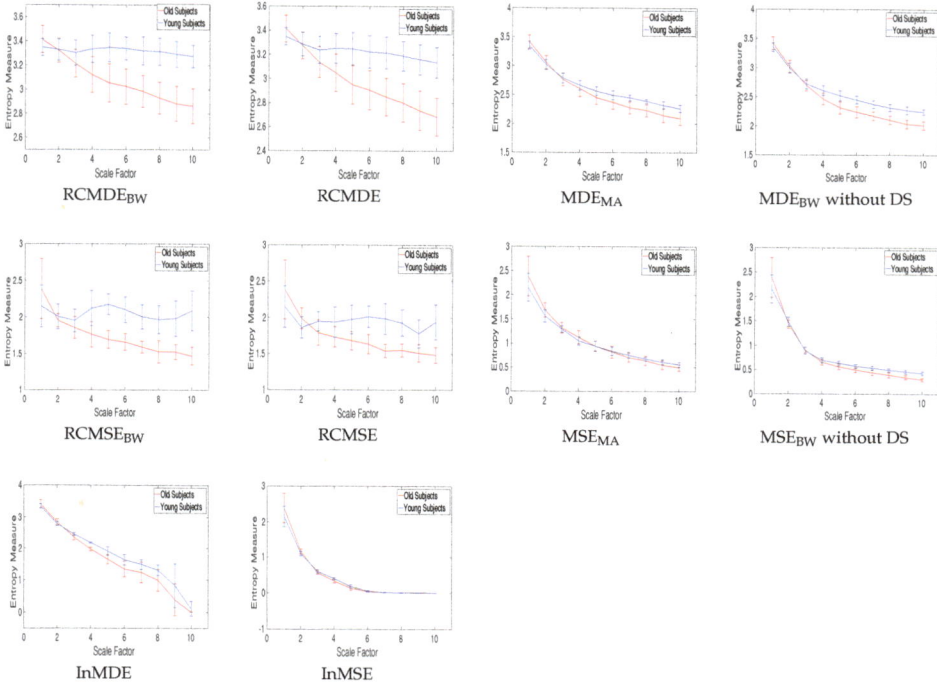

**Figure 13.** Mean value and SD of results obtained by the complexity measures computed from the young and old subjects' stride interval recordings.

**Table 6.** CV values obtained by the complexity measures for the stride interval recordings for young and old subjects.

| Signals | $RCMDE_{BW}$ | RCMDE | $MDE_{BW}$ without DS | $RCMSE_{BW}$ |
|---|---|---|---|---|
| Young subjects | 0.0355 | 0.0410 | 0.0334 | 0.0644 |
| Old subjects | 0.0517 | 0.0540 | 0.0449 | 0.0723 |

## 5. Time Delay, Downsampling, and Nyquist Frequency

According the previous complexity-based approaches [2,3,13,15], the time delay was equal to 1 in this study. Nevertheless, if the sampling frequency is considerably larger than the highest frequency component of a signal, the first minimum or zero crossing of the autocorrelation function or mutual information can be used for the selection of an appropriate time delay [39].

Alternatively, a signal may be downsampled before calculating the complexity-based entropy approaches to adjust its highest frequency component to its Nyquist frequency ($f_s/2$) [40]. Accordingly, when the coarse-graining process starts, the low-pass filtering will affect the highest frequency component of the signal at low temporal scale factors. It is worth noting that if the main frequency components of the signal are considerably lower than its highest frequency component (e.g., the signal o - please see Figure 7), the filtering process may make only a little change in the amplitude values of the signal at even large scales.

## 6. Future Work

Wavelet transform, which is a powerful filter bank broadly used for analysis of non-stationary recordings, can be employed to decompose a signal to several series with specific frequency bands [41]. Accordingly, the wavelet-based filter bank could be used as a complexity approach. VMD can also be used as an alternative to EMD in InMSE and InMDE. VMD, unlike EMD, provides a solution to the decomposition problem that is theoretically well founded and more robust to noise than EMD [16]. A recent development in the field has tried to generalize multivariate and univariate multiscale algorithms to a family of statistics by using different moments (e.g., variance, skewness, and kurtosis) in the univariate and multivariate coarse-graining process [25,42–44]. It is recommended to compare these techniques in the context of signal processing and to investigate their interpretations. As the existing univariate and even multivariate coarse-graining processes filter only series in each channel separately [38,43,45], there is a need to propose new multivariate filters dealing with the spatial and time domains at the same time.

## 7. Conclusions

In summary, we have compared existing and newly proposed coarse-graining approaches for univariate multiscale entropy estimation. Our results indicate that, as expected due to the filter bank properties of the EMD [33] in comparison with moving average and Butterworth filtering, the cut-off frequencies at each temporal scale $\tau$ of the former are considerably smaller than those for the latter. Therefore, InMSE and our developed InMDE have entropy values very close to 0 for relatively low values of temporal scales due to the exponential, rather than linear, dependency of the bandwidth at each scale. We also inspected the effect of the downsampling in the coarse-graining process in the entropy values, showing that it may lead to increased or decreased values of entropy depending on the sampling frequency of the time series.

Our results confirmed previous reports indicating that, when dealing with short or noisy signals, the refined composite approach [14,25] may improve the stability of entropy results. On the other hand, for long signals with relatively low levels of noise, the refined composite method makes little difference in the quality of the entropy estimation at the expense of a considerable additional computational cost. In any case, the use of dispersion entropy over sample entropy in the estimations led to more stable results based on CV values and ensured that the entropy values were defined at all temporal scales.

Finally, the profiles obtained by the multiscale techniques with and without downsampling led to similar findings (e.g., pink noise is more complex than white noise based on all the complexity methods) although the specific values of entropy may differ depending on the coarse-graining used. This suggests that downsampling within the coarse-graining procedure may not be needed to quantify the complexity of signals, especially for short ones. In fact, these kinds of techniques still eliminate the fast temporal scales to deal with progressively slower time scales as $\tau$ increases and take into account multiple time scales inherent in time series.

On the whole, it is expected that these findings contribute to the ongoing discussion regarding the development of stable, fast, and less sensitive-to-noise complexity approaches appropriate for either short or long time series. We recommend that future studies explicitly justify their choices for coarse-graining procedure in the light of the characteristics of the signals under analysis and the hypothesis of the study, and that they discuss their findings on the light of the behaviour of the selected entropy metric and coarse-graining procedure.

**Author Contributions:** Hamed Azami and Javier Escudero conceived and designed the methodology. Hamed Azami was responsible for analysing and writing the paper. Both the authors contributed critically to revise the results and discussed them and have read and approved the final manuscript.

**Conflicts of Interest:** The authors declare no conflict of interest.

## Appendix. Matlab Codes used in this Article

The Matlab codes of DispEn and MDE are available at https://datashare.is.ed.ac.uk/handle/10283/2637. The codes of SampEn and MSE can be found at https://physionet.org/physiotools/matlab/wfdb-app-matlab/. The code of EMD is also available at http://perso.ens-lyon.fr/patrick.flandrin/emd.html. For the Butterworth filter, we used the functions "butter" and "filter" in Matlab R2015a.

## References

1.  Yang, A.C.; Tsai, S.J. Is mental illness complex? From behavior to brain. *Prog. Neuro-Psychopharmacol. Biol. Psychiatry* **2013**, *45*, 253–257.
2.  Costa, M.; Goldberger, A.L.; Peng, C.K. Multiscale entropy analysis of complex physiologic time series. *Phys. Rev. Lett.* **2002**, *89*, 068102.
3.  Costa, M.; Goldberger, A.L.; Peng, C.K. Multiscale entropy analysis of biological signals. *Phys. Rev. E* **2005**, *71*, 021906.
4.  Bar-Yam, Y. *Dynamics of Complex Systems*; Addison-Wesley Reading: Boston, MA, USA, 1997.
5.  Fogedby, H.C. On the phase space approach to complexity. *J. Stat. Phys.* **1992**, *69*, 411–425.
6.  Rostaghi, M.; Azami, H. Dispersion entropy: A measure for time series analysis. *IEEE Signal Process. Lett.* **2016**, *23*, 610–614.
7.  Zhang, Y.C. Complexity and 1/f noise. A phase space approach. *J. Phys. I* **1991**, *1*, 971–977.
8.  Silva, L.E.V.; Cabella, B.C.T.; da Costa Neves, U.P.; Junior, L.O.M. Multiscale entropy-based methods for heart rate variability complexity analysis. *Phys. A Stat. Mech. Appl.* **2015**, *422*, 143–152.
9.  Goldberger, A.L.; Peng, C.K.; Lipsitz, L.A. What is physiologic complexity and how does it change with aging and disease? *Neurobiol. Aging* **2002**, *23*, 23–26.
10. Hayano, J.; Yamasaki, F.; Sakata, S.; Okada, A.; Mukai, S.; Fujinami, T. Spectral characteristics of ventricular response to atrial fibrillation. *Am. J. Physiol. Heart Circ. Physiol.* **1997**, *273*, H2811–H2816.
11. Valencia, J.F.; Porta, A.; Vallverdu, M.; Claria, F.; Baranowski, R.; Orlowska-Baranowska, E.; Caminal, P. Refined multiscale entropy: Application to 24-h holter recordings of heart period variability in healthy and aortic stenosis subjects. *IEEE Trans. Biomed. Eng.* **2009**, *56*, 2202–2213.
12. Humeau-Heurtier, A. The multiscale entropy algorithm and its variants: A review. *Entropy* **2015**, *17*, 3110–3123.
13. Azami, H.; Rostaghi, M.; Abasolo, D.; Escudero, J. Refined Composite Multiscale Dispersion Entropy and its Application to Biomedical Signals. *IEEE Trans. Biomed. Eng.* **2017**, *64*, 2872–2879.
14. Wu, S.D.; Wu, C.W.; Lin, S.G.; Lee, K.Y.; Peng, C.K. Analysis of complex time series using refined composite multiscale entropy. *Phys. Lett. A* **2014**, *378*, 1369–1374.
15. Amoud, H.; Snoussi, H.; Hewson, D.; Doussot, M.; Duchêne, J. Intrinsic mode entropy for nonlinear discriminant analysis. *IEEE Signal Process. Lett.* **2007**, *14*, 297–300.
16. Dragomiretskiy, K.; Zosso, D. Variational mode decomposition. *IEEE Trans. Signal Process.* **2014**, *62*, 531–544.
17. Unser, M.; Aldroubi, A.; Eden, M. B-spline signal processing. I. Theory. *IEEE Trans. Signal Process.* **1993**, *41*, 821–833.
18. Fliege, N.J. *Multirate Digital Signal Processing*; John Wiley: Hoboken, NJ, USA, 1994.
19. Oppenheim, A.V. *Discrete-Time Signal Processing*; Pearson Education India: Delhi, India, 1999.
20. Castiglioni, P.; Coruzzi, P.; Bini, M.; Parati, G.; Faini, A. Multiscale sample entropy of cardiovascular signals: Does the choice between fixed-or varying-tolerance among scales influence its evaluation and interpretation? *Entropy* **2017**, *19*, 590.
21. Richman, J.S.; Moorman, J.R. Physiological time-series analysis using approximate entropy and sample entropy. *Am. J. Physiol. Heart Circ. Physiol.* **2000**, *278*, H2039–H2049.
22. Chen, W.; Wang, Z.; Xie, H.; Yu, W. Characterization of surface EMG signal based on fuzzy entropy. *Neural Syst. Rehabil. Eng. IEEE Trans.* **2007**, *15*, 266–272.
23. Wu, S.D.; Wu, C.W.; Humeau-Heurtier, A. Refined scale-dependent permutation entropy to analyze systems complexity. *Phys. A Stat. Mech. Appl.* **2016**, *450*, 454–461.

24. Humeau-Heurtier, A.; Wu, C.W.; Wu, S.D.; Mahé, G.; Abraham, P. Refined Multiscale Hilbert–Huang Spectral Entropy and Its Application to Central and Peripheral Cardiovascular Data. *IEEE Trans. Biomed. Eng.* **2016**, *63*, 2405–2415.

25. Azami, H.; Fernández, A.; Escudero, J. Refined multiscale fuzzy entropy based on standard deviation for biomedical signal analysis. *Med. Biol. Eng. Comput.* **2017**, *55*, 2037–2052.

26. Baker, G.L.; Gollub, J.P. *Chaotic dynamics: an introduction*; Cambridge University Press: Cambridge, UK, 1996.

27. Thuraisingham, R.A.; Gottwald, G.A. On multiscale entropy analysis for physiological data. *Phys. A Stat. Mech. Appl.* **2006**, *366*, 323–332.

28. Andrzejak, R.G.; Schindler, K.; Rummel, C. Nonrandomness, nonlinear dependence, and nonstationarity of electroencephalographic recordings from epilepsy patients. *Phys. Rev. E* **2012**, *86*, 046206.

29. Gait in Aging and Disease Database. Available online: https://www.physionet.org/physiobank/database/gaitdb (accessed on 17 February 2018).

30. The Bern-Barcelona EEG database. Available online: http://ntsa.upf.edu/downloads/andrzejak-rg-schindler-k-rummel-c-2012-nonrandomness-nonlinear-dependence-and (accessed on 17 February 2018).

31. Hausdorff, J.M.; Purdon, P.L.; Peng, C.; Ladin, Z.; Wei, J.Y.; Goldberger, A.L. Fractal dynamics of human gait: stability of long-range correlations in stride interval fluctuations. *J. Appl. Physiol.* **1996**, *80*, 1448–1457.

32. Wu, Z.; Huang, N.E. A study of the characteristics of white noise using the empirical mode decomposition method. *Proc. R. Soc. Lond. A Math. Phys. Eng. Sci.* **2004**, *460*, pp. 1597–1611.

33. Flandrin, P.; Rilling, G.; Goncalves, P. Empirical mode decomposition as a filter bank. *IEEE Signal Process. Lett.* **2004**, *11*, 112–114.

34. Huang, N.E.; Shen, Z.; Long, S.R.; Wu, M.C.; Shih, H.H.; Zheng, Q.; Yen, N.C.; Tung, C.C.; Liu, H.H. The empirical mode decomposition and the Hilbert spectrum for nonlinear and non-stationary time series analysis. *Proc. R. Soc. Lond. A Math. Phys. Eng. Sci.* **1998**, *454*, pp. 903–995.

35. Gow, B.J.; Peng, C.K.; Wayne, P.M.; Ahn, A.C. Multiscale entropy analysis of center-of-pressure dynamics in human postural control: methodological considerations. *Entropy* **2015**, *17*, 7926–7947.

36. Sharma, R.; Pachori, R.B.; Acharya, U.R. Application of entropy measures on intrinsic mode functions for the automated identification of focal electroencephalogram signals. *Entropy* **2015**, *17*, 669–691.

37. Nemati, S.; Edwards, B.A.; Lee, J.; Pittman-Polletta, B.; Butler, J.P.; Malhotra, A. Respiration and heart rate complexity: effects of age and gender assessed by band-limited transfer entropy. *Respir. Physiol. Neurobiol.* **2013**, *189*, 27–33.

38. Ahmed, M.U.; Mandic, D.P. Multivariate multiscale entropy: A tool for complexity analysis of multichannel data. *Phys. Rev. E* **2011**, *84*, 061918.

39. Kaffashi, F.; Foglyano, R.; Wilson, C.G.; Loparo, K.A. The effect of time delay on approximate & sample entropy calculations. *Phys. D Nonlinear Phenom.* **2008**, *237*, 3069–3074.

40. Berger, S.; Schneider, G.; Kochs, E.F.; Jordan, D. Permutation Entropy: Too Complex a Measure for EEG Time Series? *Entropy* **2017**, *19*, 692.

41. Strang, G.; Nguyen, T. *Wavelets Filter Banks*; Wellesley: Cambridge, UK, 1996.

42. Costa, M.D.; Goldberger, A.L. Generalized multiscale entropy analysis: application to quantifying the complex volatility of human heartbeat time series. *Entropy* **2015**, *17*, 1197–1203.

43. Azami, H.; Escudero, J. Refined composite multivariate generalized multiscale fuzzy entropy: A tool for complexity analysis of multichannel signals. *Phys. A Stat. Mech. Appl.* **2017**, *465*, 261–276.

44. Xu, M.; Shang, P. Analysis of financial time series using multiscale entropy based on skewness and kurtosis. *Phys. A Stat. Mech. Appl.* **2018**, *490*, 1543–1550.

45. Azami, H.; Fernández, A.; Escudero, J. Multivariate Multiscale Dispersion Entropy of Biomedical Times Series. *arXiv* **2017**, arXiv:1704.03947.

*entropy*

**MDPI**

*Article*

# Centered and Averaged Fuzzy Entropy to Improve Fuzzy Entropy Precision

**Jean-Marc Girault** [1] and **Anne Humeau-Heurtier** [2,*]

[1]   Groupe ESEO, 49000 Angers, France; jmgirault@univ-tours.fr
[2]   Laboratoire Angevin de Recherche en Ingénierie des Systèmes (LARIS), Univ Angers, 49000 Angers, France
*   Correspondence: anne.humeau@univ-angers.fr; Tel.: +33-(0)2-44-68-75-87

Received: 16 March 2018; Accepted: 13 April 2018; Published: 15 April 2018

**Abstract:** Several entropy measures are now widely used to analyze real-world time series. Among them, we can cite approximate entropy, sample entropy and fuzzy entropy (FuzzyEn), the latter one being probably the most efficient among the three. However, FuzzyEn precision depends on the number of samples in the data under study. The longer the signal, the better it is. Nevertheless, long signals are often difficult to obtain in real applications. This is why we herein propose a new FuzzyEn that presents better precision than the standard FuzzyEn. This is performed by increasing the number of samples used in the computation of the entropy measure, without changing the length of the time series. Thus, for the comparisons of the patterns, the mean value is no longer a constraint. Moreover, translated patterns are not the only ones considered: reflected, inversed, and glide-reflected patterns are also taken into account. The new measure (so-called centered and averaged FuzzyEn) is applied to synthetic and biomedical signals. The results show that the centered and averaged FuzzyEn leads to more precise results than the standard FuzzyEn: the relative percentile range is reduced compared to the standard sample entropy and fuzzy entropy measures. The centered and averaged FuzzyEn could now be used in other applications to compare its performances to those of other already-existing entropy measures.

**Keywords:** entropy; fuzzy entropy; sample entropy; irregularity; fetal heart rate; time series; symmetrical m-patterns

## 1. Introduction

   Approximate entropy (ApEn) and sample entropy (SampEn) algorithms are now widely used to quantify the irregularity of experimental time series [1,2]. They both rely on the evaluation of vectors' similarity. However, in both ApEn and SampEn, the vectors' similarity is based on the Heaviside function, a function that has rigid boundaries. Thus, the contributions of samples inside the boundary are treated equally, but the samples outside the boundary are left out. However, in the real world, boundaries between classes may be ambiguous: it is often difficult to determine if an input pattern belongs totally to a class. To overcome this lack of reality in ApEn and SampEn algorithms, Chen et al. proposed the fuzzy entropy (FuzzyEn) algorithm [3]. In the latter case, the vectors' similarity is defined by the soft and continuous boundaries of a fuzzy function. Since its introduction, it has been reported that FuzzyEn leads to better performance than ApEn or SampEn [4–6]. FuzzyEn presents a stronger relative consistency and shows less dependence on data length than ApEn and SampEn [3].

   Nevertheless, the number of samples in a signal still plays a role in the precision of FuzzyEn: the shorter the signal, the lower the number of vectors, and thus, the lower the precision of FuzzyEn (i.e., the larger the standard deviation). Therefore, to obtain more precise entropy values, the longer the signal, the better it is. In practical situations (real data), this may be a challenge. Indeed, it is often difficult to have long recordings, particularly in the biomedical field where patients may have difficulty to stay still or to cooperate.

This is why we herein propose a new fuzzy entropy measure that presents better precision than the traditional FuzzyEn measure. This is performed by increasing the number of samples used in the computation, without changing the length of the time series.

The paper is organized as follows. The original algorithm of FuzzyEn is first detailed in Section 2; then the new entropy measure is described. The synthetic and biomedical data (fetal heart rate time series) used in our work are introduced in Section 3. In Section 4, we first present, analyze, and discuss the results obtained with the synthetic data. We then describe and interpret the results obtained with the biomedical time series. We finally end with the conclusion.

## 2. Standard Fuzzy Entropy and the New Entropy Measure

In this section, we recall the FuzzyEn concept based on the use of a membership function. For this purpose, the generalized Gaussian membership function is used since it allows the derivation of both the rectangular function used in the calculation of SampEn and the standard Gaussian function used in the calculation of FuzzyEn.

### 2.1. Fuzzy Entropy Algorithm

For a given discrete time series $\mathbf{X} = \{x(1), x(2), \ldots, x(N)\}$ of length $N$, the algorithm to compute FuzzyEn relies on the following steps [1]:

1. Split $\mathbf{X}$ into a series of subsequences $\mathbf{X}_m(i)$ of length $m$ starting at $x(i)$: $\mathbf{X}_m(i) = \{x(i), x(i+1), \ldots, x(i+m-1)\}, 1 \leq i \leq N - m + 1$.
2. For each vector $\mathbf{X}_m(i)$, compute the similarity degree $D_{ij}^m$ of its neighboring vector $\mathbf{X}_m(j)$ using a similarity function as:

$$D_{ij}^m = \mu_p(d[\mathbf{X}_m(i), \mathbf{X}_m(j)], r), \tag{1}$$

where the membership function $\mu_p$ reported in Figure 1 is defined $\forall d \geq 0$ as:

$$\mu_p(d, r) = \exp(-(d/r)^p), \tag{2}$$

and where the distance function $d$ is the maximum absolute difference $d[\mathbf{X}_m(i), \mathbf{X}_m(j)] = \max_{0 \leq k \leq m-1}(|x(i+k) - x(j+k)|)$. For $p = 2$, we have the Gaussian function, and for $p = \infty$, we have the rectangular function.

3. For each $i$ ($1 \leq i \leq N - m + 1$), compute $\phi_i^m$ as:

$$\phi_i^m(r) = \frac{1}{N - m - 1} \sum_{j=1, j \neq i}^{N-m} D_{ij}^m. \tag{3}$$

4. Construct $\varphi^m$ and $\varphi^{m+1}$ as:

$$\varphi^m(r) = \frac{1}{N - m} \sum_{i=1}^{N-m} \phi_i^m(r), \tag{4}$$

$$\varphi^{m+1}(r) = \frac{1}{N - m} \sum_{i=1}^{N-m} \varphi_i^{m+1}(r). \tag{5}$$

5. Fuzzy entropy is then calculated as:

$$FuzzyEn(m, r) = \lim_{N \to \infty} \ln \left[ \frac{\varphi^m(r)}{\varphi^{m+1}(r)} \right], \tag{6}$$

which, for finite datasets, can be estimated by the statistic:

$$FuzzyEn(m,r,N) = \ln\left[\frac{\varphi^m(r)}{\varphi^{m+1}(r)}\right].$$ (7)

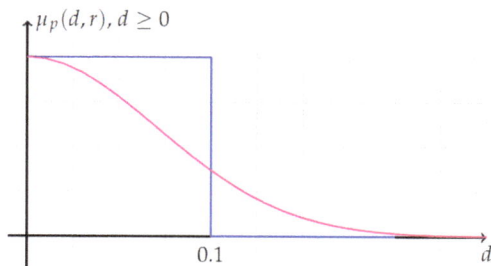

**Figure 1.** Membership functions $\mu_p(d,r) = \exp(-(d/r)^p)$ with $r = 0.1$. Gaussian function (blue) with $p = 2$; rectangular function (magenta) with $p = \infty$, for $d \geq 0$.

As shown in Figure 2, the 2-pattern '1' has only one similar 2-pattern among the 27 possible 2-patterns in the time series. From the time series reported in Figure 2, the total number of similar 2-patterns is 12: ('1','15'), ('5','21'), ('7','19'), ('8','20'), ('13','24'), ('14','25').

As for ApEn and SampEn, the statistical stability of the FuzzyEn estimation depends on the length $N$ of the time series as reported in Equation (7). To decrease this length-dependency, several strategies can be proposed.

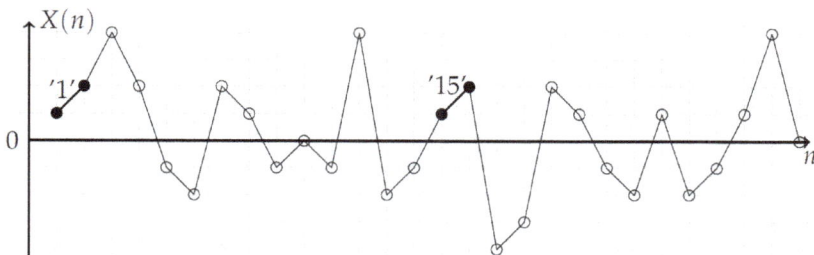

**Figure 2.** Stochastic time series where 2-patterns are pointed out. Each number corresponds to the place of the corresponding segment. No-centered 2-patterns are considered. The two 2-patterns '1' and '15' (black bullets) have the same mean value; they are similar. The total number of similar 2-patterns is 12: ('1','15'), ('5','21'), ('7','19'), ('8','20'), ('13','24'), ('14','25').

### 2.2. New Approaches

As mentioned above, from a fixed number of samples $N$ in the time series, a way to improve the statistical stability of the entropy measurement consists in artificially increasing the number of similar $m$-patterns taken into account in the entropy calculation. To do so, three different ways are proposed:

1.  The first approach is inspired by [3,7]. In the latter studies, the interest in centering each $m$-pattern has been shown. In this case, instead of limiting the search of $m$-patterns with the same mean value, any pattern can be taken into account. Therefore, the number of similar patterns drastically increases.

    Therefore, in the first approach, a centered $m$-pattern $\mathbf{X}c_m(j)$ is compared to a reference centered $m$-pattern $\mathbf{X}c_m(i)$. The similarity degree is calculated with $\mathbf{X}c_m(i) = \{x(i), x(i+1), \ldots, x(i+$

$m - 1)\} - x_0(i)$, where $1 \leq i \leq N - m + 1$ and $x_0(i) = \frac{1}{m} \sum_{j=0}^{j=m-1} x(i + j)$, through a similarity function:

$$Dc_{ij}^m = \mu_p(d[Xc_m(i), Xc_m(j)], r), \tag{8}$$

with the same membership function as the one reported in Equation (2). The centered fuzzy entropy FuzzyEn$_c$ is thus defined as:

$$FuzzyEn_c(m, r, N) = \ln \left[ \frac{\varphi_c^m(r)}{\varphi_c^{m+1}(r)} \right], \tag{9}$$

with $\varphi_c^m(r) = \frac{1}{N-m} \sum_{i=1}^{N-m} \phi_{ci}^m(r)$ and with $\phi_{ci}^m(r) = \frac{1}{N-m-1} \sum_{j=1, j \neq i}^{N-m} Dc_{ij}^m$.

As shown in Figure 3a, removing the mean value of 2-patterns increases the number of centered similar 2-patterns since the number of centered 2-patterns similar to '1' is six compared to one when the centering approach is not used. From Figure 3b, the total number of centered similar 2-patterns is 25: ('1','9','13','15','17','24'), ('2','14','25'), ('3','8','20'), ('4','23'), ('5','7','10','19','21'), ('11','18'), ('12','16'), ('22','26'). The total number of similar centered 2-patterns is much larger than no-centered 2-patterns.

2.  The second approach is inspired by [8], where transformed patterns are compared to reference patterns. Thus, in the second approach, a transformed $m$-pattern $\Gamma_k[X_m(j)]$ (see below) is compared to a reference $m$-pattern $X_m(i)$. The similarity degree is calculated with the same membership function as the one reported in Equation (2):

$$^kD_{ij}^m = \mu_p(d[X_m(i), \Gamma_k[X_m(j)]], r). \tag{10}$$

Four types of $\Gamma_k[X_m(j)]$ operations with $k = \{T, R, I, G\}$ are evaluated:

- $\Gamma_T[X_m(j)] = X_m(j + n)$ corresponds to a translation of $n$ samples, $k = T$;
- $\Gamma_R[X_m(j)] = X_m(-j + n)$ corresponds to a reflection at the position $n$, $k = R$;
- $\Gamma_I[X_m(j)] = -X_m(-j + n)$ corresponds to an inversion at the position $n$, $k = I$;
- $\Gamma_G[X_m(j)] = -X_m(j + n)$ corresponds to a glide reflection of $n$ samples, $k = G$.

At first sight, any type of operation could be used. However, from our point of view, only isometries (translation **T**, reflection **R**, inversion **I** and glide reflection **G**) are suitable. This statement is supported by the recent work reported in [8] where the concept of symmetry was placed back on stage in the study of time series. Indeed, in [8], it was shown that the concept of recurrences could be generalized by taking into account the symmetry properties of $m$-patterns. As entropy can be derived from the recurrence concept (the recurrence plot [9] is defined as $RP = (N - m + 1) \sum D_{ij}$ with $\mu_\infty(d, r)$), from [8], four new kinds of entropy ($ApEn_T$, $ApEn_R$, $ApEn_I$, $ApEn_G$ or $SampEn_T$, $SampEn_R$, $SampEn_I$, $SampEn_G$ or $FuzzyEn_T$, $FuzzyEn_R$, $FuzzyEn_I$, $FuzzyEn_G$) can be proposed. Finally, as our ultimate goal is to increase the precision of FuzzyEn, it is more appropriate here to calculate the mean value of the four new fuzzy entropies. In this case, the averaged fuzzy entropy FuzzyEn$_a$ is defined as:

$$FuzzyEn_a(m, r, N) = \frac{(FuzzyEn_T + FuzzyEn_R + FuzzyEn_I + FuzzyEn_G)}{4},$$

with:

$$FuzzyEn_k(m,r,N) = \ln\left[\frac{\varphi_k^m(r)}{\varphi_k^{m+1}(r)}\right],$$

with $k = \{T,R,I,G\}$ for $\varphi_k^m(r) = \frac{1}{N-m}\sum_{i=1}^{N-m}\phi_{ki}^m(r)$ and $\phi_{ki}^m(r) = \frac{1}{N-m-1}\sum_{j=1,j\neq i}^{N-m}{}^kD_{ij}^m$. *FuzzyEn_T* corresponds to the standard FuzzyEn measure when $m > 1$.

As shown in Figure 3b, the transformation of the 2-patterns increases the number of similar 2-patterns. From Figure 3b, for the 2-pattern ('1'), four kinds of 2-patterns can be obtained: 2-patterns with translation ('T') in black ('1','15'), 2-patterns with vertical reflection ('R') in red ('7','19'), 2-patterns with inversion ('I') in green ('13','24') and 2-patterns with glide reflection ('G') in blue ('5','21'). By considering all 2-patterns ranging from '1'–'27', the mean total number of symmetrical 2-patterns is $N_{sym} = 92$ with $N_{sym}^T = 12$, $N_{sym}^R = 30$, $N_{sym}^I = 24$, $N_{sym}^G = 26$.

3. The last approach compares a centered $m$-pattern $Xc_m(i)$ to a transformed centered $m$-pattern $\Gamma_k[Xc_m(j)]$. In this case, the centered and averaged fuzzy entropy FuzzyEn_ca is defined as:

$$FuzzyEn_{ca}(m,r,N) = \frac{(FuzzyEn_{cT} + FuzzyEn_{cR} + FuzzyEn_{cI} + FuzzyEn_{cG})}{4},$$

with:

$$FuzzyEn_{ck}(m,r,N) = \ln\left[\frac{\varphi_{ck}^m(r)}{\varphi_{ck}^{m+1}(r)}\right],$$

with $k = \{T,R,I,G\}$ for $\varphi_{ck}^m(r) = \frac{1}{N-m}\sum_{i=1}^{N-m}\phi_{cki}^m(r)$ and $\phi_{cki}^m(r) = \frac{1}{N-m-1}\sum_{j=1,j\neq i}^{N-m}{}^kDc_{ij}^m$. ${}^kDc_{ij}^m$ is defined as ${}^kDc_{ij}^m = \mu_p(d[Xc_m(i),\Gamma_k[Xc_m(j)]],r)$.

As shown in Figure 3, one can observe that the combination of the centering and averaging operations globally increases the number of $m$-patterns taken into account in the calculation of the entropy measure. Furthermore, a centered $m$-pattern transformed by an inversion ('I') is similar to a centered $m$-pattern transformed by a translation ('T'). The same remark applies for glide and vertical reflection transformations of centered $m$-patterns.

From Figure 3c, regarding the 2-pattern ('1'), two kinds of centered 2-patterns can be obtained: 2-patterns ('T','I') in black ('1','9','13','15','17','24') and 2-patterns ('R','G') in blue ('5','7','10','19','21'). By considering all 2-patterns ranging from '1'–'27', the mean total number of symmetrical 2-patterns is $Nc_{sym} = 312$ with $Nc_{sym}^T = 86$, $Nc_{sym}^R = 70$, $Nc_{sym}^I = 86$ and $Nc_{sym}^G = 70$.

The novelty of our method therefore relies on two main points: (i) the mean value of the patterns is no longer a constraint in the computation as the patterns are centered; (ii) translated patterns, but also reflected, inversed, and glide-reflected patterns are taken into account (in the standard sample and fuzzy entropy measures, only translated patterns are considered). Therefore, for a given number of samples $N$ in the time series, we managed to increase the number of similar $m$-patterns taken into account in the entropy calculation. In what follows, the new entropy measure will be applied to synthetic $1/f^\beta$ time series and biomedical datasets. Its precision will be compared to the one of the standard FuzzyEn.

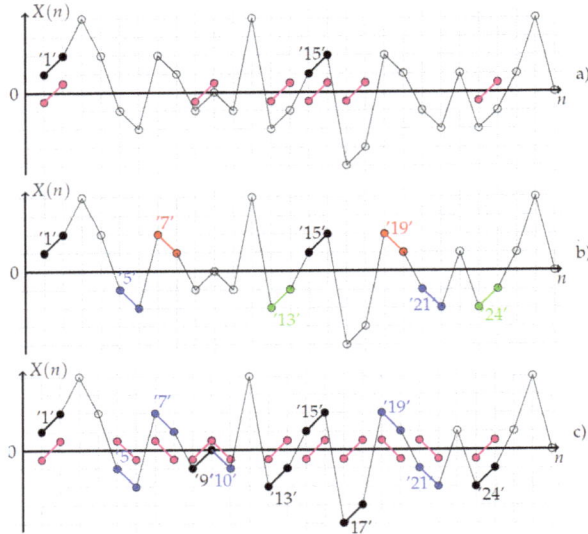

**Figure 3.** Stochastic time series with different types of 2-patterns. (**a**) Centered 2-patterns are considered. Centered 2-patterns similar to '1' are represented with magenta bullets; there are six patterns similar to '1'. The total number of centered similar 2-patterns is 25: ('1','9','13','15','17','24'), ('2','14','25'), ('3','8','20'), ('4','23'), ('5','7','10','19','21'), ('11','18'), ('12','16'), ('22','26'). The total number of similar centered 2-patterns is much larger than that of no-centered 2-patterns. (**b**) Regarding the 2-pattern ('1'), four kinds of 2-patterns can be obtained: 2-patterns with translation ('T') in black ('1','15'), 2-patterns with vertical reflection ('R') in red ('7', '19'), 2-patterns with inversion ('I') in green ('13','24'), 2-patterns with glide reflection ('G') in blue ('5','21'). By considering all 2-patterns ranging from '1'–'27', the mean total number of symmetrical 2-patterns is $N_{sym} = 92$ with $N_{sym}^{T} = 12$, $N_{sym}^{R} = 30$, $N_{sym}^{I} = 24$, $N_{sym}^{G} = 26$. (**c**) Regarding the 2-pattern ('1'), two kinds of centered 2-patterns can be obtained: 2-patterns ('T','I') in black ('1','9','13','15','17','24'), 2-patterns ('R','G') in blue ('5','7','10','19','21'). By considering all 2-patterns ranging from '1'–'27', the mean total number of symmetrical 2-patterns is $Nc_{sym} = 312$ with $Nc_{sym}^{T} = 86$, $Nc_{sym}^{R} = 70$, $Nc_{sym}^{I} = 86$ and $Nc_{sym}^{G} = 70$.

## 3. Data Processed

### 3.1. Synthetic Signals

In order to analyze the new fuzzy entropy measures and to compare their performances with the ones of the standard FuzzyEn, we used $1/f^{\beta}$ time series, with different $\beta$ values: $\beta$ varied from $-1$ to 2 in steps of 0.2. For $\beta > 0$, the $1/f^{\beta}$ signals are persistent processes with long-term correlations [10]. However, for $\beta < 0$, the $1/f^{\beta}$ signals are anti-persistent processes with short-term anti-correlations [10]. From a theoretical point of view, the higher the value of $\beta$, the larger the number of correlations in the time series and, therefore, the larger the number of similar samples used in the computation of FuzzyEn. For each $\beta$ value, 50 time series were simulated.

### 3.2. Biomedical Data

The new descriptors mentioned above were also applied to biomedical data and more precisely to fetal heart rate (FHR) time series. The latter were acquired using a homemade pulse Doppler system co-developed with Altaïs Technologies (Tours, France). This Doppler fetal monitor transmits ultrasound waves of 2.25 MHz for an acoustic power limited to 1 mW/cm$^2$ (for more details, see [11]). It was developed to measure both the FHR and fetal movements (pseudo-breathing, limb movements).

The study was approved by the Ethics Committee of the Clinical Investigation Centre for Innovative Technology of Tours (CIC-IT 806 CHRUof Tours). Before acquisition, the consent of each parent was obtained. All parents were over eighteen years of age, and pregnancies were single. After locating the fetal heart with an echographic scanner, 18 Doppler recordings of 30 min each were acquired at CHRU Bretonneau Tours, France. This corresponds to approximately 3600 heart beats for each recording. In order to constitute homogeneous groups without spurious data, gestations complicated by other kinds of disorders (hypertension, diabetes) were discarded. Two groups of fetuses were selected: normal and those with severe intra-uterine growth retardation (IUGR). The severe IUGR group included nine fetuses delivered prematurely by cesarean section. The normal group included nine fetuses without disorders, delivered at term by spontaneous labor. For this clinical protocol, the gestational ages of fetuses ranged from 30–34 weeks.

In what follows, the 30 min of data were processed, but also segments of 10 min and 20 min. Our goal was thus to compare the results obtained as the data length decreases. Moreover, in order to compare the results obtained between normal and IUGR groups, a Mann–Whitney test was used. A $p$-value strictly less than 0.05 was considered to define statistical significance.

## 4. Results and Discussion

In all that follows, the value of $r$ is set at $0.1 \times$ the standard deviation of the time series.

### 4.1. Results for the Synthetic Signals

In order to validate our hypothesis (that is, the greater the number of similar $m$-patterns taken into account in the computation, the more precise the entropy measure), we started by counting the number of similar $m$-patterns from 50 synthetic time series.

From $1/f^\beta$ noises generated with $N = 5000$ samples with $\beta$ ranging from $-1$ to 2, the median of the mean number $MN$ of similar 3-patterns and the median of the mean number $MN_{ca}$ of centered and averaged similar 3-patterns were evaluated and are reported in Table 1. As expected, the higher the sample correlation in the time series, the higher the value of $\beta$ and the higher the number of similar 3-patterns. Indeed, from Table 1, when $\beta$ increases from 0 to 2, $MN$ goes from 1 to 162. When symmetrical properties and the centering operation are taken into account, $MN_{ca}$ goes from 21 to 9278 for $\beta$ ranging from 0 to 2. From this, it can be claimed that the averaging and the centering operations increase the number of similar patterns. Furthermore, whatever the $m$-value, we obtain rising trends as $\beta$ increases (data not shown).

In order to evaluate the performance of our new approaches, for a fixed $m$-value and for 50 $1/f^\beta$ time series with different $\beta$ values, different measures have been computed: the medians $MFuzzyEn$, $MFuzzyEn_c$, $MFuzzyEn_a$, $MFuzzyEn_{ca}$ and the percentiles at 75% and 25% $PFuzzyEn(75)$, $PFuzzyEn(25)$, $PFuzzyEn_c(75)$, $PFuzzyEn_c(25)$, $PFuzzyEn_a(75)$, $PFuzzyEn_a(25)$, $PFuzzyEn_{ca}(75)$, $PFuzzyEn_{ca}(25)$ have been compared.

To quantitatively evaluate the gain brought by our new approaches in comparison with FuzzyEn, two kinds of statistics have been evaluated: percentile ranges and relative percentile ranges. The following percentile ranges have thus been computed:

- $R_F = PFuzzyEn(75) - PFuzzyEn(25)$;
- $R_{Fc} = PFuzzyEn_c(75) - PFuzzyEn_c(25)$;
- $R_{Fa} = PFuzzyEn_a(75) - PFuzzyEn_a(25)$;
- $R_{Fca} = PFuzzyEn_{ca}(75) - PFuzzyEn_{ca}(25)$.

Finally, from the percentile ranges, the following relative percentile ranges have been evaluated:

- $(R_F - R_{Fc})/R_{Fc}$;
- $(R_F - R_{Fa})/R_{Fa}$;
- $(R_F - R_{Fca})/R_{Fca}$.

The global results are presented in Tables A1–A3 reported in the Appendix and are shown in Figure 4. We observe from the tables that SampEn leads to worse results than FuzzyEn, as already shown by others. Moreover, we observe that the new approach leads to results that show a reduced percentile range compared to the standard fuzzy entropy measure. Its precision is therefore better than the other entropy measures. However, our work also has some drawbacks: the gain provided by the method depends on the signal properties. The gain differs with $\beta$ values.

**Table 1.** For the calculation of *FuzzyEn* and *FuzzyEn$_{ca}$*, the median of the mean number *MN* of similar 3-patterns and the median of the mean number of centered and averaged *MN$_{ca}$* of similar 3-patterns obtained from $1/f^\beta$ noises ($N = 5000$ samples) with $\beta$ ranging from $-1$ to $2$. $MN_{ca} = (MN_{ca}^T + MN_{ca}^R + MN_{ca}^I + MN_{ca}^G)$, where $MN_{ca}^k$ is the median of the number of centered symmetric similar 3-patterns obtained in the calculation of *FuzzyEn$_{ca}$*, $k = \{'T', 'R', 'I', 'G'\}$. For the computation, $m = 3$ and $r = 0.1\times$ standard deviation of the time series.

| $\beta$ | $-1$ | $-0.8$ | $-0.6$ | $-0.4$ | $-0.2$ | 0 | 0.2 | 0.4 | 0.6 | 0.8 | 1.0 | 1.2 | 1.4 | 1.6 | 1.8 | 2.0 |
|---|---|---|---|---|---|---|---|---|---|---|---|---|---|---|---|---|
| MN | 0.73 | 0.69 | 0.65 | 0.62 | 0.6 | 0.63 | 0.63 | 0.67 | 0.80 | 1.07 | 1.76 | 3.64 | 9.24 | 26.47 | 71.68 | 162.38 |
| MN$_{ca}$ | 16.71 | 17.03 | 17.48 | 18.13 | 19.18 | 20.75 | 23.19 | 27.35 | 35.71 | 53.15 | 93.73 | 206.09 | 540.00 | 1580.67 | 4317.40 | 9277.86 |

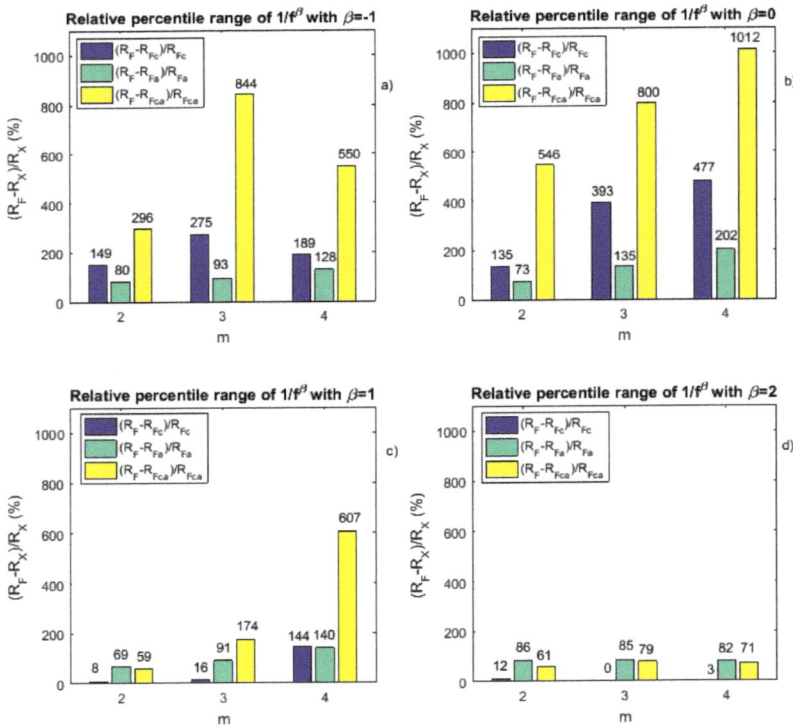

**Figure 4.** Relative percentile ranges derived from Tables A1–A3 reported in the Appendix. (**a**) For $\beta = -1$, relative percentile range values obtained for different *m*-values: for the centered fuzzy entropy compared to the fuzzy entropy (($R_F - R_{Fc})/R_{Fc}$), for the averaged fuzzy entropy compared to the fuzzy entropy (($R_F - R_{Fa})/R_{Fa}$) and for the centered and averaged fuzzy entropy compared to the fuzzy entropy (($R_F - R_{Fca})/R_{Fca}$); (**b–d**) similar to (**a**), but for $\beta = 0$, $\beta = 1$ and $\beta = 2$, respectively.

## 4.2. Results for the Fetal Heart Rate Time Series

The results obtained from FHR time series for $m = 2$ are presented in Figure 5 for data lengths of 10 min, 20 min, and 30 min. For the three data lengths, we observe that the normal fetuses show a significantly higher entropy value than the pathological fetuses. This is true for the two entropy measures: $FuzzyEn_{ca}$ and the standard $FuzzyEn$. This means that FHR time series are more irregular for the normal fetuses than for the pathological ones. We also observe that the $p$-value between the two groups decreases as the data length increases. Therefore, the longer the data, the better the separation between the two groups. However, we note that, whatever the length studied, the $p$-value is lower for $FuzzyEn_{ca}$ than for the standard $FuzzyEn$. Our new entropy measure is therefore more interesting for this classification purpose than the standard $FuzzyEn$. Other data may now be processed; see, e.g., [12–14].

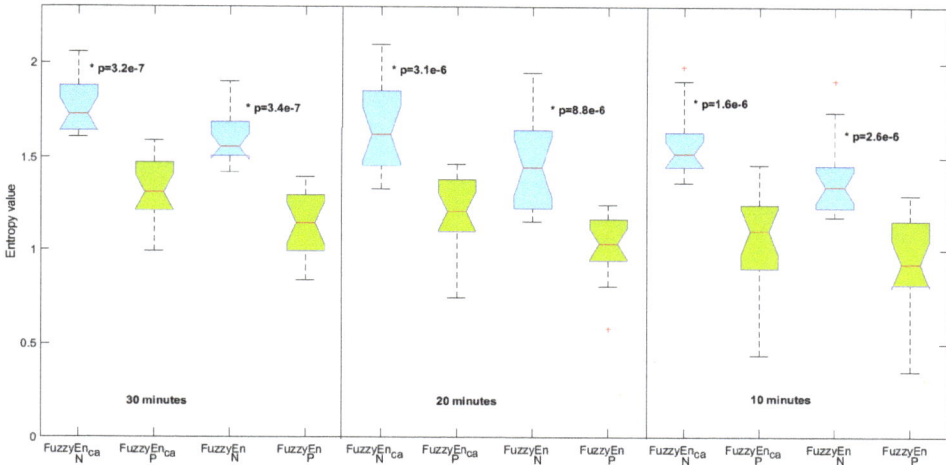

**Figure 5.** Centered and averaged fuzzy entropy ($FuzzyEn_{ca}$) and standard fuzzy entropy ($FuzzyEn$) for normal (N) in blue and pathological fetuses (P) in green with $m = 2$. The results for three data lengths are shown. * means statistically significant between the two groups.

## 5. Conclusions

A new entropy measure, $FuzzyEn_{ca}$, is proposed to improve the precision of the standard FuzzyEn. The new measure relies on centering and averaging approaches that lead to a larger number of similar patterns used in the computation of the entropy algorithm. This is performed by removing the constraint of the mean value in the comparison of the patterns. Moreover, translated patterns are not the only ones considered: reflected, inversed, and glide-reflected patterns are also taken into account. The results obtained on $1/f^{\beta}$ time series reveal that $FuzzyEn_{ca}$ shows a greater precision than FuzzyEn. Moreover, when applied to FHR time series acquired from normal and pathological fetuses, $FuzzyEn_{ca}$ leads to a better discrimination between the two groups than the standard FuzzyEn. These findings could allow one to obtain entropy-based relevant information by processing shorter datasets (we could obtain the same precision as the standard FuzzyEn, but with less data). This is particularly interesting for the biomedical field. $FuzzyEn_{ca}$ now has to be applied to other datasets, and its performance has to be compared to those of other already-existing entropy measures.

**Acknowledgments:** This study was supported financially by the "Agence Nationale de la Recherche" (Project ANR-07-TECSAN-023, Surfoetus), France. Furthermore, the authors would like to thank the Clinical Investigation Centre for Innovative Technology of Tours (CIC-IT 1415 CHRU of Tours) and F. Perrotin's team in the Obstetrics Department for their support in recording the signals.

*Entropy* **2018**, *20*, 287

**Author Contributions:** Jean-Marc Girault and Anne Humeau-Heurtier conceived and designed the study. Jean-Marc Girault and Anne Humeau-Heurtier analyzed the synthetic data. Jean-Marc Girault analyzed the FHR time series. Jean-Marc Girault and Anne Humeau-Heurtier wrote the paper. Both authors have read and approved the final manuscript.

**Conflicts of Interest:** The authors declare no conflict of interest.

## Appendix A

Results reported in Tables A1–A3 show performances that differ with the $\beta$ values. We observe that the higher the $\beta$ value, the lower the gain obtained in terms of relative percentile range. This is probably due to the level of correlation between samples in the time series.

**Table A1.** Results obtained for $1/f^\beta$ time series, for $m = 2$, $N = 5000$ samples and for different $\beta$ values.

| $\beta$ value | −1 | −0.8 | −0.6 | −0.4 | −0.2 | 0 | 0.2 | 0.4 | 0.6 | 0.8 | 1 | 1.2 | 1.4 | 1.6 | 1.8 | 2 |
|---|---|---|---|---|---|---|---|---|---|---|---|---|---|---|---|---|
| $MSampEn$ | 3.46 | 3.5 | 3.53 | 3.54 | 3.57 | 3.59 | 3.58 | 3.54 | 3.45 | 3.29 | 3.04 | 2.68 | 2.21 | 1.68 | 1.13 | 0.64 |
| $MFuzzyEn$ | 3.17 | 3.21 | 3.25 | 3.26 | 3.28 | 3.28 | 3.28 | 3.25 | 3.16 | 3.00 | 2.76 | 2.40 | 1.93 | 1.42 | 0.93 | 0.54 |
| $MFuzzyEn_c$ | 3.58 | 3.58 | 3.58 | 3.58 | 3.56 | 3.53 | 3.49 | 3.42 | 3.30 | 3.11 | 2.84 | 2.46 | 1.99 | 1.46 | 0.94 | 0.51 |
| $MFuzzyEn_a$ | 3.14 | 3.17 | 3.2 | 3.22 | 3.24 | 3.24 | 3.23 | 3.20 | 3.12 | 2.97 | 2.73 | 2.37 | 1.90 | 1.40 | 0.93 | 0.57 |
| $MFuzzyEn_{ca}$ | 3.57 | 3.58 | 3.58 | 3.57 | 3.56 | 3.53 | 3.48 | 3.41 | 3.30 | 3.11 | 2.83 | 2.44 | 1.96 | 1.44 | 0.94 | 0.53 |
| $RSampEn$ | 0.05 | 0.08 | 0.07 | 0.08 | 0.09 | 0.08 | 0.08 | 0.10 | 0.08 | 0.05 | 0.08 | 0.11 | 0.16 | 0.24 | 0.25 | 0.23 |
| $RFuzzyEn$ | 0.03 | 0.04 | 0.03 | 0.04 | 0.04 | 0.04 | 0.04 | 0.05 | 0.04 | 0.04 | 0.08 | 0.11 | 0.17 | 0.22 | 0.22 | 0.18 |
| $RFuzzyEn_c$ | 0.01 | 0.01 | 0.01 | 0.01 | 0.01 | 0.02 | 0.02 | 0.01 | 0.02 | 0.03 | 0.07 | 0.11 | 0.18 | 0.23 | 0.23 | 0.20 |
| $RFuzzyEn_a$ | 0.02 | 0.03 | 0.02 | 0.02 | 0.02 | 0.02 | 0.02 | 0.03 | 0.03 | 0.02 | 0.04 | 0.08 | 0.11 | 0.13 | 0.13 | 0.10 |
| $RFuzzyEn_{ca}$ | 0.01 | 0.01 | 0.01 | 0.01 | 0.01 | 0.01 | 0.01 | 0.01 | 0.01 | 0.02 | 0.05 | 0.08 | 0.11 | 0.14 | 0.14 | 0.11 |
| $(RSampEn - RFuzzyEn)/RFuzzyEn$ | 0.63 | 0.75 | 1.24 | 1.09 | 1.17 | 1.04 | 1.10 | 0.99 | 0.90 | 0.43 | 0.12 | 0.00 | −0.05 | 0.10 | 0.16 | 0.29 |
| $(RSampEn - RFuzzyEn_c)/RFuzzyEn_c$ | 3.08 | 5.77 | 5.6 | 7.06 | 6.11 | 3.80 | 4.27 | 5.85 | 3.17 | 0.68 | 0.21 | −0.05 | −0.12 | 0.08 | 0.09 | 0.14 |
| $(RSampEn - RFuzzyEn_a)/RFuzzyEn_a$ | 1.95 | 2.09 | 3.26 | 3.34 | 4.38 | 2.53 | 2.78 | 2.72 | 2.03 | 1.33 | 0.89 | 0.40 | 0.50 | 0.89 | 1.03 | 1.39 |
| $(RSampEn - RFuzzyEn_{ca})/RFuzzyEn_{ca}$ | 5.47 | 8.67 | 9.1 | 9.69 | 11.3 | 12.20 | 8.96 | 8.57 | 5.55 | 1.59 | 0.77 | 0.36 | 0.43 | 0.74 | 0.86 | 1.07 |
| $(RFuzzyEn - RFuzzyEn_c)/RFuzzyEn_c$ | 1.49 | 2.86 | 1.95 | 2.87 | 2.27 | 1.35 | 1.51 | 2.44 | 1.19 | 0.18 | 0.08 | 0.05 | 0.07 | 0.02 | 0.07 | 0.12 |
| $(RFuzzyEn - RFuzzyEn_a)/RFuzzyEn_a$ | 0.8 | 0.76 | 0.91 | 1.08 | 1.48 | 0.73 | 0.80 | 0.87 | 0.59 | 0.63 | 0.69 | 0.40 | 0.58 | 0.72 | 0.75 | 0.86 |
| $(RFuzzyEn - RFuzzyEn_{ca})/RFuzzyEn_{ca}$ | 2.96 | 4.51 | 3.52 | 4.13 | 4.66 | 5.46 | 3.74 | 3.81 | 2.44 | 0.81 | 0.59 | 0.36 | 0.51 | 0.59 | 0.60 | 0.61 |

**Table A2.** Same as Table A1, but for $m = 3$.

| $\beta$ value | −1 | −0.8 | −0.6 | −0.4 | −0.2 | 0 | 0.2 | 0.4 | 0.6 | 0.8 | 1 | 1.2 | 1.4 | 1.6 | 1.8 | 2 |
|---|---|---|---|---|---|---|---|---|---|---|---|---|---|---|---|---|
| $MSampEn$ | 3.47 | 3.51 | 3.48 | 3.54 | 3.56 | 3.57 | 3.54 | 3.58 | 3.43 | 3.27 | 3.01 | 2.65 | 2.18 | 1.66 | 1.13 | 0.64 |
| $MFuzzyEn$ | 3.02 | 3.1 | 3.12 | 3.15 | 3.2 | 3.18 | 3.16 | 3.15 | 3.04 | 2.88 | 2.61 | 2.26 | 1.80 | 1.30 | 0.84 | 0.49 |
| $MFuzzyEn_c$ | 3.28 | 3.27 | 3.28 | 3.29 | 3.27 | 3.26 | 3.22 | 3.16 | 3.05 | 2.87 | 2.59 | 2.22 | 1.77 | 1.26 | 0.80 | 0.44 |
| $MFuzzyEn_a$ | 2.99 | 3.02 | 3.04 | 3.07 | 3.11 | 3.10 | 3.10 | 3.05 | 2.97 | 2.82 | 2.58 | 2.22 | 1.77 | 1.29 | 0.84 | 0.51 |
| $MFuzzyEn_{ca}$ | 3.21 | 3.22 | 3.23 | 3.24 | 3.23 | 3.22 | 3.18 | 3.12 | 3.02 | 2.85 | 2.59 | 2.21 | 1.74 | 1.25 | 0.80 | 0.47 |
| $RSampEn$ | 0.51 | 0.49 | 0.43 | 0.5 | 0.66 | 0.58 | 0.59 | 0.39 | 0.40 | 0.35 | 0.19 | 0.18 | 0.18 | 0.21 | 0.25 | 0.23 |
| $RFuzzyEn$ | 0.17 | 0.16 | 0.13 | 0.17 | 0.23 | 0.17 | 0.18 | 0.14 | 0.17 | 0.11 | 0.09 | 0.11 | 0.17 | 0.19 | 0.20 | 0.16 |
| $RFuzzyEn_c$ | 0.05 | 0.04 | 0.03 | 0.03 | 0.05 | 0.03 | 0.04 | 0.04 | 0.03 | 0.05 | 0.08 | 0.11 | 0.17 | 0.20 | 0.20 | 0.16 |
| $RFuzzyEn_a$ | 0.09 | 0.07 | 0.08 | 0.07 | 0.06 | 0.07 | 0.08 | 0.07 | 0.06 | 0.06 | 0.05 | 0.07 | 0.10 | 0.13 | 0.12 | 0.09 |
| $RFuzzyEn_{ca}$ | 0.02 | 0.02 | 0.02 | 0.02 | 0.02 | 0.02 | 0.02 | 0.02 | 0.02 | 0.02 | 0.03 | 0.08 | 0.11 | 0.13 | 0.12 | 0.09 |
| $(RSampEn - RFuzzyEn)/RFuzzyEn$ | 1.95 | 2.13 | 2.46 | 1.86 | 1.94 | 2.48 | 2.23 | 1.78 | 1.38 | 2.21 | 1.15 | 0.54 | 0.08 | 0.08 | 0.25 | 0.42 |
| $(RSampEn - RFuzzyEn_c)/RFuzzyEn_c$ | 10.08 | 11.6 | 11.8 | 14.54 | 12.6 | 16.19 | 14.51 | 8.33 | 11.03 | 5.97 | 1.49 | 0.58 | 0.05 | 0.03 | 0.25 | 0.41 |
| $(RSampEn - RFuzzyEn_a)/RFuzzyEn_a$ | 4.68 | 5.55 | 4.51 | 5.74 | 9.74 | 7.20 | 6.06 | 4.93 | 6.25 | 4.90 | 3.12 | 1.60 | 0.74 | 0.64 | 1.17 | 1.62 |
| $(RSampEn - RFuzzyEn_{ca})/RFuzzyEn_{ca}$ | 26.88 | 22.74 | 17.41 | 28.43 | 33.81 | 30.33 | 30.39 | 21.08 | 23.78 | 16.41 | 4.91 | 1.34 | 0.59 | 0.57 | 1.09 | 1.55 |
| $(RFuzzyEn - RFuzzyEn_c)/RFuzzyEn_c$ | 2.75 | 3.03 | 2.7 | 4.43 | 3.63 | 3.93 | 3.80 | 2.36 | 4.05 | 1.17 | 0.16 | 0.03 | 0.03 | 0.05 | 0.00 | 0.00 |
| $(RFuzzyEn - RFuzzyEn_a)/RFuzzyEn_a$ | 0.92 | 1.1 | 0.59 | 1.36 | 2.65 | 1.35 | 1.18 | 1.13 | 2.05 | 0.84 | 0.91 | 0.69 | 0.61 | 0.52 | 0.74 | 0.85 |
| $(RFuzzyEn - RFuzzyEn_{ca})/RFuzzyEn_{ca}$ | 8.44 | 6.59 | 4.32 | 9.28 | 10.84 | 8.00 | 8.71 | 6.94 | 9.41 | 4.42 | 1.74 | 0.52 | 0.47 | 0.45 | 0.68 | 0.79 |

*Entropy* **2018**, *20*, 287

**Table A3.** Same as Table A1, but for $m = 4$. "-" means that an undefined value is obtained due the absence of similar $m$-patterns in the time series.

| $\beta$ value | -1 | -0.8 | -0.6 | -0.4 | -0.2 | 0 | 0.2 | 0.4 | 0.6 | 0.8 | 1 | 1.2 | 1.4 | 1.6 | 1.8 | 2 |
|---|---|---|---|---|---|---|---|---|---|---|---|---|---|---|---|---|
| $MSampEn$ | - | - | - | - | - | - | - | - | - | - | 2.95 | 2.71 | 2.19 | 1.59 | 1.13 | 0.64 |
| $MFuzzyEn$ | 3.09 | 3.02 | 2.95 | 3.08 | 3.01 | 3.26 | 3.17 | 3.13 | 2.99 | 2.75 | 2.50 | 2.13 | 1.70 | 1.17 | 0.78 | 0.45 |
| $MFuzzyEn_c$ | 3.15 | 3.15 | 3.14 | 3.15 | 3.19 | 3.14 | 3.13 | 3.08 | 2.95 | 2.79 | 2.51 | 2.11 | 1.70 | 1.16 | 0.76 | 0.43 |
| $MFuzzyEn_a$ | 2.73 | 2.73 | 2.84 | 2.77 | 2.81 | 2.92 | 2.84 | 2.87 | 2.72 | 2.55 | 2.35 | 2.05 | 1.65 | 1.19 | 0.78 | 0.47 |
| $MFuzzyEn_{ca}$ | 3.08 | 3.11 | 3.12 | 3.11 | 3.11 | 3.12 | 3.09 | 3.04 | 2.93 | 2.76 | 2.51 | 2.14 | 1.67 | 1.19 | 0.76 | 0.44 |
| $RSampEn$ | - | - | - | - | - | - | - | - | - | - | 1.09 | 0.49 | 0.26 | 0.28 | 0.26 | 0.22 |
| $RFuzzyEn$ | 0.61 | 0.39 | 0.71 | 0.59 | 0.58 | 0.80 | 0.63 | 0.58 | 0.29 | 0.40 | 0.22 | 0.18 | 0.16 | 0.21 | 0.19 | 0.15 |
| $RFuzzyEn_c$ | 0.21 | 0.16 | 0.16 | 0.15 | 0.12 | 0.14 | 0.12 | 0.11 | 0.08 | 0.06 | 0.09 | 0.14 | 0.16 | 0.23 | 0.19 | 0.15 |
| $RFuzzyEn_a$ | 0.27 | 0.19 | 0.29 | 0.28 | 0.29 | 0.27 | 0.29 | 0.26 | 0.24 | 0.19 | 0.09 | 0.10 | 0.10 | 0.10 | 0.11 | 0.08 |
| $RFuzzyEn_{ca}$ | 0.09 | 0.1 | 0.1 | 0.08 | 0.07 | 0.07 | 0.05 | 0.06 | 0.04 | 0.03 | 0.03 | 0.06 | 0.11 | 0.10 | 0.12 | 0.09 |
| $(RSampEn - RFuzzyEn)/RFuzzyEn$ | - | - | - | - | - | - | - | - | - | - | 4.03 | 1.75 | 0.64 | 0.31 | 0.38 | 0.49 |
| $(RSampEn - RFuzzyEn_c)/RFuzzyEn_c$ | - | - | - | - | - | - | - | - | - | - | 11.29 | 2.60 | 0.59 | 0.22 | 0.40 | 0.44 |
| $(RSampEn - RFuzzyEn_a)/RFuzzyEn_a$ | - | - | - | - | - | - | - | - | - | - | 11.06 | 4.13 | 1.54 | 1.93 | 1.44 | 1.70 |
| $(RSampEn - RFuzzyEn_{ca})/RFuzzyEn_{ca}$ | - | - | - | - | - | - | - | - | - | - | 34.58 | 6.91 | 1.33 | 1.92 | 1.27 | 1.54 |
| $(RFuzzyEn - RFuzzyEn_c)/RFuzzyEn_c$ | 1.89 | 1.37 | 3.38 | 2.97 | 3.94 | 4.77 | 4.26 | 4.10 | 2.76 | 6.28 | 1.44 | 0.31 | 0.03 | 0.07 | 0.01 | 0.03 |
| $(RFuzzyEn - RFuzzyEn_a)/RFuzzyEn_a$ | 1.28 | 1.03 | 1.49 | 1.13 | 1 | 2.02 | 1.16 | 1.26 | 0.23 | 1.07 | 1.40 | 0.86 | 0.55 | 1.24 | 0.77 | 0.82 |
| $(RFuzzyEn - RFuzzyEn_{ca})/RFuzzyEn_{ca}$ | 5.5 | 2.79 | 6.16 | 6.22 | 6.89 | 10.12 | 11.25 | 8.05 | 6.25 | 12.67 | 6.07 | 1.87 | 0.42 | 1.23 | 0.64 | 0.71 |

# References

1.  Pincus, S.M. Approximate entropy as a measure of system complexity. *Proc. Natl. Acad. Sci. USA* **1991**, *88*, 2297–2301.
2.  Richman, J.S.; Moorman, J.R. Physiological time-series analysis using approximate entropy and sample entropy. *Am. J. Physiol.-Heart Circ. Physiol.* **2000**, *278*, H2039–H2049.
3.  Chen, W.; Zhuang, J.; Yu, W.; Wang, Z. Measuring complexity using FuzzyEn, ApEn, and SampEn. *Med. Eng. Phys.* **2009** *31*, 61–68.
4.  Hu, J. An approach to EEG-based gender recognition using entropy measurement methods. *Knowl.-Based Syst.* **2018**, *140*, 134–141.
5.  Tibdewal, M.N.; Dey, H.R.; Mahadevappa, M.; Ray, A.; Malokar, M. Multiple entropies performance measure for detection and localization of multi-channel epileptic EEG. *Biomed. Signal Process. Control* **2017**, *38*, 158–167.
6.  Hu, J.; Wang, P. Noise robustness analysis of performance for EEG-based driver fatigue detection using different entropy feature sets. *Entropy* **2017**, *19*, 385.
7.  Liu, C.; Li, K.; Zhao, L.; Liu, F.; Zheng, D.; Liu, C.; Liu, S. Analysis of heart rate variability using Fuzzy measure entropy. *Comput. Biol. Med.* **2013**, *43*, 100–108.
8.  Girault, J.-M. Recurrence and symmetry of time series: Application to transition detection. *Chaos Solitons Fractals* **2015**, *77*, 11–28.
9.  Eckmann, J.P.; Oliffson Kamphorts, S.; Ruelle, D. Recurrence plots of dynamical systems. *Europhys. Lett.* **1987**, *4*, 973–977.
10. Tarnopolski, M. On the relationship between the Hurst exponent, the ratio of the mean square successive difference to the variance, and the number of turning points. *Phys. A* **2016**, *461*, 662–673.
11. Voicu, I.; Menigot, S.; Kouamé, D.; Girault, J.-M. New estimators and guidelines for better use of fetal heart rate estimators with Doppler ultrasound devices. *Comput. Math. Methods Med.* **2014**, *2014*, 784862.
12. Fang, Y.; Zhou, D.; Li, K.; Liu, H. Interface Prostheses With Classifier-Feedback-Based Usér Training. *IEEE Trans. Biomed. Eng.* **2017**, *64*, 2575–2583.
13. Zhou, D.; Fang, Y.; Botzheim, J.; Kubota, N.; Liu, H. Bacterial memetic algorithm based feature selection for surface EMG based hand motion recognition in long-term use. In *2016 IEEE Symposium Series on Computational Intelligence (SSCI)*; IEEE: Piscataway Township, NJ, USA, 2016; pp. 1–7.
14. Humeau-Heurtier, A.; Mahé, G.; Durand, S.; Abraham, P. Multiscale entropy study of medical laser speckle contrast images. *IEEE Trans. Biomed. Eng.* **2013**, *60*, 872–879.

MDPI

St. Alban-Anlage 66

4052 Basel

Switzerland

Tel. +41 61 683 77 34

Fax +41 61 302 89 18

www.mdpi.com

*Entropy* Editorial Office

E-mail: entropy@mdpi.com

www.mdpi.com/journal/entropy